TEACHER'S · RESOURCE · MANUAL

TAPESTRY

A · MULTICULTURAL · ANTHOLOGY

Alan C. Purves
General Editor

Globe Book Company
A Division of Simon & Schuster
Englewood Cliffs, New Jersey

ESL Consultant: Jacqueline Kiraithe de Córdoba, Ph.D.
Coordinator of Spanish and TESOL Areas
Department of Foreign Languages and Literatures
California State University at Fullerton

Executive Editor: Virginia Seeley
Senior Editor: Jasper Jones
Art Director: Nancy Sharkey
Production Manager: Winston Sukhnanand
Desktop Specialists: Matt Zuch, Danielle Hollomon
Marketing Manager: Elmer Ildefonso
Cover Design: BB&K Design, Inc.
Cover Illustration: John Jinks

ISBN: 0–835–90518–7

Printed in the United States of America
4 5 6 7 8 9 10   96 95

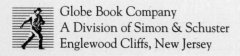
Globe Book Company
A Division of Simon & Schuster
Englewood Cliffs, New Jersey

Table of Contents

Contents: Cultural Groups

Contents: Genres

Contents: Chronology of Selection Settings

Program Rationale and Philosophy

Tapestry is a collection of stories, poems, plays, and documents written by or about the millions of people in the United States who have often been forgotten in the official histories and anthologies. Some of them are the peoples who were here before the European arrivals on the islands and continents we call the Americas—the people whom we call the First Nations. Others are those who came as immigrants but not as conquerors. Some came as enslaved people, and some came to make up the labor forces that built the railroads or toiled in the factories. They came from Europe and its surrounding islands, from Central Asia and the Indian subcontinent, from China and Japan and other parts of Asia, from the Caribbean and Central America.

Many came under appalling conditions. Some were forced to come by people in their homeland who were only too happy to drive them out. Once they got here, they worked long hours in conditions that we would think torturous. Many who worked twelve-hour days were young children. They lived in cramped and dirty conditions. Unfortunately, many recent immigrants still do.

They survived. They are the ancestors of most of us who live in the United States today. Many of them became the poets, novelists, and dramatists who are familiar to us today, but others are not so well known. *Tapestry* includes both, but there is a definite emphasis on unheard voices, those writers and speakers who have been overlooked in traditional literature anthologies. This exclusion may be due to the often disturbing or dissenting messages they convey. Also, there has been a tendency among anthologists to favor literature by European and European American authors.

This book has been prepared with an inclusive approach. Unheard voices have been sought out, but with no compromise in the quality of the selections. The editors have acknowledged the conflict and dissent among cultures in the United States but have emphasized the commonality and humanity shared by these different cultures. *Tapestry* is a book of the literature that accompanied the making of this country we live in.

It is also your book and your students' book. We want *Tapestry* to be a book in which students will find works that are enjoyable to read and at the same time challenging. We want it to be a book in which they meet new writers and hear new voices. But as your students read the selections and talk about them in class or write about them, we want them to find their voices as readers and writers and as members of the society that is the United States.

In making the above statements about the selections we have chosen, we have also established an approach to teaching literature that acknowledges the importance of the reader, the writer and culture. We have operated on the assumption that the readers of the text (you and your students) are important in determining the value of what has been read. We also know that the writers and artists, the people who wrote down or told those pieces—and the cultures and traditions of the artists—are important and worth learning about.

The selections in the student book offer not only an opportunity for exercise in self-exploration by the students but also an opportunity for inquiry into what it means to write from the wellspring of a particular culture. Together with your students discover what is means to be a member of a disfranchised or enslaved culture, a group seeking its place in a democracy, a community trying to maintain its traditional culture within a pluralistic society, or a band of artists seeking recognition as writers worthy to stand with the "great masters." As they come to accept alternative perspectives, your students will also be increasing their understanding of themselves as readers and as human beings.

To help you, we have created a Student Edition that arranges the selections thematically and historically. We have also included background information on the author and the culture surrounding the selection so that the students may begin to appreciate the work and its author and form their own responses to both. We call this "Reading the Author in a Cultural Context," an approach that helps students remember the human sources of any piece of writing and the fact that other human beings are the readers. We stress this because we think it important to be consistent in the teaching of literature and writing. The writer and the writer's world shape what is written. That's what we teach in composition classes as we guide students through the writing process. It is just as important in the teaching of reading and literature. Students should understand the significance of the author's culture and personal life in his or her writing. This understanding can be enhanced with a consideration of the literary genre and techniques the author has chosen to write about this experience and to tell his or her story.

To aid in this approach we have included model lessons as well as questions and exercises that direct the students towards expanding their reading of the writer and the work the writer has created. We have also raised questions that point the students toward a consideration of the commonalities among the cultures that make up the United States as well as their uniqueness. This approach is one of the more problematic in teaching literature today. Some would have us focus on the differences among cultures and the uniqueness of each; others would have us focus on the common elements that bind all cultures. In response, we do both.

Such a position is not just a compromise. It is a considered attempt to understand the tension that exists in the fabric of any nation, a tension that at times acts to bind it together. We have called this volume *Tapestry* in order to illustrate the tension. A tapestry is made up of threads of different colors and textures that make individual patterns woven into the base fabric. Taken as a whole, there is a unified picture, but the picture would not exist without the pieces. A tapestry can be examined for the contribution of the threads. It must also be seen as a combination that has its own unity.

So it is with the selections in this volume and the cultures from which they come. They have their individual strengths and their individual characteristics. At the same time, there are common elements that hold them together and allow us to see universal human characteristics across cultures as well as the features that unite these groups within some framework that exists in the United States.

The *Tapestry Teacher's Resource Manual* amplifies the approach and the view we have set forth. It contains a variety of ways to explore texts and writers and readers. It suggests ways of helping students use their imaginations to enter into the cultures they encounter through the selections, and it presents ways of assessing what the students have learned. What follows in this book should not be seen as a prescription, but as a set of suggestions from fellow teachers. You know your students best. Feel free to adapt what is in these pages to your specific situation. Above all, have an interesting time with your students in seeing what makes up the tapestry and what picture finally emerges from it.

SCOPE AND SEQUENCE

	Literary Skills	Critical Thinking Skills	Vocabulary Skills
UNIT 1: ORIGINS AND CEREMONIES SE pp. 2–5 TRM pp. 2–3	Identifying different uses of storytelling	Comparing and contrasting cultural ceremonies and rituals; synthesizing: selecting a cultural symbol for own culture; making connections among time line events	
THEME 1: "The Stars" SE pp. 7–14 TRM pp. 7–9	Identifying elements of a quest tale	Comparing and contrasting Tewa goals and values and own; applying lessons of story to own life	Scanning for key words and phrases; using context clues
"The Sheep of San Cristóbal" SE pp. 15–23 TRM pp. 10–12	Identifying elements of a moral tale	Drawing analogies among historical traditions represented in the tale; comparing and contrasting moral values in tale and own; summarizing key points of discussion	Scanning for religious words and phrases; using context clues
from *Things Fall Apart* SE pp. 24–28 TRM pp. 13–15	Recognizing elements of fantasy and realism in a folk tale	Analyzing communities and community values; applying story's messages to own life; evaluating use of fantasy in the story	Scanning for adjectives; using context clues
"In the Land of Small Dragon: A Vietnamese Folk Tale" SE pp. 29–35 TRM pp. 16–18	Identifying elements of narrative poetry; recognizing use of proverbs to convey values and beliefs; identifying and interpreting metaphors in the poem	Comparing and contrasting value placed on beauty and value placed on intelligence, role of father in story and own ideas about parental roles; applying knowledge of an old fairy tale to a similar tale from another culture; evaluating proverbs in terms of the Vietnamese culture's values and beliefs; summarizing underlying message of the tale	Comparing meanings of descriptive words; using context clues
"The Man Who Had No Story" SE pp. 38–42 TRM pp. 20–22	Evaluating the effects of repetition in a story	Drawing conclusions about the appeal of imaginary beings; comparing and contrasting movie characters and story characters	Defining unfamiliar words to understand theme of story; using context clues
THEME 2: from *Bless Me, Ultima* SE pp. 45–55 TRM pp. 26–28	Analyzing the elements of foreshadowing, dialogue, and sensory details	Comparing and contrasting main character's conflict and own, Antonio's father and Antonio's mother; applying feelings about story to own thoughts of exploring ancestry	Matching Spanish terms with their English meanings

SCOPE AND SEQUENCE

Communication Skills	Multicultural Skills	Cross-Curricular Skills	
Discussing traditional storytelling vs. modern media, personal symbols	Recognizing symbolism to denote beliefs and heritage	Social Studies/Music: African American folk tales and music—origins and importance	**UNIT 1: ORIGINS AND CEREMONIES** SE pp. 2–5 TRM pp. 2–3
Writing about beliefs and values; discussing symbolism	Appreciating storytelling traditions among Native Americans; values, customs, and beliefs of the Tewa	Geography: Importance of place for Native Americans	**THEME 1: "The Stars"** SE pp. 7–14 TRM pp. 7–9
Writing about life goals; discussing themes of wealth and penance	Noting the importance of religion in tradition of Spanish-speaking peoples	Social Studies: History of mestizos	**"The Sheep of San Cristóbal"** SE pp. 15–23 TRM pp. 10–12
Writing about and discussing good vs. evil; discussing community values	Becoming aware of different purposes of oral storytelling in African culture	Government: The necessity for laws	**from Things Fall Apart** SE pp. 24–28 TRM pp. 13–15
Discussing and writing about the source of beauty	Acknowledging the importance of storytelling in Vietnam and the insights it provides	Family Life: Familial roles in different cultures	**"In the Land of Small Dragon: A Vietnamese Folk Tale"** SE pp. 29–35 TRM pp. 16–18
Discussing imaginary beings; writing about community; listening for effects of repetition	Recognizing the importance of Irish storytelling to the culture	History: Conflict between Ireland and England	**"The Man Who Had No Story"** SE pp. 38–42 TRM pp. 20–22
Writing about family values as reflective of cultural heritage; writing about family goals for students	Identifying conflicts between traditional and new ways in Hispanic culture	Social Studies: History of rancheros and vaqueros	**THEME 2: from Bless Me, Ultima** SE pp. 45–55 TRM pp. 26–28

	Literary Skills	Critical Thinking Skills	Vocabulary Skills
from *Roots* SE pp. 56–61 TRM pp. 29–31	Distinguishing between fact and fiction	Comparing and contrasting ancient African ideas about justice and U. S. ideas about justice; summarizing main points of selection	Understanding emotional connotations of words; using context clues
"To Da-duh, In Memoriam" SE pp. 62–74 TRM pp. 32–34	Identifying external and internal conflict; analyzing point of view; identifying character and place	Comparing and contrasting New York City and Barbados; analyzing the effects of the main character's experiences on her perceptions, sources of conflict among characters	Scanning for descriptive words; using context clues
from *The Way to Rainy Mountain* SE pp. 75–82 TRM pp. 35–37	Identifying elements of fiction; analyzing round and flat characters	Classifying details about characters and places described in the selection	Classifying words as naming or descriptive; using context clues
"i yearn" ["We Who Carry the Endless Seasons"] SE pp. 83–85 TRM pp. 38–40	Evaluating the effect of different line lengths in free verse; analyzing poets' use of language and imagery	Analyzing levels of tolerance toward immigrants; summarizing poems; evaluating importance of poets' ancestral and personal history	Understanding connotations of the word immigrant; identifying words as Spanish or Filipino; using context clues
THEME 3: "Seventeen Syllables" SE pp. 88–98 TRM pp. 44–46	Identifying story elements; understanding haiku; analyzing story climax and cause and effect relationships within the story	Evaluating pros and cons of an arranged marriage; synthesizing: suggesting other examples of the effect of cultural background on growth	Scanning for words related to a specific occurrence; using context clues
from *Humaweepi, the Warrior Priest* SE pp. 99–106 TRM pp. 47–49	Identifying story elements; evaluating author's presentation of theme	Synthesizing: making generalizations about growth and change based on the selection; comparing and contrasting character's experiences with own	Scanning for color words; using context clues
"Cante Ishta—The Eye of the Heart" SE pp. 107–114 TRM pp. 50–52	Identifying elements of autobiography; analyzing author's use of personification	Analyzing the effect of events described by author; applying lessons learned from selection to own life	Understanding associations among key terms; using context clues
"Black Hair" "ALONE/ december/night" SE pp. 115–117 TRM pp. 53–55	Evaluating language and imagery of poems	Comparing and contrasting the poets' ways of expressing themselves; evaluating themes of growth and change in the poems	Understanding how unusual word associations are used to convey images and feelings

Communication Skills	Multicultural Skills	Cross-Curricular Skills	
Discussing concepts of justice	Developing an awareness of customs and traditions that are related to one African village's system of justice	Social Studies: Kingdoms of West Africa	**from *Roots*** SE pp. 56–61 TRM pp. 29–31
Writing about and discussing the search for ancestral roots; discussing conflict resolution in the story	Noting similarities and differences of African American society and black society of Barbados	Geography: Location and physical features of Barbados and New York City	**"To Da-duh, In Memoriam"** SE pp. 62–74 TRM pp. 32–34
Discussing memories and their impact on people; writing about what is most important in own lives now; discussing Native American relationship with nature; writing about and discussing personal symbols of heritage	Recognizing importance of the written word to Native Americans; some customs and values of Kiowa people	Social Studies: Impact on Native Americans of killing off buffalo Geography: Importance of place for Native Americans	**from *The Way to Rainy Mountain*** SE pp. 75–82 TRM pp. 35–37
Listening to poems; writing about and discussing treatment of immigrants	Understanding the experiences of immigrants as related to their roots; importance of ethnic neighborhoods	Geography: Locating the Philippines on a globe or map Social Studies: U.S. industry and government views on immigration	**"i yearn" ["We Who Carry the Endless Seasons"]** SE pp. 83–85 TRM pp. 38–40
Writing about and discussing conflicts between immigrants and their U.S.-born children; writing character sketches	Becoming aware of conflicts between Issei and Nisei; some Japanese traditions	Social Studies: Reasons for Japanese immigration	**THEME 3: "Seventeen Syllables"** SE pp. 88–98 TRM pp. 44–46
Discussing sources of continuity and stability in community of selection	Developing an awareness of some Native American rites of passage and traditions	Environmental Science: Survival in the wilderness	**from *Humaweepi, the Warrior Priest*** SE pp. 99–106 TRM pp. 47–49
Discussing ancestral roots as sources of conflict in families; writing about physical discomfort sometimes associated with rituals	Noting importance of cultural traditions, specifically some Native American traditions	Social Studies: Emergence and goals of AIM	**"Cante Ishta—The Eye of the Heart"** SE pp. 107–114 TRM pp. 50–52
Discussing economic opportunity in the United States; listening to the poems	Recognizing pride in cultural heritage despite discrim-ination; some experiences of Mexican Americans	Social Studies: Economic standing of Mexican Americans	**"Black Hair" "ALONE/ december/night"** SE pp. 115–117 TRM pp. 53–55

	Literary Skills	Critical Thinking Skills	Vocabulary Skills
UNIT 2: ARRIVAL AND SETTLEMENT SE pp. 120–123 TRM pp. 58–59	Discussing value of first-hand accounts as historical documents Understanding importance of origin stories of Native Americans	Analyzing 1891 quote on immigration; comparing and contrasting different cultures' views on arrival and settlement; synthesizing: predicting own reactions to being an immigrant, possible literary topics for unit themes	
THEME 4: "Wasichus in the Hills" SE pp. 125–130 TRM pp. 63–65	Analyzing an autobiography for fact and opinion; identifying and exploring author's use of similes; understanding personal narrative as a way of preserving culture	Synthesizing: expressing own viewpoint about compensation for land taken from Oglala Lakota; comparing and contrasting values of whites and Native Americans; applying personal views on relocation to reading of selection	Determining origins of people and place names; matching names with their significance
from "The Council of the Great Peace" SE pp. 131–137 TRM pp. 66–68	Identifying a primary source as a written version of an oral tradition; analyzing the document to find certain elements	Summarizing basic beliefs on which the United States is founded; comparing and contrasting different political beliefs and values; evaluating accuracy of selection	Scanning for words related to government; using context clues
"The Indians' Night Promises to Be Dark" SE pp. 138–143 TRM pp. 69–71	Understanding the elements of a speech; identifying simile and metaphor in the speech, as well as other powerful use of language and images	Applying knowledge of audience and translation difficulties to reading of speech; evaluating effectiveness of the speech; synthesizing: creating own questions to ask Chief Seattle, predicting outcomes of relocation to reservations	Identifying words and phrases that are opposites; using context clues
"They Are Taking the Holy Elements from Mother Earth" SE pp. 146–150 TRM pp. 73–75	Analyzing author's purpose for writing; identifying elements of an essay; recognizing use of cause and effect in an essay	Evaluating suitability of topics to essay form; applying knowledge of some Native American beliefs to draw conclusions about the impact of modern society on the Native American way of life	Scanning for words and phrases associated with strip mining; using context clues; identifying words that express anger
THEME 5: "The Drinking Gourd" "Steal Away" SE pp. 153–156 TRM pp. 79–81	Identifying repetition and refrains; evaluating effectiveness of refrains	Applying knowledge of conditions of slavery to interpret the spirituals; synthesizing: predicting tone of the spirituals; summarizing previous knowledge about slavery	Identifying repeated words and phrases and telling why they are important
"The Slave Auction" "The Slave Ship" SE pp. 157–163 TRM pp. 82–84	Identifying elements of autobiography; analyzing selections for use of factual recall and emotional interpretation	Synthesizing: making generalizations about feelings of people involved in slavery; analyzing quotations by Sojourner Truth and Frederick Douglass; summarizing emotions of people involved in slave trading	Scanning for words related to travel and slavery; understanding negative connotations of some words

Communication Skills	Multicultural Skills	Cross-Curricular Skills	
Discussing roles, needs, and struggles of newcomers; writing about family roots	Learning about experiences of diverse cultural groups upon arrival in the Americas	Social Studies: Europeans' arrival in the Americas, Africans' arrival in the Americas, U.S. laws governing immigration beginning in 1882, economic reasons for immigration	**UNIT 2: ARRIVAL AND SETTLEMENT** SE pp. 120–123 TRM pp. 58–59
Discussing whether or not Oglala Lakota should be compensated for land taken from them, impact of forced relocation; writing about the value of gold; writing and reading own similes	Recognizing visions as a part of Native American life; some Lakota traditions and experiences	Social Studies: Impact on Lakota of discovery of gold in the Black Hills Geography: Locating the Black Hills on a map	**THEME 4: "Wasichus in the Hills"** SE pp. 125–130 TRM pp. 63–65
Discussing importance of nature to Native Americans; writing and sharing orally beliefs on founding principles of U.S. government	Identifying some of the values and ideals of the Five Nations as revealed through the document	Social Studies: Government of the Iroquois	**from "The Council of the Great Peace"** SE pp. 131–137 TRM pp. 66–68
Listening to the selection; discussing Chief Seattle's decision	Comparing and contrasting Native American beliefs about nature, faith, and the afterlife and those of white settlers; some traditions, values, and beliefs of Pacific Northwest Native Americans	Social Studies: Political conflicts caused by opposing concepts of land Speech: Importance of remembering purpose and audience	**"The Indians' Night Promises to Be Dark"** SE pp. 138–143 TRM pp. 69–71
Discussing technology, the environment and Native American beliefs about the land; reading to "hear" the author's voice	Becoming aware of some Navajo beliefs and traditions; ability of modern Native Americans to continue their traditions	Environmental Science: Effects of strip mining on health of animals and people	**"They Are Taking the Holy Elements from Mother Earth"** SE pp. 146–150 TRM pp. 73–75
Listening to some spirituals; writing about slavery; discussing religion of enslaved African Americans	Recognizing importance of music, dance, and religion to enslaved African Americans; oral tradition of spirituals	Music: Social and political role of music in the culture of enslaved African Americans Social Studies: History of slavery worldwide	**THEME 5: "The Drinking Gourd" "Steal Away"** SE pp. 153–156 TRM pp. 79–81
Writing and role-playing about emotional aspect of slavery; discussing effectiveness of autobiography as source of information	Appreciating diversity of cultures represented by enslaved African Americans; how enslaved African Americans maintained cultural identity in the face of oppression	Social Studies: How Africans became enslaved; the Middle Passage; the abolitionist movement	**"The Slave Auction" "The Slave Ship"** SE pp. 157–163 TRM pp. 82–84

	Literary Skills	Critical Thinking Skills	Vocabulary Skills
"The Slave who Dared to Feel like a Man." SE pp. 164–171 TRM pp. 85–87	Identifying elements of personal narrative; appreciating the valuable role of slave narratives in American literature	Comparing and contrasting attitudes and actions of different family members in the selection, first-hand accounts and researched historical presentations; sequencing events from the selection	Scanning for words related to slavery and freedom; using context clues
"The People Could Fly" "Runagate Runagate" SE pp. 172–179 TRM pp. 88–90	Identifying elements of a folk tale, including themes, key events, and turning points in chronology of the folk tale; analyzing realistic and fantastic elements in the folk tale	Comparing and contrasting people's responses to oppression; synthesizing: suggesting characteristics that make survival of extreme hardship possible	Scanning for words related to escape and flight; using context clues
THEME 6: "Lali" SE pp. 182–189 TRM pp. 94–96	Identifying setting and flashback in the selection; analyzing the main character's conflict	Comparing and contrasting rural and urban life; analyzing consequences of Lali's move, author's use of flashback	Scanning for words that describe urban and rural settings; using context clues
from Picture Bride SE pp. 190–198 TRM pp. 97–99	Evaluating development of character in the selection	Comparing and contrasting different characters; synthesizing: forming own opinion about prearranged marriage; analyzing characters in terms of their wishes for Hana	Scanning for words important to characterization; using context clues
"I Leave South Africa" SE pp. 199–206 TRM pp. 100–102	Identifying use of irony in the selection; analyzing character's inner and outer conflicts	Synthesizing: forming own opinions about integration; comparing and contrasting different characters' backgrounds and views, main character's expectations and the realities he encounters	Scanning for words that indicate characters' views about integration; matching antonyms
"Ellis Island" SE pp. 207–209 TRM pp. 103–105	Identifying figurative language used in the poem; analyzing conflicts expressed in the poem	Evaluating meaning of immigration trends; comparing and contrasting two perspectives on immigration	Classifying words as naming or describing geographical places; using vocabulary words to complete a paragraph about Ellis Island
"Those Who Don't" "No Speak English" "The Three Sisters" SE pp. 212–217 TRM pp. 107–109	Identifying and evaluating the tone of each selection	Comparing and contrasting author's views on neighborhoods and own views; evaluating bias in people's perceptions of neighborhoods	Referring to footnotes to learn meanings of Spanish terms; matching Spanish words with English meanings; scanning for words that help to set tone

Communication Skills	Multicultural Skills	Cross-Curricular Skills	
Discussing use of violence to escape slavery; responding to oral reading of the selection; discussing impact of slavery on enslaved African Americans' families	Becoming aware of the struggle of enslaved African Americans to learn to read and write, and to escape enslavement	Social Studies: Fugitive Slave Act; the abolitionist movement	**"The Slave who Dared to Feel like a Man."** SE pp. 164–171 TRM pp. 85–87
Listening to and discussing quote by John C. Calhoun; listening to an oral reading of the poem; writing a poem on the theme of escape and freedom; discussing humans' ability to survive extreme hardship	Recognizing conditions for enslaved African Americans on Southern plantations; role of African folk tales in the culture of enslaved African Americans	Social Studies: Language and literary barriers confronting enslaved African Americans	**"The People Could Fly" "Runagate Runagate"** SE pp. 172–179 TRM pp. 88–90
Discussing sources of discrimination against newcomers, economic hardships suffered by immigrants	Identifying some Puerto Rican values and traditions; differences between generations of mainland Puerto Ricans	Social Studies: Brief history of Puerto Rico Government: Politics involved in immigration Family Life: Dynamics of main characters' marriage	**THEME 6: "Lali"** SE pp. 182–189 TRM pp. 94–96
Listening to selection excerpts; discussing feelings on prearranged marriage; writing about and discussing the different characters	Learning about some Japanese traditions; the experiences of Japanese Americans during the immigration process and as newcomers	Social Studies: Internment camps during World War II Family Life: What makes a marriage solid	**from *Picture Bride*** SE pp. 190–198 TRM pp. 97–99
Discussing issue of integration, Mark's father's beliefs	Developing an awareness of differing views among African Americans regarding integration, the Black Muslim movement	Social Studies: Apartheid in South Africa, integration in the United States	**"I Leave South Africa"** SE pp. 199–206 TRM pp. 100–102
Listening to an oral reading of the poem; discussing poem's images, symbolism of Ellis Island	Noting different beliefs on the importance of land held by Native Americans and by European immigrants	Social Studies: Handling of immigrants through Ellis Island	**"Ellis Island"** SE pp. 207–209 TRM pp. 103–105
Discussing neighborhoods, tone of selections	Recognizing some Mexican American traditions; bias and stereotypes regarding Mexican Americans; importance of ethnic neighborhoods	Social Studies: Trends in Hispanic immigration, brief history of Mexican control over present-day United States	**"Those Who Don't" "No Speak English" "The Three Sisters"** SE pp. 212–217 TRM pp. 107–109

	Literary Skills	Critical Thinking Skills	Vocabulary Skills
UNIT 3: STRUGGLE AND RECOGNITION SE pp. 220–223 TRM pp. 112–113		Synthesizing: making generalizations about teenage social groups; summarizing knowledge about cultural groups; evaluating U.S. society's efforts to achieve cultural equality from a pessimistic or optimistic viewpoint	
THEME 7: "I See the Promised Land" SE pp. 225–231 TRM pp. 117–119	Identifying literary elements in the speech; analyzing allusions in the speech	Synthesizing: suggesting alternatives to violence, creating personal definitions of freedom; evaluating the level of King's faith in the United States; comparing and contrasting violent and nonviolent means of protest	Scanning for words related to freedom and equality; using context clues
"See for yourself, listen for yourself, think for yourself" SE pp. 232–237 TRM pp. 120–122	Identifying elements of effective persuasive speeches	Comparing and contrasting Malcolm X's views and U.S. policies on integration, own beliefs on basic human rights	Scanning for words related to nonviolence and violence and analyzing their emotional appeal; using context clues
from *China Men* "Immigration Blues" SE pp. 238–246 TRM pp. 123–125	Analyzing author's development of character; identifying examples of indirect characterization	Synthesizing: making generalizations about why Chinese laborers continued working under harsh conditions; comparing and contrasting direct and indirect characterization	Identifying words as descriptive of harsh labor; using context clues
"Napa, California" "The Circuit" SE pp. 247–257 TRM pp. 126–128	Identifying importance of setting in the selections	Comparing and contrasting characters' lives and values and own; synthesizing: predicting how constant moves would affect one's life, how it would feel to work as the characters described in the selections	Identifying words as important to describing setting; using context clues
"Speech of Sojourner Truth" SE pp. 260–263 TRM pp. 130–132	Analyzing author's tone; identifying elements that make the speech effective	Analyzing reasons for discrimination against women in the work place; comparing and contrasting women's concerns in the past and present, women's rights movement and civil rights movement; applying knowledge of 19th-century United States to reading of selection	Scanning for words and phrases with religious connotations; matching religious names with the meanings they are intended to convey

Communication Skills	Multicultural Skills	Cross-Curricular Skills	
Discussing teenagers' particular prejudices, ways of making society more accepting of diversity; orally presenting viewpoints on U.S. society's ability in achieving cultural equality	Appreciating the multitude of cultural groups within the United States; the struggles of cultural groups in the face of prejudice and discrimination	Social Studies: History of exclusionary groups in the United States, efforts to achieve cultural equality	**UNIT 3: STRUGGLE AND RECOGNITION** SE pp. 220–223 TRM pp. 112–113
Discussing violence and its alternatives, effectiveness of King's use of biblical allusions	Developing an awareness of African Americans' struggle for equal rights	Social Studies: Mahatma Gandhi's beliefs, key events of the Civil Rights Movement, importance of Dr. Martin Luther King, Jr., in the Civil Rights Movement	**THEME 7: "I See the Promised Land"** SE pp. 225–231 TRM pp. 117–119
Listening to and discussing a quote by Malcolm X's teacher; writing about human beings rights; discussing what makes a speaker effective	Learning about African American movements for equality	Social Studies: The Black Muslim movement	**"See for yourself, listen for yourself, think for yourself"** SE pp. 232–237 TRM pp. 120–122
Discussing ethnocentrism, types of jobs available to immigrants; listening to oral readings of poems; writing characterization of self	Recognizing hardships experienced by Chinese immigrants and some cultural traditions and beliefs that helped them to endure	Government: Effect of discriminatory laws on Chinese immigrants Social Studies: Typical jobs taken by Chinese immigrants during the late 1800s and early 1900s	**from *China Men* "Immigration Blues"** SE pp. 238–246 TRM pp. 123–125
Discussing importance of unions for Mexican American workers, impact of migrant lifestyle, sources of hope; writing a description of own home and discussing importance of setting	Recognizing hardships experienced by Mexican American migrant farm workers; importance of preserving traditional Mexican American values in order to cope with hardships	Social Studies: The bracero program, the UFW Union and importance of César Chávez	**"Napa, California" "The Circuit"** SE pp. 247–257 TRM pp. 126–128
Discussing disparity of compensation and executive positions for working women and men, human rights; writing rhetorical questions	Noting efforts on the part of abolitionists to liberate enslaved African Americans; universal elements of struggles for equal rights	Social Studies: The abolitionist movement and corresponding women's rights movement	**"Speech of Sojourner Truth"** SE pp. 260–263 TRM pp. 130–132

	Literary Skills	Critical Thinking Skills	Vocabulary Skills
from I am Joaquín SE pp. 264–269 TRM pp. 133–135	Evaluating effectiveness of repetition and parallelism in the poem	Synthesizing: expressing own beliefs about U.S. opportunities for success; comparing and contrasting author's view of his people and the images received through U.S. media; analyzing importance of author's struggles and roots to him	Identifying verbs in the poem and discussing their emotional appeal; discussing meanings of Spanish names in the poem; matching Spanish names to the people they describe
"Free to Go" SE pp. 269–275 TRM pp. 136–138	Analyzing irony in the selection	Evaluating bias against Japanese Americans during World War II; synthesizing: developing own opinion about prejudice toward Japanese Americans; analyzing impact of prejudice on family in selection	Discussing vocabulary words related to concepts of prejudice and discrimination; using context clues
"In the American Storm" SE pp. 276–282 TRM pp. 139–141	Analyzing the author's tone	Comparing and contrasting views on desirability of immigration, hardships and benefits resulting from immigration; analyzing main character's thoughts	Scanning for and classifying words related to labor; using context clues
THEME 8: **from Their Eyes Were Watching God** SE pp. 285–288 TRM pp. 145–147	Identifying and appreciating the author's use of dialect; analyzing author's use of dialogue in developing character	Analyzing author's use of language as means of conveying cultural background; synthesizing: determining special word usages for conveying own heritage; comparing and contrasting Janie's experiences and own	Scanning for unfamiliar word usage in dialogue; defining dialect examples from the selection
"Refugee Ship" "address" "Letter to America" SE pp. 289–293 TRM pp. 148–150	Identifying similes in a poem; analyzing the mood of the poems and the poets' techniques for creating mood	Analyzing cultural make-up of own community; synthesizing: forming own opinions about hiring practices; comparing and contrasting the different poems	Scanning for words revealing cultural differences; using context clues
from Black Boy SE pp. 294–299 TRM pp. 151–153	Identifying the elements of autobiography	Analyzing the author's attitudes as revealed in the autobiography, existence of segregation in U.S. society; synthesizing: predicting effects of cultural discrimination on young children; comparing and contrasting author's experiences and own; evaluating autobiography as source of knowledge	Categorizing words as descriptive of characters' attitudes; using context clues

Communication Skills	Multicultural Skills	Cross-Curricular Skills	
Discussing role models and images of success in U.S. society, racial stereotypes in the media	Recognizing Mexican American history as a unique combination of influences; value placed by members of Chicano movement on tradition and heritage	Social Studies: The Chicano movement, changing demographics of Mexican American populace	**from *I am Joaquín*** SE pp. 264–269 TRM pp. 133–135
Listening to an oral reading of the selection; discussing examples of prejudicial actions, reasons for father's actions in the selection; writing ironic sentences	Developing an awareness of prejudice and discrimination faced by Japanese Americans	Social Studies: Internment of Japanese Americans	**"Free to Go"** SE pp. 269–275 TRM pp. 136–138
Listening to and discussing the inscription engraved on the Statue of Liberty; discussing realism of view that the United States is a land of opportunity	Learning about hardships experienced by European immigrants in the early 1900s	Social Studies: Patterns of immigration	**"In the American Storm"** SE pp. 276–282 TRM pp. 139–141
Discussing a James Baldwin quotation; listening to an oral reading of the selection	Developing an awareness of the complexity and richness of Southern African American society	Social Studies: The sharecropping system Art, Music, and Drama: The Harlem Renaissance	**THEME 8: from *Their Eyes Were Watching God*** SE pp. 285–288 TRM pp. 145–147
Discussing hiring practices as related to cultural discrimination; writing poems about cultural differences	Becoming aware of hardships and isolation experienced by Mexican Americans; ethnic pride; intermingling of different cultures forming the Mexican American heritage	Social Studies: The Chicano Movement Drama: El Teatro Campessino	**"Refugee Ship" "address" "Letter to America"** SE pp. 289–293 TRM pp. 148–150
Writing a remembrance of becoming aware of own racial identity	Developing an awareness of prejudice encountered by African Americans through reading fiction	Social Studies: Actions of the NAACP	**from *Black Boy*** SE pp. 294–299 TRM pp. 151–153

	Literary Skills	Critical Thinking Skills	Vocabulary Skills
from "Four Directions" "Saying Yes" SE pp. 300–312 TRM pp. 154–156	Identifying the elements of plot; analyzing author's use of conflict to develop characters	Synthesizing: predicting character's traits; comparing and contrasting views on cultural heritage	Understanding importance of word choice in developing character; using context clues
from Stelmark: A Family Recollection SE pp. 313–320 TRM pp. 157–159	Analyzing author's presentation of main ideas; discriminating between main ideas and supporting details	Comparing and contrasting characters; synthesizing: predicting outcomes; evaluating different ways of supporting main ideas	Scanning for names of foods; matching words and definitions
from Fences SE pp. 321–330 TRM pp. 160–162	Analyzing playwright's use of dialogue to develop character	Comparing and contrasting parents' and own views about the future; analyzing characters' attitudes; evaluating importance and quality of dialogue in movies	Scanning for sports-related words and examples of dialect; using vocabulary words to complete a selection summary; using sports terms in non-sports-related ways
THEME 9: "America: The Multinational Society" SE pp. 333–339 TRM pp. 166–168	Identifying author's use of cause and effect in the essay; appreciating the author's distinctive personal style	Applying own educational experiences to understand the author's points; comparing and contrasting author's views and own; evaluating bias detailed by author	Matching words with cultures of origin; using context clues
"For My People" "Let America Be America Again" SE pp. 340–347 TRM pp. 169–171	Analyzing poets' use of cataloging	Analyzing poll results about discrimination in the work place, poets' perspectives on how to achieve racial equality; evaluating U.S. progress in achieving racial equality, universal appeal of poems	Identifying words with harsh and pleasant connotations; using context clues
"Petey and Yotsee and Mario" SE pp. 350–355 TRM pp. 173–175	Evaluating effectiveness of first-person point of view in the story; identifying different points of view; describing elements of the short story	Synthesizing: predicting personal reactions to being the "outsider"; summarizing own cultural identity; evaluating appropriateness of different plots to short story genre	Scanning selection for context clues
"AmeRícan" "Ending Poem" "Instructions for joining a new society" SE pp. 356–362 TRM pp. 176–178	Identifying poets' use of alliteration; discriminating between initial and internal alliteration	Analyzing causes of changes in the barrios; comparing and contrasting poets' cultural identities; evaluating use of alliteration in popular music	Recognizing words that reveal the cultural blending in the United States; explaining why the poets invented some new words
from Seven Arrows SE pp. 363–367 TRM pp. 179–181	Analyzing author's use of symbolism in the allegory	Defining success, status, happiness, and commitment to community; evaluating applicability of Native American beliefs in today's world	Scanning for animal or nature references; explaining significance of terms used in the selection

Communication Skills	Multicultural Skills	Cross-Curricular Skills	
Writing about and discussing family's response to marrying outside of one's culture; listening to poem; discussing generational differences	Recognizing intergenerational conflicts among Chinese Americans; some Chinese traditions;	Social Studies: Patterns of Chinese immigration	**from "Four Directions" "Saying Yes"** SE pp. 300–312 TRM pp. 154–156
Discussing fast food in the United States, main ideas of favorite stories; writing lists of food names and food descriptions that invoke personal cultural heritage,	Noting importance of ethnic neighborhoods for immigrants; some Greek traditions and values	Social Studies: Emergence of cultural pluralism Home Economics: Foods as a link to cultural heritage	**from *Stelmark: A Family Recollection*** SE pp. 313–320 TRM pp. 157–159
Listening to an oral reading of the play; discussing importance of sports in U.S. society, different perspectives of generations	Reading about the African American experience in the 20th century; sports as a means for diverse cultural groups to achieve "success"	Physical Education: Professional African American athletes in the United States	**from *Fences*** SE pp. 321–330 TRM pp. 160–162
Discussing importance of understanding cause-and-effect relationships	Recognizing Western European bias in U.S. society; common bond of humanity among all cultures; examples of multiculturalism	Social Studies: Debate in U.S. education over Eurocentrism	**THEME 9: "America: The Multinational Society"** SE pp. 333–339 TRM pp. 166–168
Discussing affirmative action programs; writing descriptions of poets' lists	Celebrating humanity's great diversity of people, beliefs, and cultures; African Americans' status as we approach the year 2000	Social Studies: Affirmative action programs	**"For My People" "Let America Be America Again"** SE pp. 340–347 TRM pp. 169–171
Writing a paragraph about cultural identity; discussing level of ease in different neighborhoods, types of plots particularly suited to short stories	Recognizing struggles of Eastern European Jewish immigrants; some Jewish traditions; the sense of community in ethnically diverse neighborhoods	Social Studies: History of persecution of Jews worldwide	**"Petey and Yotsee and Mario"** SE pp. 350–355 TRM pp. 173–175
Listening to a quotation by a New York City police officer regarding an East Harlem barrio; writing a paragraph on how to break down communication and cultural barriers; discussing use of alliteration in popular music	Appreciating the mixing of diverse cultures with mainstream U.S. culture; some cultural traditions and values of Spanish-speaking immigrants from Caribbean Sea islands	Social Studies: Patterns of Puerto Rican immigration, the Castro regime in Cuba	**"AmeRícan" "Ending Poem" "Instructions for joining a new society"** SE pp. 356–362 TRM pp. 176–178
Discussing materialism in U.S. society, importance of animals to Native Americans, community service as a requirement for graduation	Recognizing oral tradition among Native Americans; the importance of the Medicine Wheel in bringing together all people	Social Studies: Changes experienced by Native Americans over the past three centuries	**from *Seven Arrows*** SE pp. 363–367 TRM pp. 179–181

A Multicultural View of Education

A multicultural view of education approaches classroom curriculum in such a way that the diversity of the world's and the United States's populations is celebrated. The approach recognizes the fact that, although each person is unique, every person has a sense of belonging to a group with its own particular history and its own sets of customs, ceremonies, and beliefs. The group may be identified by ancestry, by religion, by its place of origin on the globe, by common interests and customs, or by some combination of these. The group has its own ways of communicating among its members, both through the language that is used and the physical expressions that surround that language.

To many, multiculturalism in the United States acknowledges that the diverse groups making up this country have come here from different areas and countries, under different conditions, and had different experiences upon their arrival. Many of them are cut off from their ancestral homeland yet continue to feel an emotional attachment to it. Many of them have sought to forge a new identity.

The various attempts by many multiculturalists to name these groups in appropriate ways represents the growing awareness of and respect for the complexity of groups' identities. For example, we used to speak of Italian-Americans or Afro-Cuban-Americans. Although these terms were effective in acknowledging the roots of different groups, the hyphen suggested a second-class citizenship, as if some group exists that is "American American." We now use a term like "American of Asian descent, or Asian American without the hyphen," but even these labels have problems. Some people think of themselves simultaneously as members of several cultures; we don't have a good term that reflects that fact.

As another example, the peoples who inhabited the land before the coming of Columbus were named Indians by error. Recently, they have been called Native Americans, but many of those whose ancestors came here three centuries ago also think of themselves as native Americans with a lowercase *n*. The currently popular term *First Nations* is acceptable to many people, yet it possibly suggests that the people are not "American." Some prefer the title American Indian.

As these examples show, categorizing people is tricky and can lead to hurt as well as pride. Currently no answer to this dilemma exists, but the fact that labeling is recognized as an issue shows the level of concern that characterizes multiculturalism.

Multiculturalism addresses other tough issues. It does not cover up the problems and issues of a society that is made up of peoples who have come here from nearly every area of the world. Rather, the approach has as its main aim the fostering of an understanding of the problems that develop when human beings move from place to place. In particular, multiculturalism aims to stimulate the students' awareness and imagination, to help them sense what it is like to think and feel and believe as someone else might. It is an approach that helps us to understand others and to recognize and value the otherness, yet sameness, of people.

What is a multicultural approach to literature?

A multicultural approach to literature challenges the view of literature instruction that has been in place in the United States for the past 60 years. This traditional view starts from the belief that the literary text is detached from its author and its culture. In contrast, the multicultural approach begins with this premise: Literary works come from writers who inhabit cultural contexts, and these contexts shape their writing.

Viewed this way, every text is both an individual aesthetic object and a cultural document, a part of the legacy of an individual and of a group. To understand a text, then, the reader must understand—as fully as possible—the writer and the writer's culture. Experiencing the text as the writer intended it requires that it be read in the light of what one critic called the "spirit of the age" or of the culture. Such reading affirms in principle the distinction between the "meaning" of a text, that which refers to the text in cultural and authorial context, and the "significance" of the text, that which we, distinct from that culture, make of it.

How does this approach affect the student?

A multicultural approach starts from the premise that, like other pieces of art, literary texts have creators who inhabit and help to create and also perpetuate the various cultures of the world. It also assumes that each student is also a member of a culture, sharing or rebelling against beliefs, ideals, prejudices, and perspectives that are shaped by that culture. This benefits the student both by broadening his or her world of knowledge and by emphasizing the significance of the student's interaction with the text. Students see that each text comes from a context that is rich and complex; the text can be understood best within that context. As students read the text, they build upon that context and use it to help them read.

When this approach is used, students are not asked to rely on their own wits to come up with a clever interpretation of a text. Rather, they are given information to help them become more fully aware of the writer and his or her world. They can use this information to build a context for the reading that makes sense. Individual interpretation occurs as the students view and interpret this context based on their own world. In this way students are helped to make connections among texts, the cultures of the writers, and their own cultures. In doing so, they both build their personal canons and examine the mosaic of cultures that constitutes our world.

Similarly, a multicultural approach to literature does not focus solely on the students' immediate, emotional responses to the texts. Instead of always asking, "How did the story make you feel?" or "What is the hidden meaning of the poem?" the multicultural approach adds questions about *where* and *when* the text originates and from *whom*. For example, when students read a poem, they should find out something about the poet and ask themselves how that knowledge brings them a newer or deeper understanding of the text. The course focus is not "Reading Literature;" rather, it is "Reading Writers." The gender and ethnicity of the writer are always examined in terms of how they may affect the writer's and the reader's perspectives on the subject. The reader's gender and ethnicity must also be considered, for as the conditions of the writers' lives affect what and how they write, so do the conditions of the readers' own lives affect how they read and respond.

How does a multicultural approach affect the curriculum?

To answer this question, let us look specifically at *this* text as a curriculum. The *Tapestry* Student Edition is a collection of U.S. literature, much of it written by people whose voices rarely have been heard in schools and colleges until recently. The text celebrates American writers whose roots differ from those of the established English American or European groups who brought with them the traditions of Greece, Rome, and the Western European Renaissance. Most of the writers featured in this volume are from traditions primarily other than these—traditions from the First Nations and from the African, Hispanic, Eastern European, and Asian cultures. The curriculum recognizes that these cultures helped build this country and its arts and literature into what they are today. A truly multicultural reading of these texts demands that you and your students read the works with the underlying purpose of celebrating the writers and the cultures they represent.

What is important in a multicultural curriculum is to provide a broad variety of texts from around the United States without creating a myopic view of even our own country. The view of this curriculum holds that all works of literature are to be held as equally valid, and that it is not the role of literary criticism to rank them. In one of his last writings, the critic Northrop Frye defined literature:

. . .*where the organizing principles are myth, that is, story or narrative, and metaphor, that is, figured language. Here we are in a completely liberal world, the world of the free movement of the spirit. If we read a story there is no pressure to believe in it or act upon it; if we encounter metaphors in poetry, we need not worry about their factual absurdity. Literature incorporates our ideological concerns, but it devotes itself mainly to the primary ones, in both physical and spiritual forms: its fictions show human beings in the primary throes of surviving, loving, prospering, and fighting with the frustrations that block those things. It is at once a world of relaxation, where even the most terrible tragedies are still called plays, and a world of far greater intensity than ordinary life affords. In short it does everything that can be done for people except transform them. It creates a world that the spirit can live in, but it does not make us spiritual beings.*
(Frye, 1991, p. 16)

In taking such a definition, we see that an individual work is a part of the totality of myth; at the same time it is situated in the world from which it came. This *double vision* is the focus of the *Tapestry* curriculum.

In transforming this double vision to the curriculum, we must recognize the pressures of the world in which we and our students reside. For example, to ensure variety in regard to level of difficulty, time periods represented, and emotional appeal, we must exercise some restrictions during the selection process. Similarly, many voices and genres must be represented, which means we are confronted with the problem of how to group the selections.

Typically, selections are grouped by culture, by chronology, or by genre. However, we, the general editor and editors at Globe, did not choose these organizations. We believe that although cultures differ, overarching themes—what Frye calls myths—hold them together. These are themes that define the culture as a culture, that record the struggle of a culture to become recognized by the mainstream society, and that illustrate attempts to overcome prejudice, ignorance, and fear. Such thematic groupings help students learn to look at texts from different perspectives.

First, in *Tapestry* each selection can be viewed as an individual piece of the larger universe of myth and metaphor. Each selection also can be viewed in terms of authorship, as being written by a person who has a real life and a real background. Each selection also must be viewed in terms of how it expresses the author's culture—whether it be a mainstream culture or not, whether it be a culture of ethnicity, gender, political or social position, or physical differences. For example, with a thematic approach, we can read Margaret Walker as an American, African American, woman, professor, and outspoken supporter of the social liberation of African American women. Knowing these facets of the poet—instead of knowing only one identifying trait such as gender—helps the reader understand the voice of "For My People."

We, at Globe, realize that the reader also must recognize the genre of a selection. The reader must understand how the elements of literature can be manipulated by writers to make different genres as effective as possible in telling their stories. Therefore, in *Tapestry* we include both literary skill instruction and genre background within our themes. For example, Margaret Walker's poem "For My People" is identified as free verse. We discuss the calculated use of many different techniques to achieve structure in free verse, suggesting that one of these techniques, cataloging, is used very effectively in Margaret Walker's poem. Cataloging helps the poet to substantiate the claims she makes. Walker uses another technique, repetition, to emphasize her theme and call attention to her main points. If students did not have this information, they might miss important aspects of the poem.

How does a multicultural approach affect teaching?

In order to celebrate the cultures of our world, we must go beyond having courses or units on these cultures in which the texts are treated as if the writer is nonexistent and as if the reader can look only at his or her personal response to the text. We must not expect the naive reader to understand cultural differences if we continue to treat all texts as contemporary, genderless, and mainstream.

To look at texts as the works of human beings who have a past and a culture is to see literature, ourselves, and our cultural whole. We must acknowledge that texts build upon other texts and that text emerges from and reshapes a culture or a subculture. To adopt such a view as part of our teaching strategy may help our students become serious, appreciative readers of texts springing from many different cultures.

Specifically, we must call our students' attention not just to the words on the page—as if they are permanent and hold "the truth"—nor to the students' immediate, emotional responses to the texts. Both are important, but a multicultural approach must also raise certain questions about each text—where it comes from and particularly from whom it comes. This means that references must be made to the world of the writer and the culture from which the writer and the text have come. Students should be encouraged to use outside information as they read. They should be instructed to research background information on the author, to read some of the author's other works, and to consider the beliefs, assumptions, and practices of the culture from which the author comes.

Students also should be encouraged to explore their own cultures and the culture of their community. For example, they should explore what it means to be who they are—what does it mean to be a Native American teenage girl whose great or even great-great grandparents were sent to live on a reservation, or an urban teenage boy whose parents came from the Philippines and speak no English? Students need to spend time thinking about their own roots, their cultural allegiances, and those of their families, for these personal facts are also to be celebrated. One activity you should have your students pursue as they read through the various themes is to interview their parents and older relatives about the issues. For example, what do family members consider to be their origins and ceremonies? What do they remember or recall hearing about their family's experiences arriving and settling in the United States? The information they learn from such questions is not to be shared with the class; it is personal information that will enrich each student's sense of self.

In short, the most important part of teaching a multicultural curriculum might very well be the continuing role of opening wider and wider the door to the world. The *Tapestry* program will help you to keep students busy exploring themselves, for with self-recognition comes the ability to recognize others. In addition, it will help you to keep students always aware of the whole world outside of themselves. It will help you lead them to as many resources as you can to help them remain curious and excited about their explorations. *Tapestry* asks many open-ended questions. Remember to accept and value all responses, and encourage sharing in an atmosphere that emphasizes respect for all.

How does a multicultural curriculum affect the teacher?

To teach a multicultural program, you have to admit that you are a person of a certain gender and age, from a certain heritage, and with certain training that affects your own reading of the texts. You as a reader and a teacher have a cultural heritage that gives you a unique perspective, too. However, since you have but one perspective, yours (along with everyone else's in the world) cannot be viewed as the one correct perspective. With this realization come some difficult admissions: teachers usually don't know everything; teachers may have prejudices; you yourself are a reader from one culture who is reading texts from another culture.

For the same reasons, the *Tapestry Teacher's Resource Manual* cannot be viewed as an authoritative document. We don't know your culture or the culture of your students, but we hope to help instill in you a true sense of exploration.

Certainly we are not suggesting that you abandon the role of teacher and become just another of your students. Truly, your role of teacher and guide may be more important with this curriculum than with any other. You know some things the students do not know and you can help them. You know about literature. You have read more, and you have more knowledge of the diversity of the world. You are older than your students and have experienced much more. You are also a professional who knows how to elicit informed responses from students, to challenge them, to set them on projects that they will find interesting and particularly challenging. You know what questions to ask more than you know what the answers are. With all of these important functions, we will try to help you in the lessons in this Teacher's Manual.

How does multicultural education affect the classroom?

As you implement a multicultural curriculum, you can view your class as a microcosm of the world. How is your class a small culture? What are the culture's ceremonies and rituals? What are its rites of passage? Who are the leaders, and who are the subordinate ones? Does your class have a single culture, or are there several?

How the class begins to view itself as the result of multicultural instruction should represent the students' understanding of themselves as being unique even as they are part of the class group. Class members may bond together more tightly as being members of the class culture while also taking pride in individual differences. Class activities will change to reflect this shift in identity; there will be more talk—in pairs and small groups as well as in whole class discussion. There will be sharing. Learning is a joint venture, and students learn from one another. The class will be involved in team projects, group papers, and other forms of cooperative learning. After all, a culture is a form of cooperation.

Students also will have a greater comfort level when engaging in activities based on role-playing, simulation gaming, and dramatization. These are among the best ways to explore imaginatively the other cultures of our society, to try to understand their ideas, beliefs, and values.

A multicultural classroom depends also on frequent writing. Writing is a good way to explore a culture and other roles. The writing will not take the form of critical formal essays, but of response journals, imitations, and poems and stories that are composed in response to those that are read. Some of the writing will be extensive. Some will be reports on library searches and interviews of parents and other "elders" of the community.

Finally, there will be reflection, *much* reflection. Students should think about themselves and their values as they think about the values expressed in what they are reading. They should think about their world and their society. This is, after all, what multiculturalism is all about.

Multicultural Applications in the Features and Organization of Tapestry

Tapestry has been designed in such a way that the diversity of the U.S. population is represented in the literature and in the authorship. The literature and the Student Edition materials reflect the notion that each person has a unique set of customs and beliefs based on ancestry; however, the commonalities in the experiences of all peoples and groups becomes evident through the thematic organization.

The literature in *Tapestry*

- reflects the diverse and rich cultural mix in the U.S. population;
- represents the typically unheard voices in U.S. literature;
- is organized into three major thematic units and nine subthemes;
- illustrates multiple perspectives on issues of major importance to the cultures it represents;
- tells the stories of the movement of peoples from place to place.

The apparatus in the Student Edition of *Tapestry*

- Unit Opener pages provide a historical background including time lines that provide a context with which to read the literature and encourage students to think across discipline areas;
- Prereading pages provide extensive background information on the author's culture and the conditions under which the author lived;

- Motivation questions emphasize an awareness of where, when, and from whom the text originates;
- Postreading pages provide questions that broaden a student's world of knowledge and ask the student to look at the text through his or her own set of cultural beliefs and traditions;
- Journal writing activities provide numerous opportunities for students to respond to the text;
- Author's Craft emphasizes literary elements used by writers to develop the genres and to tell their stories;
- Unit Closure pages highlight writing an essay using the stages of the writing process, model cooperative/collaborative discussion of the literature in the unit, and encourage problem solving.

The *Tapestry* Student Edition special feature

- "Model Lesson" illustrates with sidenotes one student's critical reading of the text (a critical reading involves an awareness of the author's culture and the time period the piece was written, the author's perspective on the theme, elements of the genre used by the author, and the student's understanding of and responses to the text);
- "Making Connections" highlights the cultures, traditions, and heritages of the five groups represented in the text;
- "Arts of the Tapestry" features 16 pages of four-color fine art by artists from the different cultural groups that give different perspectives on the themes and encourages students to think across curriculum areas;
- "Student Handbook" contains a Listing of Literary Terms with definitions and a Writing Process section that outlines the stages in the writing process as well as student essays that model the three writing process activities on the Unit Closure pages.

The *Tapestry* Teacher's Resource Manual

- Alternate Tables of Contents assist teachers in presenting the literature in ways that best meet their individual curriculum needs — by theme, by cultural group, by genre, or by chronology of the time period during which the selection took place.
- Staff Development provides background information about the fundamentals of a multicultural curriculum; specific information about how *Tapestry* implements a multicultural curriculum; assessment tips to help teachers with

both formal and informal assessment; and model lessons by teachers who are currently engaged in teaching multicultural literature classrooms.

- ESL/LEP activities, designed and written as an integral part of the lesson plans, are appropriate for any student who may experience difficulty with concepts that require knowledge of varying cultural, regional, or experiential elements.

- Unit Introductions and theme Preview lessons include an interdisciplinary approach that outlines plans for teachers from many different subject areas to work together to present the viewpoints in *Tapestry*.

Assessment Tips

In a multicultural curriculum, there are frequent opportunities to gather evidence for assessment. The instruments for gathering evidence are varied: quizzes, papers, exams, records of reading, tapes or videotapes of discussion, performances of dramatic readings, films, games, art work and music, feedback from the students, or observations by you and others who come into the classroom. The point is not how many different ways of gathering evidence there are, but that no one kind is better than any other.

Two important facts about assessment should be remembered whenever assessment tools are designed.

Assessment must be ongoing; it cannot occur only at isolated times during study.

Effective assessment not only helps you to look at how the students are doing, but it also helps you to evaluate how effective you are as the teacher of the course. Consider, for example, the following eight questions that assessment should allow you to answer. (Note that four are about the students, and four are about the course.)

1. Have the students completed all the required work?
2. Has the work completed by students been of value to them?
3. Has the work completed by students been of the quality I expect?
4. Has the work completed by students shown any change over time? Are these the changes that I wanted to see?
5. How well has the program I have presented worked?
6. Has the program accomplished what I wanted it to for all students?
7. How well have the students received the program?
8. What changes could I make to improve the program's effectiveness?

What assessment tools are especially effective for multicultural programs?

Suggested below are just some means of assessment that are perhaps especially pertinent to a multicultural curriculum in general, and to *Tapestry* specifically.

Portfolios

We recommend the creation of student and class portfolios—carefully presented and reasoned collections of the best work that the students have done. The work should show what and how they have learned about other cultures, about their own cultures, and about the literature in *Tapestry* because it explores points of view of cultures through authorship. A class portfolio can include papers, excerpts from student journals, reading logs, tape recordings, videotapes, or perhaps a record of a game or simulation.

Having students keep portfolios has been tried by many teachers, and the time involved in tracking and evaluating a portfolio assignment is no more than would be invested in making up and scoring a multiple-choice test. Yet the results are so very valuable compared to those obtained from more traditional and sometimes more limited assessment methods. Portfolios give a comprehensive picture of what the students have learned. Portfolios allow you to study students' progress throughout the course, offering many observation points on which to base your assessment of their work. (For suggestions on how to grade portfolio assignments, see "How can I grade student work?" on page xxxv.)

By completing a class or individual portfolio, students also participate in the task of assessment. They are given the opportunity to show you the very best of what they have learned. By talking to them about what you think ought to be learned and demonstrated in the portfolio, you help the students to set their own goals and measure their own progress. In essence, you are telling students that you are interested in them as people—in their lives in the classroom, in their growth as students. You can make this point even clearer by requesting that the students include a piece of reflective writing or a reflective tape. The reflection should consider why the various pieces are included in the portfolio and what the student thinks the pieces say about his or her growth and development as a participant in a multicultural literature program.

Tapestry suggests a variety of projects that might go in the portfolio, both in the Teacher's Manual and the Student Edition. For example, each selection in the Student Edition is preceded and followed by instructions for journal writing. By having students include these writings in their portfolios, you can easily see who is keeping up with the reading while also noting how the students are responding to the selections. The Teacher's Manual also suggests frequent journal writing as a means of getting students to respond to the text. For example, students set personal goals before each unit, often create word banks to help with selection vocabulary, and write personal responses to selections, themes, and units.

In addition, the suggestions in certain Teacher's Manual sections often would lend themselves to more complex portfolio projects. For example, many resources outside of the Student Edition are suggested in the Unit Introductions, Theme Previews and selection lessons found in the *Tapestry Teacher's Resource Manual*. Students could create projects based on these outside resources, ranging from research papers to oral presentations to artistic interpretations.

You will find that portfolios work for all students, primarily because they give each of them a voice in his or her work, an opportunity to show that he or she can read and write and perform and learn. One of the best portfolio projects we know of was begun by Eliot Wiggington in 1966. Each year, Wiggington's ninth- and tenth-grade students, from a poor rural area in southern Appalachia, create a literary magazine called *Foxfire*. The magazine, which features many different genres, is rooted in the Appalachian environment and culture. (Periodically, the magazines are put together in book form, also called *Foxfire*, and published by Anchor Press/Doubleday, Garden City, New York. These volumes are usually available in most libraries. The students also do TV shows for their community cable station, record albums of traditional music, and operate a furniture-making business, all under the umbrella of The Foxfire Fund, Inc.) Through *Foxfire* the students have shown themselves as writers and transcribers of the rural culture around them. It is a testimony to their ability as readers and writers. These are not all college-bound students, but they have proved themselves through this cooperative portfolio.

Examinations

A unit or semester examination is not an evil, but a real challenge for you and your class—a way that you and they can sum up what has been happening. An examination should be related to your goals and allow a chance for the students to show their strengths and imaginations, not just what they can recall. That is, examinations should not be limited to objective questions; they should require more from the students than a correct choice or a one-word response.

We suggest three alternatives to the traditional objective examination:

- Perhaps the best test of how well students have learned to read multicultural literature is seen in what they say or write about a new text, one that is unfamiliar to them. For this type of examination, provide the students with a previously unread poem, short story, speech, play, or other literary form that focuses on the same themes you have been studying. (The Teacher's Manual Bibliography at the end of each theme suggests additional books that are related to the theme.) Have the students read the selection and respond to it in terms of the main ideas you have explored throughout the unit.

- Another good way to measure students' gains is to see the kinds of connections they can make among the various texts they have studied over the course. For example, a unit-closing exam for Unit I of *Tapestry* might ask students to compare and contrast the different ways of expressing origins and ceremonies, to discuss the different themes of the selections within the unit, and to draw their own conclusions about the importance of origins and ceremonies in the United States today.

- A third way to assess students' progress is to see

what they now think of the various themes and issues that have been discussed. In order to respond to questions such as: "Do you think cultures should define themselves more in terms of origins, or in terms of where they are today?" students must refer to what they have read, discuss the issues they have studied, and cite specific examples to support their opinions.

Each of these approaches or a combination of them can be adapted and used in the development of a final examination or project for the students. In *Tapestry*, you will find that many end-of-unit activities in the Student Edition can be easily reworked into effective and innovative examinations of the types suggested above.

Examinations do not have to be limited to a class period or to a scheduled examination time. Many good examinations can be what are known as "take-homes." These have the quality of a summary paper or set of questions. The students have their books and notes available to them. The focus of the take-home examination is not on recall but on the interpretations, applications, and syntheses of the texts that can be made.

Although the issue of cheating could arise with the take-home format, it can be eliminated by how the examination is set up. You might choose to allow students to work together; their shared viewpoints will be very enriching for them. You might also word the project or question so that each student must choose his or her own topic. Because some students might take advantage of the take-home situation no matter how you handle it, you might find it useful to have some part of the examination controlled—an in-class essay on a new text, for example.

Another good way of handling the examination issue is to have a performance week, during which each student presents a final project summarizing the course. The "Problem Solving" activity on the end-of-unit review page lends itself toward this type of assessment. The project can take the form of a paper, an oral interpretation, a sketch or piece of art, a formal debate, and so on. The presentations should be evaluated by the rest of the class after they have developed criteria by which to judge the work. (See the section "How can I grade student work?" for suggestions on these criteria.) The presentations can also be performed for a public

audience; some teachers have even invited outside "juries" to enjoy the students' work.

Daily and weekly checks

For some students, teachers need to check whether or not they have been doing the reading. The usual method is to administer some sort of post-reading test, but in too many instances of this type of assessment, the tests become the major focus of the course. In actuality, such tests usually are not intended as anything more than gentle reminders that students have responsibilities too.

Alternatives to formal tests include good follow-up discussions and the occasional (not predictable!) assignment of brief in-class exercises. To combine these methods, try the use of one thought-provoking question that each person answers in writing. The question then serves as a take-off point for class discussion. The question does not have to be a typical test question. It might be a scale to rate a character or part of the action. For example, give the students a set of ten adjectives (*good, strong, rigid, cruel, powerless, passive, moral, flexible*, and so on) and have them rate the person or action from low to high for each quality. Sharing the ratings can serve as a springboard for discussion and other activities. The student who has not read the work will often stand out. Other examples of effective quick checks include:

- giving the students a set of pictures and asking them to select those that best illustrate the work;
- having students rate a story, play, or poem and explain why they gave it that rating;
- having students suggest alternative endings to the text.

The point is that creative post-reading exercises can both help the students get their thoughts about the text in order and help you keep track of those who seem to be falling behind. In *Tapestry*, we provide such creative exercises in both the Student Edition and the Teacher's Manual following each selection.

How can I grade student work?

The various approaches to assessment described above all need to be related to a judgment of quality. In the language arts, quality is seen both in terms of how well the students deal with the content of the course—what they have to say about what they have read and discussed—and in terms of how well they use language to express what they know. We cannot say which of these two is more important;

therefore, some sort of balance between them must be made. We propose assessing several aspects of a product rather than giving a single mark. The aspects are suggested below with criteria that might be used during rating. Certainly, any criteria that you use for assessment should be shared with students so that they can perform to the best of their abilities in these areas.

Content

How well is the material covered? Is it clear? Is it interesting? What has been included? Does what is included cover the major aspects of the topic? Does the product show an understanding of what has been read or studied? Does it show a thoughtful or mature understanding? Does it show an awareness of alternative views?

Organization

Is the presentation coherent? Does it begin and end clearly? Is there sufficient development and use of specifics?

Style and Tone

Does the presentation use language or the chosen medium appropriately to fit the purpose of the product? Is the style consistent throughout? Is there an effective mood presented? Does the writer/presenter appear to understand how the medium can affect the audience?

Mechanical and Technical Aspects

Has the writer/presenter used the medium well within the range of his or her ability? If the product is written, has the writer used grammar, syntax, spelling, and punctuation so that the message is clear? Are these aspects in keeping with what I know are the student's potential? If the product is in another medium, do the content and organization outweigh any technical problems so that they are hardly noticeable?

Notice how the suggested criteria provide guidelines but not prescriptive rules for good performance. They show that writing or other forms of performance are complex phenomena that need practice and that clearly are not matters of right versus wrong.

Criteria like these can also be used by the students themselves to evaluate the performances of other students. If you work with the class to explore these criteria, you can make a checklist that students can then use when they look at the performances of their classmates. Students also can use the checklist to evaluate their own work.

How do I arrive at final grades for my students?

Given all we have said about assessment, what about the final grade? Most teachers live within a system that requires a letter or a number grade at the end of the semester. How can all that has been done be caught in a single letter or number?

The answer is that it cannot, and yet the system demands this kind of summary rating. We have found that many students understand the system, and they know how important grades can be for their future. We have also found that the best way to handle the problem is to be clear from the beginning as to what kinds of things have to be done in order to earn an A, or a 75, or whatever grade the student is interested in for the course. In this way, the grade represents the completion of a contract. The contract involves satisfactory performance on a number of tasks done both individually and in groups. The judgment of the teacher or the jury is a part of that contract. If the various criteria are clear and the ways in which the judges will operate are spelled out, then a complex assessment can be turned into a single grade. This is not the best summary of a student's work, and we believe that a descriptive summary or checklist should be added to it. Many schools are now operating in this way and are asking that the portfolio be carried forward by the student to the next class. Through this procedure, grades are still there, but assessment is much broader than the grades themselves.

Sample Lesson Plans

The *Tapestry Teacher's Resource Manual* includes thorough lesson plans for every selection. However, as expressed in the Introductory section on pages xxviii-xxxiii, the manual is not an authoritative document on multicultural instruction. We at Globe cannot know your culture or the culture of your students; nor do we know your preferred style of instruction. We do know that *Tapestry* represents for you the opportunity to explore, expand, and enrich personal knowledge, even while encouraging the students to do the same. Perhaps more than any other curriculum you use, *Tapestry* will allow you to try different teaching techniques to achieve the responses and outcomes you desire.

For these reasons, we offer in this section the suggested lesson plans of teachers who have been involved in multicultural education in the English classroom. In some ways, these lesson plans are very different from those in the selection lessons of the *Tapestry Teacher's Resource Manual* —and also different from one another. In other ways, the lesson plans are all very similar; all call for the students to reflect on the universal themes of multiculturalism and the students' personal experiences and values.

Our hope is that your exposure to these teachers' different approaches will give you ideas, stimulate your enthusiasm about the possibilities of multicultural literature, and give you an opportunity to feel connected to other instructors in this curriculum area. You will note that each lesson plan is designed around selections actually included in the *Tapestry* Student Edition; this means you can adapt them to your own instruction.

LESSON 1
Noreen Benton, English Teacher, Grade 11
from *Fences,* by August Wilson
Notice that Noreen Benton utilizes her students as her best resource. Through question-and-answer dialogues, discussions, and shared listening, speaking, and writing activities, she elicits from her students much valuable information. This interaction forms the core of her instruction.

Preparation

Research, or assign students to research, the following topics:

- the Negro Baseball League
- Jackie Robinson
- Branch Rickey
- the controversial issue of the underrepresentation of African Americans in the management levels of professional sports today

These topics will come up during class discussion, and insight gained from research of the facts will enrich the discussions.

Motivation
Write these questions on the board:

What makes a good parent?

What are the unwritten "rules" of parenting?

(Note: If there are students in your class who live with a caregiver other than a parent, such as a grandparent or legal guardian, you may wish to include the word *caregiver* wherever appropriate in the discussion.)

Encourage students to suggest qualities as well as actions that they feel characterize a good parent. Write their suggestions on the chalkboard. Next ask:

What is "parent love"? Does this definition work for both "mother love" and "father love"? Why or why not?

Encourage students to suggest a list of unwritten but universally accepted behaviors and attitudes, or "rules," regarding parental love. Write their responses on the chalkboard. Ask questions to keep the discussion moving:

- What are parents obligated to do for their children?
- How do current television shows reflect what society says about the role of parents today?
- Is "buying things" part of a parent's responsibility toward children?

Keep the "characteristics of good parents" list and the "rules regarding parental love" list for use after the selection is read.

Preteaching

Introduce students to two themes they will encounter in the *Fences* selection:

- Parents' dreams for their children versus their children's dreams for themselves;
- Parents who make their children "pay for" the parents' mistakes.

Write the themes on the chalkboard, and encourage students to comment on them. Have them look for these themes as they read the selection.

Postreading

Direct students' attention to the lists you saved from the Motivation activity. Discuss these lists as they apply to the selection. Ask these questions:

- Is Troy a good father? Why or why not? What characteristics from the class list does he have?
- How does Toby define the rules of parental love? How does this match the class list?
- What rules does Toby suggest children have in regard to what they should do for their parents? What do you think children should do for their parents?

Also, discuss the themes outlined in the Preteaching activity, giving specific references to them in terms of the selection.

Following Up

Students who are especially interested in this selection can engage in several independent activities to continue thinking about the themes:

Writing   Have students write in their journals regarding communication in their households. What conflicts occur between generations? How do these conflicts affect communication?

Engaging the Family   Have students create a questionnaire for their parents, listing choices for things on which to spend money. As the students discuss the lists and the ways their parents prioritize the items, they can gain personal insight on the issue of what parents can and cannot buy for their children. Since this information is personal, students do not need to share their insights with the group.

Summarizing   As closure to the lesson, point out that one of the goals of multicultural education is to recognize multiple perspectives. Point out that opposing or conflicting viewpoints may be not only based on generational differences, but also on cultural differences. Ask students to look at the lists they have made and to think about whether or not any of the differences in parental expectations listed are culturally based. Students may wish to write on this issue in their journals.

LESSON 2

Marcie Belgarde, English Teacher, Grades 7 and 8

"Seventeen Syllables" by Hisaye Yamamoto

Marcie Belgarde uses writing as a means of accessing students' knowledge, feelings, and experiences. She also successfully interweaves traditional literature instruction in the lesson plan she outlines below.

Outcomes

The students will:

- infer the characters' traits and motives by what they say and think, and/or what others think or say about them;
- respond to a piece of literature and share that response;
- engage in role-playing based on inferences;
- extend the literary experience through the writing of poetry.

Prereading

Before reading the selection, introduce the concept of inference and explain how inference helps us to understand a character. Ask students to list ways they learn about characters in a story (by what they do, say, think; through descriptions of physical appearance; by what other people say about them or by how other people treat them).

Explain that sometimes readers base their understanding of characters on inferences. An inference is made when the reader reads between the lines or makes an educated guess. The reader looks at the clues an author gives and makes a supposition or inference. By making inferences about characters, readers can get involved in the reading. Give students two examples of how inferences are made:

- If I came into the room, threw your test papers down on my desk and said, "I've had it with this class," what might you infer from my actions and speech? What clues helped you make that inference?
- If I were giving a test and when I turned my back, Jack slowly eased a paper from his notebook and quickly studied it, what would you infer from Jack's actions? What clues helped you make that inference?

Explain that "Seventeen Syllables" is a story that relies on the reader to make inferences. The author allows the reader to come to his or her own conclusions.

Into the Reading

Have the students work in groups of four, assigning each group one of the characters in the story: Rosie, her mother, her father, Jesus, or Mr. Kuroda. (Some characters may need to be assigned to more than one group.) As the students read, they should focus on their assigned character. They are to note what he or she says, thinks, does, and how he or she relates to others. They will then be able to make inferences based on these clues. Write these questions on the board for students to consider as they read:

- Who is this character?
- What is important to this character?
- What does the character value?
- Who are the people that matter to this character?

- How does this character view himself or herself?
- What does this character want?
- How has this character been influenced by cultural heritage?
- Are you in any way like this character, or does he or she remind you of anyone you know?

Postreading

After the students have read the selection and considered their assigned characters, have the group members work together to create a character web. Provide each group with a large piece of paper and each group member with a different-colored marker. (This allows the teacher to assess how much each member of the group is participating.) Instruct each group to write its character's name in a square or circle in the center of the paper. Then have the group members use their responses to the questions listed above to write inferences about the character. Each group member can write his or her inferences on any area of the paper and should feel free to add to another group member's ideas. Encourage the group members to make connections among ideas. (You may want to demonstrate on an overhead projector how to create a web, using a familiar character such as Cinderella.)

After the groups have had adequate time to complete their character webs, each group should select a member to "become" its character. That person will study the web and be ready to use the information to respond to questions posed by the rest of the class. These questions are to be inference questions. For example:

- For the mother: Why do you write haiku? How did you feel when you won the contest? How did you feel about your husband burning your prize?
- For Rosie: How do you feel about your mother? father? Did you expect Jesus to kiss you? How do you feel about Jesus now?

Beyond the Reading

Teach the class how to write haiku. Share several examples. (Two readily available collections are: *Cricket Songs: Japanese Haiku* (1964) and *More Cricket Songs: Japanese Haiku* (1971), both translated by Harry Behn and published by Harcourt Brace Jovanovich). Then have each student write a haiku that is developed from his or her character's point of view and reflects on the character's experience.

LESSON 3
Joe Quattrini, English Teacher, Grades 9–12
Theme 7: The Quest for Equality

Joe Quattrini has created a panel discussion that allows students to connect the past with the present through the vehicle of text selections. Designed to be a five-day project, this theme closer activity involves communication skills, critical thinking skills, and literary analysis from a multicultural perspective. Students assume the roles of authors or characters and must have a thorough understanding of the text in order to participate in the discussion effectively. Below is an outline of the project, showing what occurs during each day of implementation.

Preparation

Make enough copies of the "Opening Dialogue Sheet" (page xli) and the "Student Instruction Sheet" (page xlii) so that each student will have one. Decide how to place the students in 10 groups, assigning each group one of the selections from Theme 7.

Day One (Student Preparation)

Distribute the "Student Instruction Sheet" and discuss it with the students. Explain that this activity will give them an opportunity to "take on" the persona of an author or character from a selection they have read and respond in character to the theme of equality. To do this, they will revisit a selection through intensive study of it, and formulate their ideas regarding how the person they have chosen would contribute to a discussion on equality.

Distribute the "Opening Dialogue Sheet" and point out that this will be the springboard for the panel discussion. As groups formulate the statements the discussion participants will make, they need to link their ideas to this modern commentary. Ask for two volunteers who are to play the part of the "Speakers" in the Opening Dialogue. Speaker 1 and Speaker 2 will read the Opening Dialogue and then participate in their designated roles during the panel discussion.

Have the students form the preassigned groups and begin work. First, they need to assign tasks within the group. One person will represent the group as Presenter in the panel discussion. (The two students who volunteered to be Speakers in the Opening Dialogue cannot take these roles.) Another will be the designated Listener, the person who will carefully follow and note the comments of each speaker during the panel discussion. (These notes will be shared by all designated Listeners during a follow-up discussion.) A third person should be appointed as Moderator, the one who keeps the group on task. Depending on group size, you might suggest other roles such as Recorder or Oral Reader.

Groups also need to decide which author or character from Theme 7 they would like to present in the panel discussion. Once this is determined, the groups can begin their intensive research on and discussion of that person and his or her views on equality. Encourage them to refer frequently to the "Student Instruction Sheet" and "Opening Dialogue Sheet" for guidance. If you would like to offer specific examples to the students, you might share some typical student notes regarding Sojourner Truth:

- would join conversation politely but forcefully;
- likes to make comparisons between rights and abilities;
- will speak from personal experience as a worker, a woman, and an African American;
- uses rhetorical questions and simple physical images;
- not educated, but intelligent;
- speaks for women's rights when others might claim that's not the issue being discussed;
- something of a mystic and a missionary;
- shows comparisons of what people want for themselves and what they think others should be satisfied with.

Day Two (Final Student Preparation)

Have the student groups continue their work. During this session, each group should reach consensus on its character's views and write the person's one-minute opening statement. Monitor the groups as they work to ensure that everyone is contributing and that students are using all the resources available—including their own journals—to complete the characterizations.

Day Three (Panel Discussion)

Explain the rules for the panel discussion:

- You are the moderator, the person in charge of keeping the discussion on track.
- Only one person can speak at a time.
- When someone on the panel wants to speak, he or she must wait to be recognized by the moderator.
- The moderator has the right to move the discussion to a new point if the topic under discussion cannot be resolved or if too much time has been taken up by a single topic.

The panel discussion begins with the Opening Dialogue. Following that, each speaker on the panel gives his or her opening statement, and then the discussion is open. If students have difficulty getting the discussion going, you might suggest that two characters with opposing viewpoints direct some comments to each other. You'll find that once discussion begins, the problem will be stopping it, not starting it.

Day Four (Follow-Up)

Have the designated Listeners share their notes regarding the panel discussion. As the discussion is revisited in this way, students can discuss their reactions. Below are some questions you can use to enliven the discussion:

- How authentic were the views represented by the speakers?
- Which speakers mostly agreed? Which speakers mostly disagreed?
- Whose views did you relate to best? Whose views seemed most in conflict with your own?
- What did we accomplish with this discussion?

Day Five (Closure)

Have students write in their journals their responses to these questions: "How did the panel discussion serve as a bridge between the past and the present? How did it serve as a bridge between your views and those represented in the selections?" Ask willing students to share the insights they gained from the discussion.

Name _____ Date _____

OPENING DIALOGUE SHEET

Speaker 1: You want equality? Your rights should be equal to the "rights" my grandfather had when he came to this country. He had the right to take only the dirtiest and lowest paying jobs. He had the right to crowd in with relatives, eight people in a room, because that's all he could afford. He had the right to be cheated and abused because he couldn't speak English. That's equality—same as for any immigrants who ever came here.

Speaker 2: But it's no longer your grandfather's day. Two generations have passed, and everything is different. My school system has students of 58 nationalities. They speak 105 different languages. It's not so simple as coming over here and moving into your ethnic neighborhood to get a start. It's different.

Speaker 1: The difference is that it's easier. Look at all the social programs we have now. It's the people who are really different. My grandfather couldn't afford to say, "That kind of work is beneath my dignity." He couldn't say, "I demand an education—and it has to be in *my* language." Those people weren't afraid to work. They had some pride. They did what they needed to do to fit in.

Speaker 2: I don't want to fit in. People now have pride, too. That's why they insist on keeping their own identities and asserting their rights as individuals, instead of settling for what's left after everyone has a piece of the pie. Men and women of all races want an equal chance.

Speaker 1: Then don't you think they should earn it, like everyone else had to— earn a place in our society?

Speaker 2: Do you think newcomers should be satisfied with what's left for them? That's supposed to be fair?

Speaker 1: Fair or not fair, that's the way it's always been.

Speaker 2: But does it have to stay that way?

STUDENT INSTRUCTION SHEET

This week you will be developing and participating in a panel discussion on the theme of equality. You will work in a group to analyze closely one author or character from an assigned selection in Theme 7. You will be figuring out that person's position on equality and determining how he or she would present that position in a panel discussion.

Your teacher will assign you to a group and will assign the group a selection. As a first step, the group will decide whom to represent in the panel discussion from the selections: the author, a character or someone else, perhaps one of the speakers or someone who has heard one of the speeches firsthand. Then you will analyze the selection, the prereading and postreading materials, and all of your journal entries or other writings about the selection as you explore the person's character.

As you explore, look for evidence to support a particular view that the author or character would present in the panel discussion. For example, in the prereading materials, look for background features that may be important. In the selection, look for other kinds of evidence to show the person's view on equality: quotations, words and phrases, definitions, favorite images, stories, and metaphors or other literary devices. In the postreading materials, refer particularly to "Going Back Into the Text" for other types of evidence. In your journal writings, examine your own views and reactions to the selection and the author or character.

When your group has enough evidence to establish a strong sense of the view your person will represent, prepare a one-minute opening statement to begin your contribution to the discussion. After each participant gives an opening statement, your participant can continue to speak from the view he or she represents as the discussion continues.

LESSON PLANS

1 Origins and Ceremonies

One way in which people connect themselves to their cultural roots is through rituals and ceremonies. Oral tradition, folktales and legends are not always far away in time or place. We are surrounded by them in our daily lives.

UNIT OBJECTIVES

Literary

- to explain how traditional storytelling conveys cultural beliefs and values
- to show how the authors use traditional storytelling to convey themes of morality, virtue, and heroism

Historical

- to explain the significance of traditional storytelling and ceremonies in world societies
- to evaluate storytelling as a means to study and pass on history

Multicultural

- to compare and contrast the beliefs, rituals, and customs embodied in different cultures' traditional storytelling
- to recognize recurring themes among diverse cultures' folktales

UNIT RESOURCES

The following resources appear in the Student Edition of *Tapestry*:

- a full-color **portfolio of theme-related art,** Humanities: Arts of the Tapestry, pages A1–A6, to build background and activate prior knowledge about the unit themes and to generate writing ideas.

- a **unit overview,** page 6, to provide historical background about cultural ceremonies and their origins, and to show how a culture can be identified and understood by these ceremonies.

- a **time line,** pages 4–5, to help students place the literary works in their historical context. You may wish to have students refer to the time line before they discuss each work.

- the **Focusing the Unit Theme** at the end of the unit, page 119, to provide a cooperative/collaborative learning project on the theme of "Origins and Ceremonies," a writing project to develop an essay of comparison and contrast, and a problem-solving activity in which students analyze one of the ceremonies or stories in the unit.

INTRODUCING UNIT 1

Providing Motivation

Before students begin to read the unit overview in the student edition, you might want to stimulate their interest in the theme of "Origins and Ceremonies" with some of the following activities.

- Suggest to students that in present-day U.S. society print and broadcast media have replaced traditional storytelling both as a source of entertainment and as a tool for learning about values and culture. Discuss the pros and cons of this change. Ask: "What, if any, advantages do you see in learning about values and cultural heritage from traditional storytelling versus television or other forms of media?"

- Have students work in groups of four. First, have each group member write a detailed description of a ceremony or ritual he or she has participated in or witnessed (suggestions: graduation, baptism, marriage, funeral). Next, have groups answer the following questions collaboratively. Assign one member of each group to record the student answers.

How did each of the ceremonies represent a "tradition"?

How were the ceremonies and rituals passed down from generation to generation?

Has this ceremony always been the same or has it changed over the years?

How are the ceremonies you have listed similar? How do they show that ceremonies for varying events have common elements?

Call on group leaders to report the answers and then have a class discussion about the findings.

- Tell students that many cultures use symbols to illustrate various aspects of their culture. These can range from flags and patriotic medallions to totem poles of the Northwest Coast Native Americans and kachina dolls of the Hopi. (If possible, show pictures of these items to the students.) Ask: "If you had to choose one symbol to embody your cultural heritage, what would you choose?" Encourage students to be creative in their responses. Discuss the diversity of U.S. culture and why many symbols exist.

- Have students work in pairs or in small groups to identify any similarites among the events listed on the time line on pages 4–5. (For example, European explorers came to lands already inhabited by other peoples, nations gained independence from other nations.) Ask groups to imagine what types of stories a group of people might develop when faced with situations such as these.

Cross-Discipline Teaching Opportunity: Unit Theme

Collaborate with the social studies and/or the music teacher to explore some traditional ceremonies and stories that have made an impact on African American culture. Remind students that in the early 1800s black codes outlawed reading and writing for African Americans in many states. Consequently, music and folk tales took on special significance as a means to pass on tradition and history in an era where learning to read and write was forbidden. Recordings of traditional folk music can be obtained from most libraries. An excellent source for information on this subject (including a thorough discussion of cultural themes and educational music) is *Deep Like the Rivers: Education in the Slave Community, 1831–1865*, by Thomas L. Webber, W.W. Norton & Company, Inc., 1978.

Setting Personal Goals for Reading the Selections in Unit 1

Have students keep a copy of the following chart in their journals or notebooks. Provide class time every few days for students to review and expand their charts. Encourage them to add topics of their own.

Origins and Ceremonies

Topic	What I Know	What I Want to Learn	What I Have Found Out
Storytelling as a way to teach children about right and wrong			
Common literary themes among diverse cultures			
The value of understanding ancestral roots			
The process of and occasional conflicts related to personal growth and change			

3

A Heritage of Traditional Stories

THEME PREVIEW

A Heritage of Traditional Stories			
Selections	Genre/Author's Craft	Literary Skills	Cultural Values & Focus
"The Stars," Pablita Velarde, pages 7–9	quest tale	quest hero; quest goal	oral tradition and heritage; humankind and nature
"The Sheep of San Cristóbal," Mexican American folk tale, retold by Terri Bueno, pages 10–12	moral tale	moral of a tale; characterization	Spanish heritage in Mexican culture; learning lessons through life trials
from *Things Fall Apart*, Chinua Achebe, pages 13–15	folktale	realism vs. fantasy	African culture in stories; learning morality from storytelling; life in what is now Nigeria before European colonization
"In the Land of Small Dragon: A Vietnamese Folktale," Dang Manh Kha, pages 16–18	narrative poem	narrative elements	storytelling as entertainment; traditional values and beliefs in duty, honor, and family
MAKING CONNECTIONS: Vietnamese Culture Linking Literature, Culture, and Theme, p. 19			
"The Man Who Had No Story," Michael James Timoney and Seámus Ó Catháin, pages 20–21	Irish tale	rhythm and repetition	storytelling tradition and culture; views of nature and life

Assessment

Assessment opportunities are indicated by a ✔ next to the activity title. For guidelines, see Teacher's Resource Manual, page xxxiii.

CROSS-DISCIPLINE
TEACHING OPPORTUNITIES

Social Studies As we confront the environmental issues threatening our planet, there has been a renewed interest in Native American ideas on nature and our relationship with the earth. *Keepers of the Earth* by Michael J. Caduto and Joseph Bruchac is an excellent source of Native American stories and environmental activities. Encourage students to apply traditional Native American views of nature and use of natural resources with current environmental problems and solutions.

Music and Art Provide students with additional in-depth material on traditional ceremonies, folk art, and their origins. Many of these traditions are being kept alive today, both in rural areas and by interested urban groups. The January 1991 issue of *National Geographic* offers an insightful, up-close look at traditional storytellers, dancers, weavers, and carvers in U.S. society today.

Geography The selections in Unit 1 offer perspectives from a variety of world cultures. On a world map locate with students Vietnam, Nigeria, Ireland, Mexico, and the Southwestern U.S. Make a classification chart to contrast these geographical regions in the following categories: weather, wildlife, and native peoples. Discuss how differences in geography might affect cultural traditions that would be reflected in storytelling. For example, the use of animals as symbols will vary from region to region. Ask students what animals they might expect to read about in a Nigerian folk tale versus a Native American story that features animals.

SUPPLEMENTING THE
CORE CURRICULUM

The Student Edition provides humanities materials related to specific cultures in Making Connections on pages 36-37 and in Humanities: Arts of the Tapestry on pages A1–A16. You may wish to assign students the following activities outside of class.

- The illustration on page A1 in Arts of the Tapestry was craeted by Pablita Velarde for her retelling of the folktale, "The Stars". The tradition of authors creating art to accompany their writing is an old one. The English writer William Blake made carefully tinted drawings which intertwined with the hand drawn letters of his poems in his *Songs of Innocence*. The association of two arts, writing and drawing, could not be closer. The contemporary poet, Margaret Burroughs, has published her poems along with her own woodblock and linoleum prints. After students read the selections in Theme 1, ask them to decide how Velarde's illustration adds a special dimension to the folktale, "The Stars." How is she specially talented for this? Do they think this illustration might exemplify the traditional arts of the American Southwest? Ask students to choose a selection in Theme 1 that appeals to them and suggest an illustration. Encourage those students who are interested to create their own illustrations for these choices

- Traditional storytelling has played and continues to play an important role in Native American cultures. Beverly Hungry Wolf has recorded in *The Ways of My Grandmothers* personal history and legends of the Blackfoot Indians. Students may use the book to present a report about storytelling, ceremonies, and rituals.

- Teaching morality through storytelling is a tradition that continues to the present day; the form now is often television or movies. Ask students to evaluate how film and television teach morality. After they have read the selections in Unit 1, have the students compare television and movie strategies used to convey morals to the storytelling techniques used in one of the selections. Students should compare plot, theme, narrative technique, and entertainment value.

- Sociologists have noted that in times of rapid social change communities often turn to the past for guidance and security. For this reason, the past is often idealized. Discuss with students what benefits traditional ceremonies and storytelling might have for a rapidly advancing technological society such as our own. Discuss the relevancy of theme titles in Unit 1 to some present-day social problems.

INTRODUCING THEME 1:
A HERITAGE OF
TRADITIONAL STORIES

Using the Theme Opener
Introduce Theme 1, A Heritage of Traditional Stories, with this quotation by Willa Cather:

There are only two or three human stories, and they go on repeating themselves as fiercely as if they had never happened before.

Explain to students that Willa Cather (1873-1947) was an American novelist and short story writer. Her work centers mostly on the pioneer spirit in Nebraska and the southwestern United States. Invite students to consider the quotation by Willa Cather and identify two literary themes that are likely to be found in all cultures. Lead a discussion on what these themes might be (search for happiness, struggle between good and evil, tales of quest, and so on). Suggest to students that in the study of traditional stories, common themes will emerge that unite all cultures in a common bond of human experience. Explain that one reason to study literature from many cultures is not only to recognize and appreciate diversity, but also to understand the similarities between all peoples and cultures.

To focus your discussion on diversity in American cultures, ask students, "Who and what is an American?" As they respond, remind students that multicultural societies are not necessarily fragmented. Different cultures offer diverse viewpoints on the same aspects of the human experience. For example, the selections in Unit 1 contain common themes of good versus evil, finding a homeland, and heroic deeds. Within each culture, however, the reader may discover a variety of viewpoints on these themes.

Have each student write a paragraph on his or her own definition of what it is to be an American. They should refer to the themes you have outlined for Unit One. Tell them that the definition must contain three components: an example of their own ancestral roots, a story they consider to be part of their heritage, and a story that they associate with being American. To stimulate their writing, have them skim the selections in Unit 1. When they have finished writing, have students work in groups of three to exchange and discuss their work. Select a group representative to summarize how the writers in each group defined *American*.

✔ Developing Concept Vocabulary

Help students to understand the nuances and implications of the terms *heritage* and *tradition* with the Venn diagram exercise below. Draw the outline on the chalkboard. Use the questions below to explore similarities and differences between the two words. (List similarities in the portion of the diagram where the circles intersect. Differences are listed under each term.) As they work, suggest to students that in a multicultural society such as our own, these terms have many different shadings of meaning.

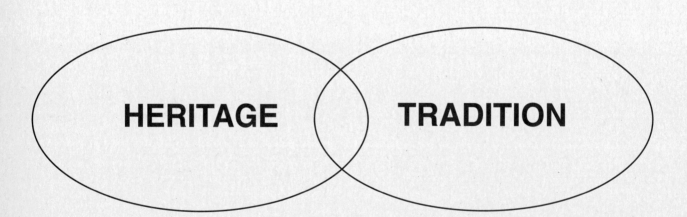

ASK:

- Do traditions change or remain the same? Does a community's heritage change or remain the same?
- What are some traditional ceremonies that many Americans practice? Are the general outlines the same while the details may vary?

- How does a common heritage help to hold a community together?
- Is a younger generation's heritage the same as its parents'?
- Will you have the same traditions as your parents?
- Is the love of a particular sport part of our heritage, or is it a tradition?

The Stars
by Pablita Velarde (pages 7–14)

OBJECTIVES

Literary
- to understand the values, customs, and beliefs the story communicates
- to identify the elements of a quest story

Historical
- to show how traditional stories can teach future generations about historical episodes
- to discuss the leadership qualities that kept Long Sash and his people together during times of hardship

Multicultural
- to compare traditional Native American storytelling to present- day methods of information dissemination
- to discuss common literary themes among diverse cultures

SELECTION RESOURCES

Use Assessment Worksheet 1 to check students' vocabulary recognition, content comprehension, and appreciation of literary skills.

 Informal Assessment Opportunity

SELECTION OVERVIEW

Native Americans did not depend on a written literature to express their views of the world. Instead, art, rituals, and the oral tradition—often intermingled—served this purpose. Symbolism is an important feature in the Native American oral tradition, with animals playing an especially important role. Animals, through their actions, help humans learn life's important lessons.

In this selection, the author describes the voyage of Long Sash and his people who, to arrive at their predestined homeland, must face a number of moral, physical, and spiritual challenges.

ENGAGING THE STUDENTS

Have students imagine that they have embarked on a long voyage to find a homeland. Now they are lost and beginning to doubt the wisdom of the trip. Ask them what they would consider in determining whether or not to continue. To help them respond, have them think of imagined times in the past. What has sustained them when things seemed hopeless?

Suggest that the search for a homeland might be compared to the search for peace, knowledge, and happiness. Explain that in the "The Stars" they will read about people who endured great hardship to arrive at a homeland. Ask them to determine the reasons for enduring these hardships.

BRIDGING THE ESL/LEP LANGUAGE GAP

The unusual narrative style and vocabulary of this piece will be difficult for some students. Expand the prereading activity with a discussion about stories, legends, or beliefs that people have about stars. (Do they know any? Why do people make up stories about the stars? What kinds of stories are they? What kind of people are in these stories?) Be sure the students understand the word constellation; provide a chart of constellations in the classroom for reference.

Photocopy and distribute a copy of the two paragraphs in the final page, beginning with "The signs in the sky will always be there to guide us." Ask them to scan these two paragraphs and highlight the following phrases:

> Long Sash
> Endless Trail
> two stars of decision
> headdress
> team of young men
> big star
> seven bright stars

Tell them to watch for these phrases as they read, and to find out how they fit into the story. After they have read the story, have them work together in groups to reconstruct orally the story, using the highlighted phrases as a guide.

☑ PRETEACHING SELECTION VOCABULARY

Have students create a word bank by writing down all the terms from "The Stars" that relate to stars, terrain, travel, and animals (for example: Orion, **Milky Way;** trees, water; **journey, destination; mole, coyote**). While they are writing, use these categories to make four columns on the chalkboard. Call on students to fill the columns with the words they have found in the text. Use the categories to discuss the questions below:

• Which of the words in the word bank do you think are symbols? Explain why.
• Which of the word bank terms are characteristic of Native American culture?
• Judging from this vocabulary activity, what do you think "The Stars" is about? Explain how you arrived at your answer.

The words printed in boldface type are tested on the selection assessment worksheet.

PREREADING

Reading the Story in Cultural Context

Help students read for purpose by asking them to consider the importance that geographical places (i.e., the land) had for Native Americans. Discuss the unique perspective Native Americans had on nature and its role in Native American culture.

Focusing on the Selection

Have students write a short paragraph on the most important belief or value that they hold and how this belief has been affected by their experiences as teenagers. Tell students they will need their paragraphs as they read "The Stars" to compare their experiences with the lessons Long Sash learns.

POSTREADING

The following activities parallel the features with the same titles in the Student Edition.

Responses to Critical Thinking Questions

Possible responses are:

1. reverence for ancestors; respect for leaders; importance of patience, tolerance, and love for one another
2. Storytelling passes on the values and beliefs of the people.
3. Guide students to think about the role of community leaders and to consider the ways a community deals with hardships and adversities.

☑ Guidelines for Writing Your Response to "The Stars"

Have students share their journal entries with a partner. Or as an alternative writing activity, begin by asking each student to write in a few sentences what he or she believes the story is about. Next, have students work in pairs and exchange their writing. Ask them to collaborate on a single interpretation of the story incorporating both viewpoints.

Guidelines for Going Back into the Text: Author's Craft

Help students understand the universal appeal of a quest tale by asking them to consider what adventures have taken place in their own lives. Point out that not all quests are as weighty as the one in "The Stars," but they can be significant just the same. Ask them to consider their own personal experiences in school, relationships, sports, or other goals they have worked to achieve. Possible answers to the questions in the Student Edition are:

1. On the surface, the goal is to find a homeland. The goal is also significant for the moral and spiritual growth experienced by the travelers.

2. The travelers in "The Stars" must overcome hunger, disease, wild animals, and fear. To overcome these obstacle they develop new ways to travel and survive and develop faith in their leader.

3. Answers will reflect students' own experiences.

☑ FOLLOW-UP DISCUSSION

Use the questions that follow to continue your discussion of "The Stars."

Possible answers are given in parentheses.

Recalling

1. Why did Long Sash and his people want to move? (to escape a harsh ruler and a life of misery)

2. How did Long Sash and his people learn to carry loads more easily? (by dragging their belongings on poles)

3. How did Long Sash know that he found the homeland? (sign of the mole and spider woman)

Interpreting

4. How would you describe the attitude of Long Sash's people toward the quest? (sometimes doubtful of the quest and occasionally losing faith in Long Sash, but willing to learn and persevere)

5. How would you explain Long Sash's momentary feelings of despair and his fear that he was losing his mind? (Possible answer: it is possible Long Sash doubted his quest and momentarily lost faith in his ability to lead.)

6. What important life skills did Long Sash and his people learn on the quest? (Possible answers: learned to adapt and change, recognize symbols, learned importance of staying together and how to find happiness out of misery)

Applying

7. Long Sash and his people learned the benefits of perseverance. Why was this an important lesson? (Guide students to understand that to escape unjust situations might require dedication, patience, and faith.)

8. If you had been one of Long Sash's followers, would you have stayed with him until reaching the homeland? Explain why or why not. (Guide students to understand the consequences of each decision.)

ENRICHMENT

To expand your discussions of traditional stories, students might benefit from viewing *Ancient Spirit, Living Word: The Oral Tradition*, available from Native American Public Broadcasting Consortium, P.O. Box 86111, 1800 N. 33rd St., Lincoln, NE, 68501, (402) 472-3522.

The Sheep of San Cristóbal

by Terri Bueno (pages 15–23)

OBJECTIVES

Literary

- to identify the elements of a moral tale
- to understand the story's message about peace and fulfillment

Historical

- to show the relationship between the mestizos and Native Americans in the Southwest
- to discuss themes of community and the socioeconomic factors

Multicultural

- to develop an awareness of Hispanic cultural values through reading the folk tale
- to discuss religious themes in literature of various cultures

SELECTION RESOURCES

Use Assessment Worksheet 2 to check students' vocabulary recognition, content comprehension, and appreciation of literary skills.

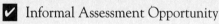 Informal Assessment Opportunity

SELECTION OVERVIEW

The migration of people from the interior of Mexico to what is now the southwestern United States began in the early 1600s. The first Spanish colonial settlement was established in Santa Fe, New Mexico, in 1610 and was at this time the farthest point north in the Spanish Empire. This region of settlement eventually came to include the present-day states of Texas, Arizona, and the coastal strip of California. The settlers themselves were mestizos—people of Spanish and Indian descent. Spaniards born in the Americas (criollos) made up only a small part of this migration to the Southwest. The religious themes of the selection (such as martyrdom and penance) reflect these groups' beliefs.

This selection tells the story of a young woman whose child is taken away. She is tormented by a wealthy neighbor with a large flock of sheep and consequently she wishes for his death. Accidentally, the neighbor dies. As penance, the woman travels the region giving away the dead man's sheep to poor people. In the end, she finds her child with the help of San Cristóbal.

ENGAGING THE STUDENTS

On the chalkboard write this question: "How do people define *wealth*?" To help students answer, ask the following questions:

- What is more important: food and shelter, or love and friendship?
- In our culture, do we equate wealth with happiness?
- How do people define "success"?

Discuss the student responses. Tell them to summarize their impressions of the discussion in a short paragraph so they may compare it with the selection.

Explain that in the "The Sheep of San Cristóbal" they will read about a woman who exchanges one kind of wealth for another; they can then add her experiences to their ideas on what is wealth.

BRIDGING THE ESL/LEP LANGUAGE GAP

The vocabulary and the religious concepts in this story may present difficulty for some students. Discuss the introductory paragraph to give students the background of the story. Tell students that the story begins by describing Felipa as "very religious." Discuss with the students what this might mean: What kind of beliefs do they think she might have? Discuss the following concepts, making sure that students understand each of them: evil/sin, penance, forgiveness, prayer, faith, guidance, blessings, and signs.

Tell students that the two main characters in the beginning of the story are Felipa and Don José. Ask them to pay attention to how the religious concepts discussed earlier relate to each of the characters. (For example, students should be able to answer questions like: "What was Felipa's/Don José's sin? What was Felipa's penance? How does faith fit into the story? Was there a blessing? When?") When the students have finished reading, have them work in pairs to answer these questions.

☑ PRETEACHING SELECTION VOCABULARY

Tell students that religion is very important in this story. Have them skim the selection and write all the words they can find that are related to religion. (Examples: Our Lady of Sorrows, **Padre, priest, indigent,** San Cristóbal, **penance, prayed**) Next, lead a guided discussion using the questions below to analyze the vocabulary words:

- What religion do you think the characters in the selection practice?
- What language do you think the characters in the selection speak?
- Based on this vocabulary activity, what do you think the message is in "The Sheep of San Cristóbal"?

The words printed in boldface type are tested on the selection assessment worksheet.

PREREADING

Reading the Story in Cultural Context

Help students read for a purpose by asking them to consider the importance religious themes have in the selection. Discuss the unique perspective of Mexican Americans who are descended from the mestizos, based on an ancestry that combines western European influences (Spain and the Roman Catholic Church) with traditional Native Mexican culture (animals, symbolism).

Focusing on the Selection

Have students each write short paragraphs that summarize their own life goals. Acknowledge that not all teenagers have completely defined their life goals, but challenge them to focus on this theme: How will they define happiness and wealth? Suggest that in the selection, happiness for the main character is related to defining what is right and wrong. Tell students to keep their writing and to compare their ideas with the lessons Felipa learns in "The Sheep of San Cristóbal."

POSTREADING

The following activities parallel the features with the same titles in the Student Edition.

Responses to Critical Thinking Questions

Possible responses are:

1. The beliefs and values are based on a religion that emphasizes confession, penance, and forgiveness. The moral is that when you do wrong, if you admit it and find a way to make up for it, your guilt will be relieved and your happiness restored.

2. The conflict between the Utes and the Spanish settlers is symbolized by the taking of Manuel. The religion and way of life depicted in the story are blends of the two cultures.

11

3. Students' views will differ according to their cultural backgrounds, although most cultures share a common value: Bad deeds should not go unnoticed and should be rectified.

☑ Guidelines for Writing Your Response to "The Sheep of San Cristóbal"

Have students share their journal entries with a partner. Or, as an alternative writing activity, ask each student to write in a few sentences what he or she believes Felipa learned about "right" and "wrong." Next, have students exchange their writing with a partner. Ask them to collaborate on a single interpretation of the story incorporating both viewpoints.

Guidelines for Going Back Into the Text: Author's Craft

Remind students that this folk tale is a moral tale. Discuss with students the kinds of moral decisions teenagers must deal with today. Ask: "In the selection the main character did 'penance' for her deed; does our society believe in penance? What is the difference between penance and punishment?" Ask students to compare and contrast modern ideas about penance and punishment with the ideas in the selection. Possible answers to the questions in the Student Edition are:

1. Students may choose to focus on the wrongful actions of Don José (harassing Felipa) or Felipa's wrongful actions (wishing Don José was dead).

2. Felipa performs penance by traveling throughout the region giving away Don José's sheep. The penance is effective because Felipa eventually finds peace.

3. Answers will reflect students' own reading and experiences.

☑ FOLLOW-UP DISCUSSION

Use the questions that follow to continue your discussion of "The Sheep of San Cristóbal." Possible answers are given in parentheses.

Recalling

1. Who took Felipa's son? (Utes)

2. What made Felipa curse Don José? (He let his sheep eat her crops.)

3. Many people in this selection gave away wealth; how many can you list? (The people in Felipa's village helped her after her son disappeared; Felipa gave away the sheep; people in the region gave Felipa food during her journey.)

Interpreting

4. How would you interpret the symbolism of Negrita the black sheep? (Sample answer: Negrita was a symbol for Felipa's lost son.)

5. What purpose did the trip serve for Felipa? Was the trip successful? (The trip was her penance. It was successful in two ways: Felipa regained her inner peace and her son was returned to her.)

6. Did Felipa fulfill her obligation of penance, even though she kept Negrita? Explain your reasoning. (Answers will vary, but students should note that Felipa did try to give Negrita away; they must decide whether she tried hard enough.)

Applying

7. Did Felipa commit a wrongful action when she cursed Don José? Use ideas on "right" and "wrong" from the selection to support your answer. (Student opinions may vary, though the tale certainly indicates that her actions were wrong and justified some form of penance.)

8. If Felipa were traveling from town to town today, how would she be perceived? (Guide students to consider how someone dressed in rags, begging for food, and traveling with only an animal for a companion would be regarded in a society dealing with homelessness.)

ENRICHMENT

Students might benefit from viewing a videotape entitled *Mexican Indian Legends*, which deals with ancient stories that teach moral lessons. The video is available from Phoenix/BFA Films, 468 Park Avenue South, New York, NY 10016, (800) 221-1274.

from *Things Fall Apart*
by Chinua Achebe (pages 24–28)

OBJECTIVES

Literary

- to identify the elements of fantasy and realism in a folk tale
- to recognize the results of the main character's weaknesses

Historical

- to compare African communities before and after European contact
- to discuss power hierarchies in traditional African villages

Multicultural

- to contrast present-day African American cultural themes with traditional African folk tales
- to discuss the use of animals among different cultures

SELECTION RESOURCES

Use Assessment Worksheet 3 to check students' vocabulary recognition, content comprehension, and appreciation of literary skills.

 Informal Assessment Opportunity

SELECTION OVERVIEW

Europeans in the 19th century, whose perspectives often were clouded by ignorance and racism, were unaware of the intricate and well developed social patterns of African societies. Europeans did not understand that African societies stressed personal relationships over material progress. In many ways, Africans had learned out of necessity to put social harmony and community values above those of the individual. These themes echo strongly in the selection from *Things Fall Apart*. The title is a reference to the fabric of social relationships.

In this selection the author describes how the birds of one village are invited to a great feast in the sky. The conniving Tortoise talks his way into an invitation and borrows feathers so that he can fly. Once the birds and Tortoise arrive at the feast, Tortoise eats and drinks most of the repast. The birds leave the feast angrily and take their feathers with them. Tortoise barely survives a long fall back to earth.

ENGAGING THE STUDENTS

On the chalkboard write this question: "Are people basically good or evil?" Allow students five minutes to write an answer to this question. When they have finished writing, explain that theologians and philosophers have pondered this question for centuries. Point out that some believe there is only a thin shell that protects our society from anarchy, and therefore laws are essential for order. Others believe that humans will "do the right thing" when moral questions arise.

Call on students to read their writing aloud and discuss their interpretations. Explain that in

Things Fall Apart they will examine the human character in an African folk tale. Encourage them to compare these ancient ideas with their own as they read.

BRIDGING THE ESL/LEP LANGUAGE GAP

This simple but amusing story provides a good opportunity for limited English speakers to participate in class discussions about this folk tale, as well as about ones they already know that have similar elements. Ask the students to consider the following questions as they read:

1. What are some things that Tortoise does that show how clever and cunning he is? What about Parrot? Does the storyteller admire these things in Tortoise and Parrot?
2. Do you know of any other stories that are about clever animals who play tricks on other characters in the story?
3. This story provides an explanation of why tortoises' shells look as they do. Do you know of any other stories that explain or answer questions people have about things they see in the world?

After the students have read the story, discuss these questions as a class. Encourage students — especially limited English speakers — to share other stories they know that are similar. Because of different cultural backgrounds, these students often know stories that other students have not heard, and so can add variety to the discussion.

☑ PRETEACHING SELECTION VOCABULARY

Have students scan the story for words that describe the main character, Tortoise. Make a list of the words on the board (for example: **cunning, delectable,** ungrateful, happy, **voluble, orator, eloquent).** After reading the selection the students may use these words to write descriptions of Tortoise.

The words printed in bold-facetype are tested on the selection assessment worksheet.

PREREADING

Reading the Story in Cultural Context

Help students read for purpose by asking them to consider the importance of the themes of social responsibility in the selection. Have them look for ways cooperation and honesty come into the story.

Focusing on the Selection

Ask students to consider their own neighborhood or community for a moment and answer the following questions:

• Would you categorize your community as "cooperative"? Why or why not? Is it possible for neighbors to help one another in large cities?
• How does our society deal with members who cheat or connive?
• As our cities become larger and families more extended, have we lost touch with community values? Has urbanization made humans insensitive to community values?

Discuss student responses and ask them to consider the issues facing their communities, neighborhoods, and cities as they read about the community in *Things Fall Apart.*

POSTREADING

The following activities parallel the features with the same titles in the Student Edition.

Responses to Critical Thinking Questions

Possible responses are:

1. Student responses should address the theme of community cooperation and responsibility.
2. Human qualities range from greed to anger to revenge. The narrator shows a degree of compassion for human frailty (Tortoise does not die), yet cautions against the problems associated with greed and dishonesty.
3. Responses will reflect students' own communities. Students will likely address the lack of community spirit in some areas and point to communal efforts and spirit in others.

☑ Guidelines for Writing Your Response to *Things Fall Apart*

Have students share their journal entries with a partner. Or, as an alternative writing activity, have students work in pairs to describe the age group for which they believe this folk tale would be most appropriate. Point out to students that besides addressing a moral issue, this selection is both humorous and entertaining. Call on each pair to explain and discuss its choice.

Guidelines for Going Back Into the Text: Author's Craft

Discuss with students the difference between "pure" fantasy and the fantasy tale from *Things Fall Apart* in which a strong moral theme is coupled with fantastic events. Discuss the limitations and strengths of a fantasy tale. Ask students to consider whether or not the use of animals (instead of humans) is more or less effective for teaching moral lessons. Possible answers to the questions in the Student Edition are:

1. The life of the animals is pure fantasy: they talk and act like humans. The feat in the sky is also fantastic, including Tortoise's ability to fly with the help of some borrowed feathers.
2. The meals and village life are drawn from real-life details. The characters all exhibit real-life human traits such as pride, greed, anger, and remorse.
3. Realism and fantasy are combined in the description of why tortoises have shells that are "pieced" together.

☑ FOLLOW-UP DISCUSSION

Use the questions that follow to continue your discussion of *Things Fall Apart*. Possible answers are given in parentheses.

Recalling

1. What condition was Tortoise in before the feast? (He was starving and thin.)
2. Where was the feast located? (in the sky)
3. According to the selection, why does Tortoise

have a shell with different sections? (When he fell, the shell broke apart, but a medicine man pieced it back together.)

Interpreting

4. What human characteristics do you think Tortoise is supposed to illustrate? (Possible answers include greed, dishonesty, deceit, selfishness.)
5. What factor do you think convinced the birds to take the tortoise along for the feast? (Possible answers: Tortoise convinced the birds that he had changed, and he impressed them with his ability to speak and the fact that he was widely traveled.)
6. What message do you think the folk tale sends by allowing Tortoise to live after his fall? (This message is one of forgiveness and compassion despite the wrongdoings of Tortoise.)

Applying

7. What "lessons" do you think this selection has for modern communities? Give an example. (Student answers should address the need for people in communities to cooperate and for individuals to be honest and take responsibility for their actions. Answers might also address the need to be aware of "smooth operators" such as Tortoise.)
8. Suppose that after Tortoise fell to earth and recovered, he asked to be included once again in a feast. If he claimed to have "changed" would you allow him to come along? Explain your reasoning. (Students should weigh the past actions of Tortoise, which included a supposed change for the better, and their own conditions for forgiveness.)

ENRICHMENT

Students might benefit from viewing a videotape that shows life in one African village. The videotape, *Africa: Chopi Village Life*, is available from AIMS Media, Inc., 9710 DeSoto Avenue, Chatsworth, CA, 93111-4409, (800) 367-2467.

In the Land of Small Dragon: A Vietnamese Folktale

told by Dang Manh Kha (pages 29–35)

OBJECTIVES

Literary

- to identify the elements of a narrative poem
- to recognize the use of proverbs to convey the values and beliefs in the poem

Historical

- to show how traditional stories provide historical information
- to discuss the social and economic conditions illustrated in the selection

Multicultural

- to compare traditional Vietnamese folk tales to present-day concepts of family and honor
- to list the cultural customs and tradition illustrated in the selection

SELECTION RESOURCES

Use Assessment Worksheet 4 to check students' vocabulary recognition, content comprehension, and appreciation of literary skills.

✔ Informal Assessment Opportunity

SELECTION OVERVIEW

For thousands of years, Southeast Asia has been a crossroads for traders and missionaries carrying goods, ideas, and beliefs. The region's variety of languages and cultures is a testament to these diverse influences. Southeast Asia has never solidified into one entity; mountains and rivers create barriers. Buddhist missionaries from India introduced Buddhism to Southeast Asia from India. Some Hindu myths were also introduced. The large palaces and temples in Southeast Asia are another indicator of India's influence. The area of present-day Vietnam diverged in development from the rest of Southeast Asia around 100 B.C., when the Chinese Han dynasty took control of the region. For the next 1,000 years the area remained under Chinese control. Despite the Chinese influence, the people of the region maintained their own language and customs. They also rebelled frequently. In A.D. 939 Vietnam became an independent kingdom and began its own period of expansion.

In "In the Land of Small Dragon," the heroine through good deeds and a strong moral fiber overcomes the evil plotting of her stepmother and sister to marry the local prince.

ENGAGING THE STUDENTS

Begin by asking students this question: "What does our society value more—physical beauty or intelligence?" As they respond, ask students to give examples to support their answers.

When the discussion is finished, give students this quickwrite: Explain this phrase from the text: "beauty comes from within and is reflected in actions." When the students have finished writing, call on volunteers to read their responses aloud. Tell students to keep their discussion and writing in mind as they read how beauty is defined in the selection.

BRIDGING THE ESL/LEP LANGUAGE GAP

The verse format may intimidate some students, but in this selection the verse is not that difficult. To help students focus on what they *do* understand, first discuss how proverbs are used in the story and point them out to the students (italicized text). Then, write the proverbs and some questions (samples below) on a separate sheet for the students to use as they read, as a guide through the narrative.

1. The main characters are described by the proverbs next to their names below; use three to five words or phrases to describe each character.

    <u>Description</u>

    T'âm: A jewel box of gold and jade
             holds only jewels of great price.

    Cám: An evil heart keeps records
             On the face of its owner.

    Father: He lived his days in justice,
             Standing strong against the wind.

    Mother: Her heart had only one door
             And only Cám could enter.

2. [A man's worth is in what he does . . .] How is this proverb shown to be true in this story?
3. [The stars looked down in pity . . .] Why did the stars have pity on Tâm? What kind of character does T'âm have?
4. [What is to be must happen . . .] and [What is written in the stars . . .] How are these two proverbs related to the lesson of the story?

✔ PRETEACHING SELECTION VOCABULARY

The two daughters described in the poem are very different. Make a list of descriptive words from the poem to indicate the differences. For example:

T'âm	Cám
elder	younger
face a golden moon	long and ugly face
dark eyes	**scowling** and
Rice paddy	discontented
Monsoon	
stepping lightly	**indolent,** slow and idle
no envy or bitterness	filled with hatred

beautiful	frowning
graceful	greedy
gentle	seeking **revenge**

Post the list for students to study before reading so that they can get a sense of character. Ask them to add to the list after reading in order to complete the characterization of the sisters.

The words printed in boldface type are tested on the selection assessment worksheet.

PREREADING

Reading the Story in Cultural Context

Help students read for purpose by asking them to consider the importance that family values play in this selection. Ask them to compare the role of the father in the selection with their own ideas about parenting and parental roles. Finally, discuss the role model this selection sets for children.

Focusing on the Selection

Have students work in pairs. Assign one student in each pair to summarize the underlying message of this selection and the other to summarize the message of the traditional Cinderella story. Afterwards, have pairs collaborate to find common values in both stories. Ask students to evaluate the identification of beauty with goodness in both cultures.

POSTREADING

The following activities parallel the features with the same titles in the Student Edition.

Responses to Critical Thinking Questions

Possible responses are:

1. The underlying theme is that beauty reflects a good heart and good deeds.
2. Sample responses: dedication of children to parents; the importance of honor; one can judge goodness from actions; people are worth what

they do, not what they say; beauty is the same as goodness.

3. Answers will reflect students' beliefs. Guide them to understand the correlation between beauty and goodness illustrated in the tale.

✔ Guidelines for Writing Your Response to "In the Land of Small Dragon"

Have students share their journal entries with a partner. Or, as an alternative writing activity, ask each student to write in a few sentences the lesson that he or she believes the story is trying to teach. Next, lead a discussion on how teenagers in particular perceive beauty. Ask: "Do you think teenagers, more than adults, judge one another on physical attractiveness? Explain your answer." Next, call on students to read their interpretations of the story, and ask: "What message does this story have for teenagers in the 20th century? Does the message of this story hold up today?"

Guidelines for Going Back Into the Text: Author's Craft

Help students to understand the impact a narrative poem can have by comparing the narrative technique used in the selection to that of many popular songs. Some similarities include: using verses, the use of four- and six-line stanzas, and the use of a refrain or chorus.

Possible answers to the questions in the Student Edition are:

1. The selection is in verse form; the selection describes characters involved in a plot; the selection revolves around a central conflict, the use of a refrain, and the use of stanzas.

2. Answers will reflect students' own preferences, but should analyze one of the elements listed above.

3. Answers will reflect students' own reading.

✔ FOLLOW-UP DISCUSSION

Use the questions that follow to continue your discussion of "In the Land of Small Dragon." Possible answers are given in parentheses.

Recalling

1. How did Cám become the number one daughter? (by stealing T'âm's catch of fish)

2. Where did the silken dress and jeweled hai come from? (from the bones of T'âm's pet fish)

3. Who helped Tâm clean the rice so that she could go to the festival? (a flock of blackbirds)

Interpreting

4. What do you think this opening phrase from the selection means: "One cannot know the whole world, But can know his own small part"? (Answers will vary, but should revolve around understanding human nature and our own selves.)

5. How would you explain T'âm's father's behavior toward both of his daughters? (His actions illustrate the dilemma parents face in trying not to show preference among children.)

6. What is the moral of this story? (Good deeds and actions will triumph, and beauty is the result of goodness.)

Applying

7. What traditions about family from this selection do you think are applicable today? (Sample answers: respect for parents, obeying the wishes of parents even in the face of injustice)

8. If you had been T'âm with a sister like Cám, how would you have dealt with her deceitfulness? What does this illustrate about your own personality? (Students will likely express an unwillingness to take the injustices as meekly and quietly as T'âm did.)

ENRICHMENT

Students might benefit from viewing any of the three videotapes in the series, *Asians in America*. The series is available from Centre Productions, Inc., 1800 30th Street, Suite 207, Boulder, CA, 80301, (800)824-1166.

MAKING CONNECTIONS VIETNAMESE CULTURE

OBJECTIVES

Overall

- to discover the connections between Vietnamese culture and the themes in "In the Land of Small Dragon: A Vietnamese Folktale"

Specific

- `to compare the cultural traditions expressed in "In the Land of Small Dragon: A Vietnamese Folktale" with those represented in a Vietnamese ceramic artwork and Vietnamese handwriting
- to identify Vietnamese cultural traditions
- to appreciate aspects of Vietnamese culture

ENGAGING THE STUDENTS

Give students a quickwrite on the following question:

> If you could pass on just one family or cultural tradition to your own children, what would it be? Why?

Before students share their writings, brainstorm a list of definitions for the word tradition.   Discuss why people find traditions important. Ask volunteers to read their work aloud. Finally, ask students to name some of the Vietnamese traditions that were included in "In the Land of Small Dragon: A Vietnamese Folktale."

BRIDGING THE ESL/LEP LANGUAGE GAP

Use the photographs on pages 36 and 37 as well as other illustrations in Theme 1 to helps students explore various aspects of cultural traditions and traditional stories, such as "The Stars" by Pablita Velarde.. Have them study each illustration, and explain orally what it suggests about aspects of traditions. Invite students to share their own traditional stories.

EXPLORING ART

Vietnamese architecture has been called landscape painting because of the value it places on achieving harmony with nature. Notice how the lines of the roof of the building pictured on page 36 are softened by being curved instead of being straight and rigid. This allows the artwork to blend with the curving lines of nature that surround the structure. In addition to being harmonious with nature, art has been a part of everyday life in Vietnam for hundreds, if not thousands, of years. Ask: "How do the ceramics shown here reflect nature as well as everyday life?" Guide students in reading the captions and discussing the artworks. Tell students that calligraphy is an art form of high status that takes years to master and that children often begin studying it before they enter school. Ask: "Why is it important that this ancient art form has become an integral part of Vietnamese American culture?" Point out that Vietnamese art is characteristically decorative and curvilinear, meaning with many curved lines. Challenge students to discover how the artworks shown here reflect these traditions (for example, curved lines of roofs and arches, decorative details on the walls).

Invite students to compare and contrast the Temple of Literature with other famous Vietnamese architectural structures, including the Doi-Son Pagoda, Ha-Loi Temple, Huong-Tich Pagoda, Kiep-Bac Temple, One Pillar Pagoda, and Thien-Mu Pagoda.

LINKING LITERATURE, HISTORY, AND THEME

Guidelines for Evaluation

Student answers should include themes of nature, and respect for knowledge, history, family, and tradition.

Students should note that the major theme of the folktale is allegiance to family. Several of the proverbs in the selection also suggest that the Vietnamese culture places value on goodness, nature, and honesty.

Guide students to understand how art and literature express cultural values and how these forms of expression can be preserved for future generations. Furthermore, students should note that art and literature are forms of expression that can be understood and appreciated by many different generations.

The Man Who Had No Story

by Michael James Timoney and Séamus Ó Catháin (pages 38–43)

OBJECTIVES

Literary

- to identify plot sections and phrases that are repeated in a story
- to evaluate the effects of repetition in a story

Historical

- to show how traditional stories can teach generations about historical peoples and places
- to discuss how traditional stories can be used to express political viewpoints

Multicultural

- to compare Irish folk tales with traditional storytelling from other cultures.
- to list the cultural traits that make Irish folk tales unique

SELECTION RESOURCES

Use Assessment Worksheet 5 to check students' vocabulary recognition, content comprehension, and appreciation of literary skills.

☑ Informal Assessment Opportunity

SELECTION OVERVIEW

The reference to "bad times" in "The Man Who Had No Story" is to the conflict between Ireland and England, which was based in part on the religious differences between Roman Catholic Ireland and the largely Protestant England. Ireland had come under English rule during the reign of Henry VIII (1509-1547). Henry VIII, Elizabeth I (reigned from 1558-1603), James I (reigned from 1603-1625), and Charles I (reigned from 1625-1649), all ruled Ireland and faced the necessity of putting down Irish rebellions. In the mid-1600s, Oliver Cromwell invaded Ireland. During the invasion, the English took lands and homes. Entire counties were given to English soldiers, and the Irish inhabitants were driven out.

In this selection the main character is transported through a number of fantastic voyages to a fairyland. These experiences provide him with a "story," and thus the man with no story finally has one to tell.

ENGAGING THE STUDENTS

Poll the class on this question: "Do people believe in ghosts? fairies? trolls?" Discuss the responses. Challenge students to find reasons why these imaginary beings hold so much fascination for many cultures. Point out that even in present-day U.S. society, people continue to be fascinated with and entertained by stories and accounts of contact with imaginary beings. Ask: "What entertainment value do stories about fairies and little people have?"

BRIDGING THE THE ESL/LEP LANGUAGE GAP

Although most of the vocabulary in this story is simple, the fantasy aspect of the narrative could be confusing to students, because the plot is unpredictable and some words are unfamiliar. Help students to identify the narrative elements with the activities described below.

Before they read, give the students a vocabulary list and quickly define for them the following words as they are used in the story: *rods* (straight, slender sticks growing on trees or bushes), *fairy tale* (a story about a human's adventures with fairies), *wake-house* (a house in which a watch over a dead person is taking place), *bow and fiddle* (a stringed instrument like a violin and the bow that is drawn across the strings to make music), *coffin* (a box in which a corpse is buried), *lance* (a spear-like weapon).

After the students have read the story, organize them into groups of two or three and ask them first to answer these questions: (1) According to the story, what are the three things Brian never did in his life? (2) At the beginning of the story, Brian says, "I'll do anything except tell a story." How does this statement come true? (3) What are the phrases that are repeated in the story?

After students have identified the narrative elements of the story by answering the questions above, ask them to retell the story to one another, using their notes and vocabulary list.

☑ PRETEACHING SELECTION VOCABULARY

On the board write these phrases from the selection: **fairy glen**, **fairy tale**, **terrible fog**, **Fenian Tale** and **wake-house**. Challenge students to guess what kind of story they are about to read using these phrases as clues. Ask the questions below to guide them:

- Is this selection a quest tale? a moral tale?
- Will this selection have a serious or light tone?
- What do you think "The Man Who Had No Story" is about? Explain your answer.

The words printed in boldface type are tested on the selection assessment worksheet.

PREREADING

Reading the Author in Cultural Context

Help students read for purpose by asking them to consider the importance economics plays in this selection. Suggest that the main character's motivation to explore the fairy glen comes partly from the economic hard times. Discuss what the author might be suggesting by this focus.

Focusing on the Selection

Tell students to write a short anecdote that illustrates a characteristic of their own neighborhood, community, or country. Stipulate that they provide some detail on everyday life in their writing. When they have finished, call on them to read their writing aloud. Discuss how the student authors conveyed a sense of their own backgrounds.

POSTREADING

The following activities parallel the features with the same titles in the Student Edition.

Responses to Critical Thinking Questions

Possible responses are:

1. Responses: animosity for the English, belief in fairies, wakes, fiddler music, dancing
2. One overall theme of the selection is that the main character is at a disadvantage until he has a story to tell. The characters he meets expect him to have a story to tell, suggesting that this talent is a long-held tradition.
3. Responses will reflect students' own cultural backgrounds.

☑ Guidelines for Writing Your Response to "The Man Who Had No Story"

Have students share their journal entries with a partner. Or, as an alternative writing activity, have students work in pairs to write a list of all the events in the selection that are not "real." Encourage students to discuss the purpose of these fantasies and how they are used to convey a theme. Discuss what they have learned about Irish folk tales.

Guidelines for Going Back Into the Text: Author's Craft

Help students understand the universal appeal of a fairy tale by asking them to consider some movies with characters that have these qualities. (Examples: *Peter Pan*, *The Wizard of Oz*) Have students compare these characters with those in the selection. Challenge students in their comparisons to find instances of repetition. Possible answers to the questions in the Student Edition are:

1. Each time that Brian gets lost he sees a light. First he meets an old couple, and next he comes to the wake.
2. Repeated phrases include: "Where there is light, there must be people," "Oh, that is something I never did in my life." The repetition gives the reader a sense of continuity and recognition.
3. Answers will reflect students' own reading.

✔ FOLLOW-UP DISCUSSION

Use the questions that follow to continue your discussion of "The Man Who Had No Story." Possible answers are given in parentheses.

✔ RECALLING

1. Why did Brian enter the fairy glen? (to cut rods)
2. What was happening in the second house Brian visited? (a wake)

Interpreting

3. What purpose does the storyteller serve by mentioning the English? (This is criticism of the English control of Ireland.)
4. What do you think the reader is supposed to learn from this tale? (Answers will vary, but guide students to consider the entertainment value of this selection as well as the emphasis on the importance of traditional storytelling.)

Applying

5. What did the selection teach you about Irish culture? (Sample answers: importance of storytelling, animosity toward the English, details of an Irish wake, belief in fairies)

ENRICHMENT

Students might benefit from viewing the videotape *Ireland: An Introduction*, which deals with the history of Ireland. The videotape is available from Centre Productions, Inc., 1800 30th Street, Suite 207, Boulder, CO, 80301, (800) 824-1166.

RESPONSES TO REVIEWING THE THEME

1. Themes on morality and lessons about life include: relationship of humans to nature, greed, honesty, dedication, duty, and honor.
2. Students should note that all selections are folk tales but with differences in form. "The Stars" is a quest tale. The story from *Things Fall Apart* and "The Sheep of San Cristóbal" are moral tales. "In the Land of Small Dragon" is a fantasy poem with a moral lesson. "The Man Who Had No Story" is a fantasy tale. The storytellers used different techniques. In "The Stars" and "The Sheep of San Cristóbal," the tone is serious. The storytellers in *Things Fall Apart* and "The Man Who Had No Story" use humor to entertain and teach.
3. Effects of the selections on students' perspectives may include: nature and animals ("The Stars"); family honor ("In the Land of Small Dragon"); community values (*Things Fall Apart* and "The Sheep of San Cristóbal").

✔ FOCUSING ON GENRE SKILLS

Dang Manh Kha uses several elements of narrative poetry effectively in "In the Land of Small Dragon: A Vietnamese Folk Tale." She tells a story in verse form which has a plot, characters, and dialogue. The italicized lines serve as a type of refrain, echoing the relationship of beauty and goodness, evil and ugliness. Select, or have students select, another narrative poem (such as Paul Laurence Dunbar's "How Lucy Backslid," written in dialect; or Gladys Cardiff's "Candelaria and the Sea Turtle,") and ask students to read the poem and identify narrative elements such as plot, characters, dialogue, and refrain.

BIBLIOGRAPHY

Books Related to the Theme:

Erdoes, Richard, and Alfonso Ortiz, eds. *American Indian Myths and Legends.* New York: Pantheon Books, 1984.

Griego, José, *Cuentos: Tales from the Hispanic Southwest.* Santa Fe: Museum of New Mexico Press, 1981.

2

Exploring Ancestral Roots

THEME PREVIEW

Exploring Ancestral Roots			
Selections	**Genre/Author's Craft**	**Literary Skills**	**Cultural Values & Focus**
from *Bless Me, Ultima* Rudolfo A. Anaya, pages 26–28		mood; sensory details	Mexican American heritage; conflict between old and new ways of life
from *Roots*, Alex Haley, pages 29–31	historical fiction	historical fact vs. fiction	African ancestral roots; studying African society through historical fiction
"To Da-duh, In Memoriam," Paule Marshall, pages 32–34	biography	internal and external conflict	conflicts between different generations; differences between African Americans and blacks living in the West Indies
from *The Way to Rainy Mountain*, N. Scott Momaday, pages 35–37	nonfiction narrative	round or flat characters	importance of place to Native Americans; discovering ancestral roots by visiting homelands
"i yearn," Ricardo Sanchez ["We Who Carry the Endless Seasons"], Virginia Cerenio, pages 38–40 biography	free verse poetry	line placement and line length	cultural identity; understanding how cultural identity links people to their heritage

Assessment

Assessment opportunities are indicated by a ✔ next to the activity title. For guidelines, see Teacher's Resource Manual, page xxxiii.

CROSS-DISCIPLINE
TEACHING OPPORTUNITIES

Social Studies Besides having political goals, many of the culture-based protest groups that evolved in and since the 1960s have also advocated the exploration and discovery of ancestral roots as a way of establishing a renewed interest in cultural identity. Have interested students research the goals and beliefs of groups such as the American Indian Movement (AIM), the Black Panthers, the United Farmworkers movement (made up largely of Chicanos), and La Raza to discover how these groups used ancestral roots to help forge their political and social agendas.

Music An exciting way to explore ancestral roots is through music. Many traditions, beliefs, and stories have been passed down from generation to generation through traditional folk songs, chants, and spirituals. Many traditional forms of music have been revitalized and rediscovered by present-day musicians. The musical group Los Lobos, for example, combines traditional Mexican American and modern musical forms. A recording of Native American music, *Desert Dance* by R. Carol Nakai, is another source of traditional music by a contemporary artist.

Geography Ask students: "If you were to go on a quest for your ancestral roots, where would you travel?" Challenge students to map out the countries and routes they would take on their own quest for their ancestral roots. Tell them to search as far back as possible for their "roots."

SUPPLEMENTING
THE CORE CURRICULUM

The Student Edition provides humanities materials related to specific cultures in Making Connections on pages 36–37 and in the art insert on pages A1–A16. You may wish to invite students to pursue the following activities outside of class.

- Have students examine the three works of art on pages A2 and A3. Tell them that works of art are often valued for what they tell us about their creators as they are for their beauty. Ask students what the Taino vessel tells them about the culture that inspired such craft. Discuss with them why the particular shape was chosen and what

significance the decorations might have. Similarly, ask them how the Mexican santo reflects something special about Mexican American culture. Finally ask them what modern art works they can think of that may give future generations information about the culture that created them. Ask them if the *Rice Eaters*, a modern work, provides any cultural information.

- A number of the selections in Theme 2 address the migration of families from rural to urban settings (*Bless Me Ultima*, "To Da-duh, In Memoriam," *The Way to Rainy Mountain*). Lead a class discussion about how the migration of people from farming villages to cities affects the development of many cultures, and how urbanization influences our society. Students who are interested in exploring this further could find valuable information in *Urban Legacy: The Story of America's Cities*, by Dianna Klebanow, Franklin L. Jonas, and Ira M. Leonard (NAL-Dutton, a division of Penguin USA, 1977).

- Tell students that American educators have been criticized in the past for ignoring the cultural diversity of American history. Discuss with students how cultural bias in the past might have prevented students from exploring and discovering their ancestry. Brainstorm with students about possible ways that they would like to infuse multicultural studies into their studies. Have them suggest specific ways they would emphasize the exploration of each individual's ancestry.

- Technological advances, especially video recorders and highly sensitive microphones, offer new ways for people to record and document their lives. Point out to students, for example, that even recordings of weddings and birthday parties could provide links between one generation and the next. Discuss with students how these technologies could be used to preserve cultural traditions and to document one's heritage. A readable study of technology's impact can be found in Robert C. Toll's *The Entertainment Machine* (Oxford University Press, 1982).

INTRODUCING THEME 2:
EXPLORING ANCESTRAL ROOTS

Using the Theme Opener

On the chalkboard write: "Who are you?" Challenge students to list all of the nouns they would use to define themselves. Guide them by giving a few examples—everyone in the class is a *teenager*, a *student*, a *son or daughter*, and so on. Call on students to find as many of these categories as they can. Point out that each student will define himself or herself in a unique way. When you have finished, survey the list and ask: "How many of you defined who you are by who your ancestors were?" Tell students that some sociologists believe ancestral roots are the most important influence on an individual's personality. One sociologist has stated, "The behavior of an individual is determined not by his racial affiliation, but by the character of his ancestral roots and his cultural environment." Ask students to discuss the importance of one's "ancestral roots" and "cultural environment."

Conclude your discussion by reading this quotation aloud:

> If you write about the things and the people you know best, you discover your roots. Even if they are new roots, fresh roots. . .they are better than. . . no roots.

Discuss what the author means by *fresh roots*. Ask: "How can you discover new roots to your own past?" Suggest to students that in *Tapestry* they may have opportunities to discover their own ancestral roots.

☑ Developing Concept Vocabulary

Illustrate the concept of *Roots* through a simple graphic representation. Draw on the board a large tree and label it *family*. Then draw many roots for the tree. Explain that the soil from which the tree springs is the *ancestral past*. Write this label on the board. Then explain that the roots are the important elements of ancestry that feed the family and help to keep it alive. Label some of the roots: *customs, values, ethnicity, beliefs*. Ask students to draw their own trees and label the roots with descriptions specific to their families, based on their responses to the following questions:

- How do ancestral roots link people to their past?
- Do cultures change? Is your set of cultural values and traditions the same as your parents'? What are some similarities? Some differences?
- How do we discover our ancestral past? What role do family members play in this discovery? How can literature help us to discover our ancestral roots?
- If you were a parent, what single idea, belief, or custom would you want your children to inherit from you?
- How can an understanding of the past give you a new perspective on the present? Provide an example.

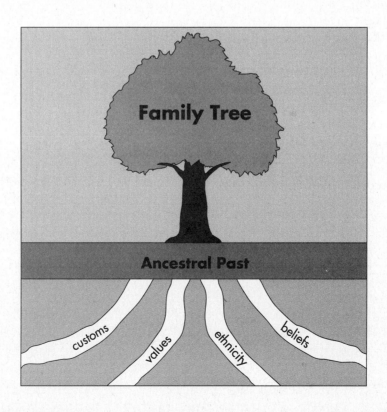

from *Bless Me, Ultima*
by Rudolfo A. Anaya (pages 45–55)

OBJECTIVES

Literary
- to recognize the elements of foreshadowing, dialogue, and sensory details
- to understand the tension in the writer created by the influence of Christian and native spiritual values

Historical
- to describe the way of life for Mexican Americans in the early 20th century
- to discuss the conflict between rural and urban Mexican American populations

Multicultural
- to understand the author's perspective on cultural changes
- to become aware of conflicts between traditional ways and new ways in Hispanic culture

SELECTION RESOURCES

Use Assessment Worksheet 6 to check student's vocabulary recognition, content comprehension, and appreciation of literary skills.

 Informal Assessment Opportunity

SELECTION OVERVIEW

The author of this selection writes about his ancestors in the present-day southwestern United States during the era of ranchos. In the early 19th century, the area was part of Mexico, and many of the ranchos were small land grants, run by vaqueros, or cowboys. Women were central to this society both as laborers and managers. This selection highlights the urban migratory pattern a century later when many of the vaqueros and their families, now citizens of the United States, moved from the ranchos into towns. As the story suggests, many families struggled with the resulting changes in their way of life. They also found themselves having to adjust to living in an area that was not predominantly Hispanic. The growing demand for Mexican labor in the 20th century brought more Mexicans northward—now as immigrants—creating even greater changes for Mexican Americans already living in the United States.

In this selection, the narrator, a young boy, describes the arrival of Ultima—a healer who had helped during the birth of the boy. The story revolves around the boy's growing perceptions of the world and the conflict he faces between two ways of life.

ENGAGING THE STUDENTS

On the board, write: "Do your parents want you to be just like them?" Tell students to consider what expectations and hopes their family holds for each one of them. Have them write a short paragraph on this topic: "How do family values and traditions reflect cultural heritage?" To start them thinking on this subject, pose the questions below:

- Do parents and children usually agree on what is best for the children?
- Is tradition still important in U.S. society?
- What can we learn about ourselves by discovering our ancestral roots?

When students have finished writing, tell them that in this selection the main character faces a conflict between two ways of life. Challenge the students to compare this conflict with those they might have in the future as they make choices in their own lives.

BRIDGING THE ESL/LEP LANGUAGE GAP

The style of the prose and the vocabulary in this selection will be difficult for many students, and they will probably need to read the selection more than once. Before the first reading, ask them to write, as they read the story the first time, everything they read in the text that tells them something about Ultima: Who is she? What is she like? Why is she so important to Antonio's parents? to Antonio?

After they have read the story once, have them work in mixed groups of both limited and proficient English speakers to list some of the conflicts they see in the story. (They may need to reread the story with this focus.) They should see elements of both the conflicts between Antonio's mother's and father's backgrounds and families and the conflict between what Antonio wants his future to be and what his mother wants for him. Ask them to discuss why Ultima is so important to Antonio in resolving these conflicts.

☑ PRETEACHING SELECTION VOCABULARY

Have students create word banks by writing all the terms in the selection that are in Spanish. (Examples: **llano, crudo, curandera, brujas,** adobe, **molino**) Using the footnotes in the Student Edition, review these terms and discuss their meanings. Use the questions below to lead a guided discussion of the author's use of these terms:

• Why do you think the author chose to use Spanish words instead of English ones in his writing?
• Which of the word bank terms are characteristic of Mexican American culture?
• Based on this vocabulary activity, what do you think *Bless Me, Ultima* is about?

The words printed in boldface type are tested on the selection assessment worksheet.

PREREADING

Responding to Literature: Modeling Active Reading

Bless Me, Ultima is annotated with the comments of an active reader. These sidenotes, prepared to promote critical reading, emphasize the cultural content of the piece, address author values, call attention to literary skills, invite personal response, and show how the selection is related to the theme. If you have time, read the entire selection aloud as a dialogue between reader and text. Encourage students to discuss and add their own responses to the ones printed in the margins. Model these skills by adding your own observations, too.

Reading the Author in Cultural Context

Help students read for a purpose by asking them to consider the cultural conflict faced by the main character in the selection. Discuss the struggle between the traditional life of Antonio's father and the hopes of his mother.

Focusing on the Selection

Ask: "Do we explore ancestral roots to find differences between the past and the present, or to find similarities?" Tell students to write a goal that they believe their family members have for them. When they have finished writing, ask: "What goals do Antonio's parents have for him?" (education, good manners, respectfulness) Ask: "How do goals for children change as the children grow older?" Discuss the students' responses.

POSTREADING

The following activities parallel the features with the same titles in the Student Edition.

Responses to Critical Thinking Questions

Possible responses are:
1. Student responses may vary, but should address the conflict Antonio feels between his attraction to the life of a vaquero and his present life in town.

2. Student responses should include information the author provides about the past, including the migration of his family and the conflict between his mother's people and his father's people.

3. Guide students to compare the family traditions in the selection with those of their own families.

☑ Guidelines for Writing Your Response to *Bless Me, Ultima*

Have students share their journal entries with a partner. Or, as an alternative writing activity, on the chalkboard make two categories: *Antonio's Father* and *Antonio's Mother*. Call on students to list the differences between the two characters. When you have finished, ask students to write responses to these questions: "Despite these differences, what bonds make Antonio's parents close? Do they agree on the arrival of Ultima?" Tell students to think of the factors that tie families together as they write their response.

Guidelines for Going Back Into the Text: Author's Craft

Remind students that Anaya wrote this novel as part of his journey into the language and traditions of his ancestry. Suggest to students that they too may some day want to discover their ancestral roots and record their findings. Ask: "How did the author of *Bless Me, Ultima* use sensory details to describe traditions, values, and culture?" Have students work in pairs to make a list of four methods (two per student) the author used to explore his ancestral roots. Call on student pairs to read their writing aloud and discuss the different interpretations. Possible answers to the questions in the Student Edition are below:

1. Sample answers: sight (looked down the slope to the green of the river), tactile (dug in dirt till his hands bled)

2. Sample answers: smells (Ultima's fragrance of herbs), sounds (listens to the sounds of activities in the house and a repeating "Hail, Mary" from his perch in the attic)

3. Student answers will reflect students' own perceptions, but should provide an example from the text of an effective description.

☑ FOLLOW-UP DISCUSSION

Use the questions that follow to continue your discussion of *Bless Me Ultima*. Possible answers are given in parentheses.

Recalling

1. Why did Antonio's family leave La Pastura? (His mother convinced his father that there was more opportunity—including the ability to attend school—in the town.)

2. Where did Antonio live? (in Guadalupe)

3. Where was Antonio working? (in a garden)

Interpreting

4. What purpose did the dream serve in the story? (The dream was of a fight between the people of his mother and of his father at his birth.)

5. In your opinion, was Antonio's father happy in Guadalupe? (Students should weigh his father's longing for his old life as a vaquero and the life he had created in Guadalupe.)

6. Was Ultima a symbol of the past or the present? Explain your answer. (Students should note that Ultima in many ways represents traditional customs in medicine and healing.)

Applying

7. Do you think the author is at ease with his past? Explain your answer. (Students should address the conflicts between life in town and life as a vaquero, and the longing for the past the author expresses through the eyes of the child.)

8. What lessons does this story hold for Americans today about family and responsibility? (Guide students to understand how values such as respect for elders and taking responsibility for those in need are applicable today as well. The underlying theme of respect for tradition embodied in the character of Ultima is not obsolete.)

ENRICHMENT

Students may benefit from viewing the videotape *Chicano from the Southwest*, which deals with a young boy's attempt to work out conflicts between his traditional family upbringing and the stresses of city life. The videotape is available from Encyclopedia Britannica, Educational Corporation, 310 S. Michigan Avenue, Chicago, IL, 60604.

from *Roots*
by Alex Haley (pages 56–61)

OBJECTIVES

Literary
- to distinguish between fact and fiction in the selection
- to understand the importance of the author's cultural heritage in his search for self

Historical
- to describe African concepts of justice
- to compare African social structures to present-day U.S. society

Multicultural
- to recognize cross-cultural themes
- to describe some African customs and family traditions

SELECTION RESOURCES

Use Assessment Worksheet 7 to check student's vocabulary recognition, content comprehension, and appreciation of literary skills.

✔ Informal Assessment Opportunity

SELECTION OVERVIEW

The present-day nation of Gambia was initially populated by the Wolof, Malinke, and Fulbe peoples in the 13th century. By the 1400s, large kingdoms had been established throughout this region of West Africa. Europeans who first began exploring and raiding the coastal lands of West Africa encountered what they assumed to be small, isolated villages. In fact, however, these coastal villages were outposts of well-organized African kingdoms and states. Europeans soon realized that the small gains from raiding these outposts were nothing compared to what could be gained from trade agreements with these rich kingdoms. Besides benefiting economically by having friendly relations with the coastal African kingdoms, Europeans also protected themselves from the military force of the West Africans. The African kingdoms of this region supported large, well-organized armies. According

to one governor of Portugal during this time, the king of the Wolof people could "put into the field an army of about 10,000 cavalry and 100,000 footmen." Guide students to understand that West Africa during this era contained a number of well-organized, politically advanced, and wealthy kingdoms.

In this selection, the author describes the observations and thoughts of Kunta as he attends a Council of Elders meeting in the village of Juffure in what is now The Gambia, West Africa. The selection offers keen insights into the social structure and cultural values of the people.

ENGAGING THE STUDENTS

On the board, write the following question: "Is there such a thing as justice?" Use the questions below to lead a guided discussion of this topic:

- Are concepts of "right" and "wrong" basically the same for all people? For example, are the exact same actions by children considered "wrong" in every family?
- How does our society decide issues of guilt and innocence? (courts and jury system)

Tell students that by reading the selection from *Roots* they will have the opportunity to witness how justice in one African village was administered centuries ago. Suggest that they might be surprised to find a number of similarities between the centuries-old system and our own justice system.

BRIDGING THE ESL/LEP LANGUAGE GAP

Before students read this selection, make sure they understand the concept of the Council of Elders. Have students role-play a Council scenario, and use words from the selection to define what the students are doing: The meeting is called a *hearing* or *session*. The people coming before the Council are the *petitioners*. They might be there to have a *dispute resolved*; to *accuse*, or *charge*, someone else of wrongdoing; or to seek *approval*, or *permission*, for an action they want to take. The Council will listen to the people *testify* and will weigh the *evidence* before deciding on the *case*.

Write the words on the board, and identify them with the action of the role-play. Encourage students to look for these words as they read the selection.

☑ PRETEACHING SELECTION VOCABULARY

Write the following words on the board and ask students what the words have in common: conflict, **decrease, succession, irate,** furious, **argumentative,** formidable, **irritable.** Discuss how the words have emotional connotations. Explain that the excerpt from *Roots* deals with a judicial system—and that anytime someone's fate is being decided, emotions will run high. Suggest that students note other emotionally charged words they encounter as they read the selection.

The words printed in boldface type are tested on the selection assessment worksheet.

PREREADING

Reading the Author in Cultural Context

Help students read for a purpose by asking them to consider the importance tradition plays in the African community described in the selection. Students should address topics such as Kunta's relationships in the community, traditional laws and how they are passed on, and the maintenance of law and order.

Focusing on the Selection

Ask: "Did the characters and places in the selection seem foreign or familiar to you?" Point out that though aspects of the culture are different, the issues discussed by the Council of Elders are very similar to those we face daily in our own society. Discuss the students' responses and use them to help answer the questions in the student text.

POSTREADING

The following activities parallel the features with the same titles in the Student Edition.

Responses to Critical Thinking Questions

Possible responses are:

1. From the selection we can gather that Haley's ancestral roots lie in a small African village with strong traditions.
2. Guide students to understand the importance of ancestral roots for African Americans and their ongoing struggle for identity and equality in the United States.
3. In general, students should see the relative equity and wisdom of the Council, including the similarities to our own judicial system.

☑ Guidelines for Writing Your Response to *Roots*

Have students share their journal entries with a partner. Or, as an alternative writing activity, ask each student to briefly summarize the selection. Next, have students exchange their writing with a partner. Ask them to collaborate on a single summary of the main points including the scene each student found most memorable.

Guidelines for Going Back Into the Text: Author's Craft

Ask: "Is reading historical fiction a good way to learn about history? How can the reader determine what is fact and what is fiction in this type of writing? What can be gained from historical fiction?" Challenge students to discuss this type of writing in terms of the purpose for reading it. Possible answers to the questions in the Student Edition are:

1. facts about how justice and government were carried out in the village, facts about how the villagers made money, facts about crime and wrongdoing among the villagers, facts about the value system of the village including views on marriage and slavery

2. details about wrongdoing and personal relationships, Kunta's personal thoughts on his own role in the community as he watched the Council

3. Answers will reflect students' own readings.

☑ FOLLOW-UP DISCUSSION

Use the questions that follow to continue your discussion of the excerpt from *Roots*. Possible answers are given in parentheses.

Recalling

1. How many elders were in the Council? (six)

2. What Council sessions attracted the most people? (sessions that dealt with matters of the people)

3. How did the village summon traveling magic men? (by drumtalk)

Interpreting

4. How would you describe Kunta's feelings about the decisions of the Council? (Kunta describes them matter-of-factly and accepts the Council's decisions.)

5. In your opinion, was the Council too harsh when the marriage permission was refused to a man who had stolen a basket long ago? (Student answers will vary, but it is likely there will be disagreement on this issue.)

6. How would you describe the relationship of master to slave in the village? (The relationship almost seems to be one of employer to employee, since the Council listened to grievances made by slaves on working conditions and wages. Slaves also married into the master's family.)

Applying

7. Consider some of the decisions made by the Council. Do you think a court today would have made the same decisions under similar circumstances? (A court today probably would have made decisions similar to those of the Council. Guide students to analyze the universality of the justice concepts dealt with by the Council.)

8. Give two ideas young Americans today could learn from reading this selection. (Some possible answers: Africa, contrary to some people's notions, had developed effective judicial and political systems before European contact; there are many similar cultural values between present-day U.S. society and earlier African societies.)

ENRICHMENT

Students might benefit from viewing any of the six videotapes in the *Roots* series. The videotapes are available from Warner Home Video, Inc., 4000 Warner Boulevard, Burbank, CA 91522-0001, (818) 954-6000.

To Da–Duh, In Memoriam

by Paule Marshall (pages 62–74)

OBJECTIVES

Literary

- to identify internal and external conflicts in a story
- to recognize the author's attitudes toward two disparate cultures

Historical

- to compare civil rights for African Americans in the United States during the 1930s with blacks living in Barbados during the same era
- to understand the impact of colonial rule in the West Indies

Multicultural

- to compare and contrast African American society in the United States with black society in Barbados
- to describe the customs and culture of the blacks of Barbados

SELECTION RESOURCES

Use Assessment Worksheet 8 to check students' vocabulary recognition, content comprehension, and appreciation of literary skills.

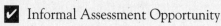 Informal Assessment Opportunity

SELECTION OVERVIEW

By the time of the Great Depression, African Americans were in the midst of transformation from an essentially rural to a mostly urban people. In 1900, nearly 90 percent of the African American population lived in the predominantly rural South. During World War I, African Americans began migrating in growing numbers to the industrial centers of the North. They were lured by the promise of jobs and a standard of living higher than they were able to have as sharecroppers in the southern United States. Though African American urban migration slowed temporarily during the Great Depression, it gained momentum again during World War II. Conversely, the island nation of Barbados remained largely agrarian, and most of its black population lived in rural areas. Many blacks living in Barbados continued to harvest tobacco, sugar cane, and cotton—living in conditions not that far removed in some ways from slavery.

In this selection, a young girl from New York City visits her grandmother in Barbados. Together they discover some ways in which their lives are different. They also discover important ways that their lives are similar as well as deep bonds that are not immediately obvious to them.

ENGAGING THE STUDENTS

Have students do a quickwrite on the following topic: "Imagine you have embarked on a long voyage to find your ancestral roots. Where would you search? What people would you want to meet? What would you want to learn?" Allow students five minutes to write their responses, then call on volunteers to read their answers aloud.

Discuss the students' writing. Explain that in this selection they will read about a girl's visit to her ancestral roots that she gains as a result. This knowledge changes not only her perceptions of the past, but her perceptions of the present as well.

BRIDGING THE ESL/LEP LANGUAGE GAP

This story is about two people who have a bond because they are family — Da-duh and the little girl — yet who are very different and who live in very different worlds: Brooklyn and Barbados. Form groups of three to four students and ask them to read small sections of the text to themselves and then to discuss each section within their groups, looking for the differences that the author sees between Brooklyn and Barbados.

Ask the students to read the following passages in the Student Edition and discuss the questions that follow them.

> . . . and her voice almost broke under the weight of her pride, "Tell me, have you got anything like these in that place where you were born?" (p. 68)

Ask: Why does the little girl mention Da-duh's pride in this passage?

> For long moments afterwards Da-duh stared at me as if I were a creature from Mars, an emissary from some world she did not know but which intrigued her and whose power she both felt and feared. (p. 70)

Ask: Why is Da-duh afraid?

> [Da-duh said:] "All right, now, tell me if you've got anything this tall in that place you're from." (p. 71)

> "I almost wished, seeing her face, that I could have said no.." (p. 71)

Ask: Why does the little girl wish she could have said no?

☑ PRETEACHING SELECTION VOCABULARY

The women in this story are all very strong characters. Inform students that they will see the different strengths of the women as they read. Suggest that these words from the selection indicate strength: **unrelenting, formidable, truculent,**

dissonant, flaunting, **roguish.** Have students scan to find the words and figure out which woman in the story each describes. Ask students to list other words they encounter as they read that indicate the strength of the girl, the mother, and the grandmother.

The words printed in boldface type are tested on the selection assessment worksheet.

PREREADING

Reading the Author in Cultural Context

Help students read for purpose by asking them to consider the importance that remembrances have for the narrator in her search for her ancestral roots. Student answers should address the changes in perception the main character experiences after her trip to Barbados, and perhaps even later as she recalls the trip as an adult.

Focusing on the Selection

Have students form pairs. Have one student in each pair imagine that he or she lives in New York City; the other should imagine that he or she lives in Barbados. Each must list the characteristics of his or her assigned location. Tell students that their lists must contain social (ways of life) characteristics as well as geographical differences. Afterwards have each pair collaborate on a single paragraph summarizing the differences between New York and Barbados. When they have finished, ask students what values separate the young child and her grandmother and which values bond them together.

POSTREADING

The following activities parallel the features with the same titles in the Student Edition.

Responses to Critical Thinking Questions

Possible responses are:

1. Sample response: The author, through the eyes of the child, expresses love for her grandmother

and her own feelings about the resoluteness and strong will of her grandmother.

2. Sample response: Paule Marshall's feelings of isolation as an African American child of West Indian immigrants surely affected the way she viewed and admired her grandmother's strength and apparent stability.

3. Responses will reflect individual cultures and experiences.

☑ Guidelines for Writing Your Response to "To Da-duh, In Memoriam"

Have students share their journal entries with a partner. Or, as an alternative writing activity, have students work in small groups to share points that struck them most deeply as they read the selection. On the chalkboard write the questions below for the groups to answer collaboratively.

- How did the trip to the island change the young girl's feelings?
- How do you think this "remembrance" reflects the author's perception of her ancestral roots?

Guidelines for Going Back Into the Text: Author's Craft

Ask students to define the conflict between Adry and her grandmother. Discuss whether or not this conflict is resolved in the story. Ask students whether they believe writing about such a conflict is a good way to understand and define it. Possible answers to the questions in the Student Edition are:

1. The little girl is torn between her desire to boast about New York to her grandmother and her emerging feelings of affection and understanding of why her grandmother also boasts about Barbados.

2. Externally she faces conflicts in school with white children who call her names, with her grandmother's sternness, and with her misgivings about the trip to the island.

3. The two conflicts are joined in the growing relationship between the child and her grandmother.

☑ FOLLOW-UP DISCUSSION

Use the questions that follow to continue your discussion of "To Da-duh, In Memoriam." Possible answers are given in parentheses.

Recalling

1. What was the main crop? (sugar cane)
2. What island landmark did Adry compare to the Empire State Building? (a large royal palm)
3. What event coincided with the death of Da-duh? (an English air raid)

Interpreting

4. What did Adry's father think of the trip? (He thought it was foolish.)
5. How would you explain the rivalry between Adry and her grandmother? (The grandmother was condescending and perhaps a bit envious of her family who moved to New York. Adry was too eager to impress her grandmother and glibly convinced that New York was superior to Barbados.)
6. How did Da-duh's attitude change after the argument about the Empire State Building? (Guide students to understand the effect this had on the grandmother's spirit and to recognize the girl's regret over having said it.)

Applying

7. What lessons does this story have for children and their grandparents today? (Guide students to understand how stubbornness and insecurity kept Adry and her grandmother at odds.
8. What does the selection tell us about cultural differences? (Guide students to understand how lack of understanding of and respect for individual backgrounds can create hard feelings.)

ENRICHMENT

Students might benefit from viewing the videotape *Barbados, A Culture in Progress.* The videotape is available from the Museum of Modern Art of Latin America, Audio-Visual Program, 1889 F Street NW, Washington, DC, 20006, (202) 458-6016.

from *The Way to Rainy Mountain*
by N. Scott Momady (pages 75–82)

OBJECTIVES

Literary

- to be aware of round and flat characters in the selection
- to identify feelings and attitudes about a search for one's ancestral roots

Historical

- to evaluate the impact of white settlement on Native American culture
- to discuss the historical setting of the selection

Multicultural

- to discuss cultural customs and values of the Kiowa
- to evaluate the relevance of past cultural traditions to contemporary society

SELECTION RESOURCES

Use Assessment Worksheet 9 to check students' vocabulary recognition, content comprehension, and appreciation of literary skills.

 Informal Assessment Opportunity

SELECTION OVERVIEW

The Kiowa are generally believed to have originated in the Montana mountains. (Some would argue that they came from the southern plains.) In Montana, the Kiowa allied with the Crows there before drifting southward to the Arkansas River, where they became allied with the Comanche. From this time onward, they occupied the plains east of what is now northern New Mexico. From the late 1700s to the mid-1800s, the peoples of the Great Plains reached the peak of power and culture. By the mid-1800s, however, the U.S. government began a long, endless war of attrition against the Plains peoples. Treaties were made and broken, and much of the Native Americans' way of life disappeared as the buffalo herds were killed off by whites and the Native Americans were forced into reservations.

In this selection, the author describes a trip back to his homeland to visit the grave of his grandmother. The trip awakens in him memories of his childhood and of the history of his people.

ENGAGING THE STUDENTS

Ask students to explain this saying: "You can never go home again." Use the questions below to guide your discussion:

- What importance do childhood memories have for people as they grow older?

- Why are people in general interested in exploring their heritages and uncovering the past?

- How effective is it to rely only on memories as a way of studying the past?

With the discussion as a backdrop, give students the following writing assignment: "Imagine that you must leave a brief written description of your life today for future generations to read. Write such a description." Give students a time limit of five minutes, to ensure the brevity of their writing. When they have finished, call on volunteers to share what they have written. Ask the students to reflect on what they chose to write about—what does this say about what is most important to them in life. Tell them to save their writing so that they can compare it to the selection they are about to read.

BRIDGING THE ESL/LEP LANGUAGE GAP

In this selection, the vocabulary and non-chronological prose will be difficult for students with limited English proficiency. Prepare them for the author's method of presenting ideas by pointing out that he writes about his grandmother using specific, personal recollections and also by telling us of the history of which she was a part. The details of his grandmother's life and character remind him of their shared heritage.

As they read, have students list two personal details that the author mentions about his grandmother and two things about the history of their people. Ask students to look for ways in which the personal memories remind the author of family history.

☑ PRETEACHING SELECTION VOCABULARY

Have students create a chart of words based on the geographical descriptions in the selection. Tell them that one column of the chart should contain the geographical landmarks mentioned in the story (Examples: mountain, **knoll,** headwaters, fork, Devil's Tower, groves, **range),** and the second column should list the adjectives used to describe the landmarks (Examples: rainy, linear, open, **sacred,** ancient, **deicide, unrelenting).** When they have finished the charts, ask the questions listed below to generate a discussion on the importance of place in the selection:

- What kinds of words does the author use to describe the places in the story?
- Does the author describe landmarks in purely physical terms, or do the descriptions go beyond physical characteristics?
- Based on the vocabulary, would you say that land has importance for the author's heritage and ancestral roots? Explain your answer.

The words printed in boldface type are tested on the selection assessment worksheet.

PREREADING

Reading the Author in Cultural Context

Help students to read for purpose by asking them to consider the importance the human relationship with nature had for Native Americans. Discuss the unique perspective Native Americans had on nature and why a search for ancestral roots would be likely to include remembrances of the land.

Focusing on the Selection

Have students write a short paragraph on the symbols and images they associate with their own past. Explain that their writing should include specific references to places, people, and things that they associate with their personal heritage and important remembrances. Afterwards, have students work in groups of three to share and discuss their writing.

POSTREADING

The following activities parallel the features with the same titles in the Student Edition.

Responses to Critical Thinking Questions

Possible responses are:

1. Sample responses: The author conveys a sense of loss, renewal, and a certain amount of bitterness. The selection is based on memories that are painful and that deal with a lost way of life.
2. Sample response: The author parallels the decline of the Kiowa with the aging process of his grandmother. He states: "I like to think of her as a child. When she was born, the Kiowas were living the last great moment of their history."
3. Responses will reflect students' own cultural group.

☑ Guidelines for Writing Your Response to *The Way To Rainy Mountain*

Have students share their journal entries with a partner. Or, as an alternative writing activity, have

students consider what they have read and then write three questions they would ask the author if they were to meet him. Encourage students in small groups to share the questions that they have written and select the best ones. Then have one member of the group take the part of the author and respond to these questions.

Guidelines for Going Back Into the Text: Author's Craft

Help students understand how the author created characters that are round and well-developed by rereading the description of his grandmother on page 79. (The paragraph begins: "Now that I can have her only in memory....") Call on students to specify methods the author uses to make this character round. Possible answers to the questions in the Student Edition are:

1. Momaday's grandmother is a round character. Though the author does not give many direct details about himself, we can sense his feelings and ideas by reading his interpretation of events. This makes him a round character as well.

2. Sample answers: childlike, mysterious, wary, religious, stern, joyful

3. Flat characters: eight children at Devil's Tower, people who came to visit his grandmother in summer

☑ FOLLOW-UP DISCUSSION

Use the questions that follow to continue your discussion of the excerpt from *The Way to Rainy Mountain*. Possible answers are given in parentheses.

Recalling

1. Why did Momaday return to Rainy Mountain? (to visit the grave of his grandmother)
2. Which Native American group gave the Kiowa many cultural traits? (the Crow)
3. How did the author's memory of his grandmother's house differ from what he

witnessed on the trip back? (The house was actually much smaller than he remembered.)

Interpreting

4. What discoveries did the author make on his trip to Rainy Mountain? (Guide students to understand how the trip reawakened childhood memories in the author.)
5. In your opinion, is the author bitter about what happened to the Kiowa? Support your answer with facts from the selection. (Student answers will vary, but the tone of the selection is more sorrowful than bitter, and no direct attacks against whites are made.)
6. Explain this phrase from the text: "Although my grandmother lived out her long life in the shadow of Rainy Mountain, the immense landscape of the continental interior lay like memory in her blood." (Deep in her conscious awareness, almost in an archetypal way, his grandmother had memories of her people's past and their struggles to survive.)

Applying

7. What do you think the author learned about himself during the trip to Rainy Mountain? (Student answers should address the author's reflections on his people and his grandmother's life, and how far he had come from this heritage.)
8. Consider what you have read, then describe your feelings about grave sites and why people visit them. (Student answers will reflect students' own opinions, but should acknowledge that the author turned his visitation into a learning experience and a chance to reflect on his ancestral roots.)

ENRICHMENT

N. Scott Momaday hosts a PBS documentary about Native Americans who are holding onto ancient customs and values. *American Indians: Winds of Change—A Matter of Promises* is available from Pacific Arts Video, 50 North La Cienega Boulevard, Beverly Hills, CA 90211, (800) 538-5856.

i yearn
by Ricardo Sánchez

[We Who Carry the Endless Seasons]

by Virginia Cerenio (pages 83–86)

OBJECTIVES

Literary

- to become aware of how the poets' word choices reflect cultural roots
- to appreciate the effect created by the use of different line lengths in free verse

Historical

- to evaluate the impact of immigration on traditional cultures
- to discuss the social and economic hardships faced by foreign immigrants

Multicultural

- to contrast the experiences of Mexican and Filipino immigrants
- to list the cultural traditions and values illustrated in the poems

SELECTION RESOURCES

Use Activity Worksheet 10 to check students' vocabulary recognition, content comprehension, and appreciation of literary skills.

 Informal Assessment Opportunity

SELECTION OVERVIEW

The immigrant experience on this continent has been an underlying theme since people started settling here. Americans have always grappled with immigration in conflicting ways. During the 19th and 20th centuries, U.S. industry often courted immigrants in order to have a supply of cheap labor, while at the same time the U.S. government (led by voters) passed and repealed numerous anti-immigration laws. In the late 20th century, the bulk of immigrants were coming from Latin America and Asia. In the decade from 1980 to 1990, the Asian population in the United States grew by 108 percent, while the Hispanic population grew by 53 percent.

In these poems the authors describe the experiences, struggles, and cultural conflicts faced by immigrants from Mexico and the Philippines.

ENGAGING THE STUDENTS

Ask students: "In general, does the word *immigrant* have a negative or a positive connotation?" Allow them two minutes to write all the words they associate with the term *immigrant*. When they have finished writing, tell them to study and evaluate their lists using the questions below as a guide:

- In general, do the terms in your list have negative or positive connotations?
- Immigrant groups often cling to cultural traditions. How do you think this affects their assimilation into American society?

• In your opinion, are Americans generally tolerant of immigrant groups? Is there any stigma attached to having a recent immigrant past?

With your discussion as a backdrop, have students reconsider their lists from the opening quickwrite and make changes.

BRIDGING THE ESL/LEP LANGUAGE GAP

Since poetry is meant to be heard, these selections provide an opportunity to focus on listening skills and on finding meaning through aural clues. Encourage students to make notes as they listen to you read the poems.

"i yearn": Ask the students which words stood out or were unfamiliar. Read the poem a second time, asking them to write the words that they want to know more about. "[We Who Carry the Endless Seasons]" Direct the students' attention to the stanza which reads, "like shadows/attached to our feet/ we cannot walk away." Discuss what this means. Are there things the students cannot —or feel they cannot — walk away from? What are they? Is this good or bad?

☑ PRETEACHING SELECTION VOCABULARY

Have students create a word bank by writing all the terms in the selection that are either Spanish or Filipino. (Examples: **chicano; infinitum; caló; assail; ¿ que tal, hermano?; barrio.**) Call on volunteers to read the terms in their word banks. Write the words on the chalkboard as they are read. Use the footnotes on pages 84–85 to go over these terms and discuss their meanings.

The words printed in boldface type are tested on the selection assessment worksheet.

PREREADING

Reading the Poets in Cultural Context

Help students read for purpose by asking them to consider the importance ethnic neighborhoods have for immigrants. Discuss with students the ethnic neighborhoods that exist within your area. Ask students to share their perceptions of these ethnic neighborhoods.

Focusing on the Selections

If you wish, have students work in pairs. Each student should take responsibility for summarizing one of the poems in the selection. Afterwards, have pairs exchange and compare their summaries. Tell them to use their summaries to answer the questions in the Student Edition. Students should support their answers with examples from the selections that illustrate how each author has conveyed his or her ancestral roots and cultural identities.

POSTREADING

The following activities parallel the features with the same titles in the Student Edition.

Responses to Critical Thinking Questions

Possible responses are:

1. "[We Who Carry the Endless Seasons]": pain of their families growing up in the United States and dealing with struggles associated with speaking a foreign language and marrying outside their ethnic group; "i yearn": longing to speak his language as it is meant to be spoken, eating his favorite foods, and the warmth of people who truly are close to him.
2. Both poets express longing for their heritage and a sense of loss at the changes they have experienced.
3. Student responses should include facing a loss of heritage and of customs that are familiar.

☑ Guidelines for Writing Your Response to "i yearn" and "[We Who Carry the Endless Seasons]"

Have students share their journal entries with a partner. Or, as an alternative writing activity, have students work in groups of four to write short descriptions of the images conveyed in an assigned verse from the poem. Each group must then collaborate on a final answer to the questions in the Student Edition.

Guidelines for Going Back Into the Text: Author's Craft

Discuss with students how free verse can lead to

different interpretations of the same words. Ask: "Why is free verse an effective method for illustrating day-to-day speech patterns? Would a poem in metrical style be as effective?" Possible answers to the questions in the Student Edition are:

1. Line lengths vary; there is no fixed metrical pattern or predictable rhythm; ideas are emphasized by breaks and word placement.
2. Authors end lines at dramatic moments and highlight feelings with short, direct lines.
3. Answers will reflect student poem choices.

✓ FOLLOW-UP DISCUSSION

Use the questions that follow to continue your discussion of "i yearn" and "[We Who Carry the Endless Seasons]." Possible answers are given.

Recalling

1. In the first selection, what desire did the mothers have for their daughters? (that their daughters marry boys from the islands)

Interpreting

2. How do you interpret this phrase: "We wear guilt for their minor sins singing lullabies in foreign tongue"? (The author laments the struggles her mother faced raising children in a land foreign to her.)
3. Do you think the authors regret being immigrants? Explain why or why not. (Guide students to understand that a longing for cultural roots does not mean that the present is not valued.)

Applying

4. What could Americans who were born in the United States learn from these poems? (Guide students to understand how empathy and compassion can arise from knowledge about other cultures and the difficulties immigrants must face in a foreign land.)

ENRICHMENT

Students might benefit from viewing a videotape that illustrates the process that immigrants must go through to become U.S. citizens. *The American*

Experience: Becoming an American is available from Phoenix/BFA Films, 468 Park Avenue South, New York, NY, 10016, (800) 221-1274.

RESPONSES TO REVIEWING THE THEME

1. Sample answer: In the excerpt from *Roots,* the author's relationship with Kunta and his ancestral past is removed in time, and based on Haley's research. Paule Marshall in "To Da-Duh, In Memoriam" has direct contact and first-hand experiences with her grandmother. While Haley's writing has a tone of interested discovery, Marshall's is more forlorn and bittersweet.
2. In *Bless Me, Ultima,* there is a conflict between two ways of life. When Antonio's family moves to town, his father's relationship with the land is lost. In *The Way to Rainy Mountain,* the author describes the close relationship his people, the Kiowa, had with the earth.
3. Sample answer: *Bless Me, Ultima* The young boy's struggle to understand his father's love for the land and his mother's hopes for a better future are themes continually played out today.

✓ FOCUSING ON GENRE SKILLS

Paule Marshall made effective use of the short story, portraying both inner conflict and conflict between characters. Marshall manages through the first person point of view to convey more than the narrator comprehends. Select another short story (such as "Blues Ain't No Mockin' Bird" by Toni C. Bambara, or "We're Very Poor" by Juan Rulfo). Have students read the story and determine if the major conflict is interior (between the main character and values and events) or exterior (between the main character and external events).

BIBLIOGRAPHY

Additional Books by Paule Marshall:

Soul Clap Hands and Sing. Washington, D.C.: Howard University Press, 1988

Books Related to the Theme:

Gates, Henry Lewis, *Bearing Witness.* New York: W.W. Norton, 1991.

3

Celebrating Growth and Change

THEME PREVIEW

Celebrating Growth and Change			
Selections	**Genre/Author's Craft**	**Literary Skills**	**Cultural Values & Focus**
"Seventeen Syllables," Hisaye Yamamoto, pages 44–46	short story	climax	cultural and generational differences; changes of adolescence
from *Humaweepi, the Warrior Priest*, Leslie Marmon Silko, pages 47–49	novel	theme	relationship between teenagers and elders; Native American culture passes knowledge to younger generations
"Cante Ishta—The Eye of the Heart," from *Lakota Woman*. Mary Crow Dog, pages 50–52	autobiography	personification	importance of ancestral roots for Native Americans today; how roots impact their lives
"Black Hair," Gary Soto "ALONE/december/ night," Victor Hernandez Cruz, pages 53–55	poetry	imagery	isolation that comes from not being accepted into mainstream American society

Assessment

Assessment opportunities are indicated by a ✔ next to the activity title. For guidelines, see Teacher's Resource Manual, page xxxiii.

CROSS-DISCIPLINE TEACHING OPPORTUNITIES

Social Studies Discuss with students how the expanding global economy has brought diverse cultures into contact with one another. Occasionally the contact creates tension. Probably the most visible clash of cultures is between the United States and Japan. Stories abound regarding how U.S. businesspeople unwittingly insult their Japanese hosts. Ignorance regarding Japanese customs lies behind such incidents. Work with the social studies teacher to compare some of the traditional aspects of Japanese culture, as described in "Seventeen Syllables," to American culture. As an example of Japanese traditions with which the average American is unfamiliar, see the excellent article on marriage in modern Japan ("Change Comes Slowly for Japanese Women") in the April 1990 *National Geographic*.

Music Teenage defiance and generational conflict are themes that transcend cultural boundaries. These themes are found in a number of the selections in Theme 3 ("Seventeen Syllables," *Humaweepi, the Warrior Priest*, "Black Hair"), and are recurrent themes in teenage popular music. Ask students to compare the themes related to generational differences that they will read about in Theme 3 to similar themes in popular music.

Geography Work with the geography teacher to explore how technology and changing settlement patterns are affecting the land worldwide. Discuss these changes in relation to some of the selections in Theme 3. For example, in *Humaweepi, the Warrior Priest*, land plays an important part in the main character's development. Discuss how these ways of life are becoming more difficult to maintain.

SUPPLEMENTING THE CORE CURRICULUM

The Student Edition provides humanities materials related to specific cultures in Making Connections on pages 36–37 and in the art insert on pages A1–A16. You may with to invite students to pursue the following activities outside of class.

- The three works of art on pages A4 and A5 are all, in different ways, expressions of growth and change. They are expressions of traditional themes in new arresting styles. Ask students to read the caption describing the drawing by Bill Traylor. Discuss with them why the artist would choose to depict a man sprouting leaves. Ask students if they think this is a positive image. Ask them to defend their views. Then ask them how it could show growth or change in a spiritual way. Continue this discussion by asking students how the other works here show growth and change in similar ways.

- Many Americans think of baseball as a strictly American pastime. However, many players in the major leagues are Latin American (like Hector Moreno in "Black Hair"). This can be directly traced to the important role of baseball in Latin American cultures. Ask interested students to research baseball in other cultures and the prevalence of Latin American players in American baseball.

- Rites of passage and ceremonial rituals are an important part of life in high school. Encourage students to research some initiation rites found in American high schools. For example, athletes receive "letters" to signify their group membership. *Down These Mean Streets* by Piri Thomas describes the initiation rites that the author experienced when he moved to a new block in Spanish Harlem, New York, in the 1950s.

INTRODUCING THEME 3: CELEBRATING GROWTH AND CHANGE

Using the Theme Opener

Challenge students to define their own perspective on personal changes. On the chalkboard write the following topics:

birthday	marriage	high school graduation
first date	baptism	leaving home

Use the questions below to lead a guided discussion on student views of growth and change:

- Is change always related to personal growth?
- Which of the topics on the chalkboard represent meaningful changes to you? Which of the topics are more trivial in your opinion?
- Which of the topics on the chalkboard do you associate with celebration? Why do you think people often like to celebrate changes?

Tell students to consider the different ways that they have interpreted change, and to keep these ideas in mind as they read about the changes celebrated in *Tapestry* Student Edition.

☑ Developing Concept Vocabulary

Help students to understand the breadth and scope of the terms *change*, *tradition*, and *ceremony*. Write these three words on the board as headings for three columns. Tell each student to write one example of each topic on a piece of paper. Suggest to them that they list examples they have witnessed or participated in. Collect the students' writing. Read the students' examples aloud, but have students decide in which category each topic belongs. Fill in the columns on the chalkboard. When the columns are filled, use the questions below to lead a guided discussion on the theme of Celebrating Growth and Change:

• How are these topics related to one another?

• How does each one relate to the theme of growth and change?

As students respond, draw connecting lines between the columns as students discover how the topics relate to one another.

CHANGE	TRADITION	CEREMONY
a move to a new location	Thanksgiving dinner	Memorial Day parade

Seventeen Syllables

by Hisaye Yamamoto (pages 88–98)

OBJECTIVES

Literary

- to appreciate the author's attitude toward differences in values
- to identify the climax of the selection

Historical

- to discuss the conditions and prejudices faced by Japanese Americans in the United States
- to understand the economic forces that brought Japanese immigrants to the United States

Multicultural

- to evaluate the impact of American culture on traditional Japanese American families
- to discuss the differences and similarities between Japanese and American cultures

SELECTION RESOURCES

Use Assessment Worksheet 11 to check students' vocabulary recognition, content comprehension, and appreciation of literary skills.

 Informal Assessment Opportunity

SELECTION OVERVIEW

Japanese immigration to California was not significant until the late 1890s. A steady flow of immigrants continued from this period into the early 20th century. Japanese immigrants and their descendants faced blatant, and often government-sponsored, acts of racism designed to exclude them from business, government, and education. In the early 1900s, for example, Japanese children in California were forced to attend separate "Oriental" schools. Initially Japanese laborers were brought to the United States to replace low-paid Chinese laborers, who were no longer available because of the Chinese Exclusion Act. Soon, however, Japanese immigrants acquired farmland of their own and became the first in the western United States to grow rice and potatoes in commercial quantities. This success resulted in a new wave of anti-Japanese legislation, including laws restricting land ownership.

In this selection, the author, through the eyes of a teenager, describes the conflict between her parents over her mother's haiku and her husband's traditional ideas on the role of a wife. This conflict is played out during the teenage girl's first romantic experience.

ENGAGING THE STUDENTS

On the board write *Generation Gap*. Then ask: "Are differences between parents and teenagers inevitable? How does peer pressure create conflict between parents and children?" Explain to students that the term on the board has been used for decades to describe struggles between teenagers and their parents.

Ask each student to write a paragraph on the singular struggles and conflicts children of immigrants to the United States might face when their parents cling to old traditions while children attempt to assimilate. Students should address these topics in their writing:

- clashes between old and new ideas
- arguments over assimilating into a new society
- a teenager's need to "fit in" and belong to a group of peers

Ask volunteers to read their writing aloud and discuss their ideas. Tell them to keep their work for a later comparison with the selection they are about to read.

BRIDGING THE ESL/LEP LANGUAGE GAP

This story may be difficult for some students because of their cultural backgrounds; the emotions and relationships may be unfamiliar to many students, and familiar to others. Encourage students who may understand the conflicts and relationships in the story to explain the dynamics of the characters and plot to their peers.

After the students have read the story, ask them to discuss, in small groups, the conflicts they see in the story. What kind of conflicts are there between Rosie and her mother? between her mother and her father? Why do they have these conflicts? How are they expressed? Are the conflicts resolved? Why or why not? Do the students have similar conflicts in their lives?

☑ PRETEACHING SELECTION VOCABULARY

One of the main themes in this selection is change. Point out to students that Rosie's first romantic encounter with Jesus is a true awakening for her. Tell students to skim pages 89–97 for vocabulary words and phrases the author uses to describe this episode. (Examples: **delectable,** beyond speech, **helplessness,** distressingly constricted, **dubious,** victim, **giddy, grave)** Call on students to read the words they have discovered, and write them on the board as the students identify them. Discuss the meanings and connotations of the adjectives. Ask students to use these vocabulary words to speculate on what they will read about in the selection.

The words printed in boldface type are tested on the selection assessment worksheet.

PREREADING

Reading the Author in Cultural Context

Ask students to consider the pros and cons of an arranged marriage. They might be interested to know that some studies have found that these kinds of marriages are more stable than marriages based on romantic love. Ask students to evaluate this cultural custom as they read the selection.

Focusing on the Selection

Have students work in pairs. Assign one student the role of Rosie's father and the other the role of her mother. Have each student write a short description of his or her assigned character, focusing on their values and views of Japanese traditions. When they have finished, each pair must collaborate on an answer to this question: "Is the father or the mother more traditional?" Call on student pairs to read their answers aloud and then discuss the questions in the student text.

POSTREADING

The following activities parallel the features with the same titles in the Student Edition.

Responses to Critical Thinking Questions

Possible responses are:

1. Both Rosie and her mother face conflicts between tradition and their changing needs. Rosie's mother is torn between her love of writing and her husband's traditional views on her role as wife. Rosie must balance the confusion caused by her mother's admonition never to marry and her newfound romance with Jesus.

2. The author most likely faced the conflicts posed by traditional Japanese culture and her assimilation into American culture.

3. Student responses should address the conflicts that can arrive when cultural heritage clashes with the demands of adolescence and the need to "fit in" and belong to a peer group.

✔ Guidelines for Writing Your Response to "Seventeen Syllables"

Have students share their journal entries with a partner. Or, as an alternative writing activity, have students form groups of three to share the scenes from the selection that they believe most resemble their own lives. Point out to students that this selection has a message for any teenager going through the discovery of romance and the problems involved with parental conflicts.

Guidelines for Going Back Into the Text: Author's Craft

Discuss with students the final scene in the story. Ask: "How does the author convey the emotional intensity of the climax? How does the dialogue convey emotions?" Discuss how the author successfully conveys the tensions of this moment. Possible answers to the questions in the Student Edition are:

1. The climax of the story is when Rosie's mother makes Rosie promise never to marry. Leading up to the climax is the conflict between Rosie's mother and her father over her writing and her role as wife.

2. Her emotional pain stems from a romantic liaison she had long ago in Japan, and the subsequent hasty marriage to her present husband. These experiences provide harsh insights for Rosie, who is feeling the first blush of teenage romance.

3. The climaxes in other stories include: Felipa's reunion with her lost son in "The Sheep of San Cristóbal," Antonio's meeting with Ultima in *Bless Me, Ultima*, and Adry's argument with her grandmother over landmarks in "To Da-duh, In Memoriam."

✔ FOLLOW-UP DISCUSSION

Use the questions that follow to continue your discussion of "Seventeen Syllables." Possible answers are given in parentheses.

Recalling

1. What do Rosie's parents do for a living? (They are tomato farmers.)

2. How did Rosie meet Jesus? (He and his family work on the farm.)

3. What is haiku? (a traditional Japanese poem with only seventeen syllables)

Interpreting

4. In your opinion, are there cultural differences between Rosie and Jesus? (Guide students to see how cultural differences never come between Jesus and Rosie, though some, such as language, race, and religion, surely exist.)

5. Would you characterize Rosie's mother as traditional or not? Explain your answer. (Students should weigh the mother's traditional traits, such as writing haiku, with her beliefs about marriage which seem to represent a rebellion against tradition.)

6. Do you think Rosie understands her mother's request that she never marry? Explain your reasoning. (Probably the mother's request confuses Rosie—this was the first time her mother had discussed her marriage with her daughter. Furthermore, Rosie's own feelings for Jesus make the situation more confusing for the young girl.)

Applying

7. How do contemporary views on marriage and women's roles differ from the ideas in this selection? (Marriages today in the United States are usually based on romantic love; women and men are generally free to choose their own spouses as well as careers and interests.)

8. What traditions illustrated in this selection continue to be important today? (respect for family and parents, appreciation for cultural heritage)

ENRICHMENT

Students may benefit from viewing a videotape that describes one Asian woman's life, from her arranged marriage to her work in a San Francisco sewing factory. *Sewing Woman* is available from the Anti-Defamation League of B'nai B'rith, Audio-Visual Department, 823 United Nations Plaza, New York, NY, 10017, (212) 490-2525.

from *Humaweepi, the Warrior Priest*
by Leslie Marmon Silko (pages 99–106)

OBJECTIVES

Literary

- to recognize the author's attitude toward change, learning, and human growth
- to understand the elements of a story

Historical

- to evaluate the impact of European farming methods and land ownership on traditional Native American ways of life
- to discuss the role of medicine men and women in traditional Native American communities

Multicultural

- to compare traditional Native American rites of passage to present-day ceremonies and rituals
- to discuss how generational differences are managed in Native American culture

SELECTION RESOURCES

Use Assessment Worksheet 12 to check students' vocabulary recognition, content comprehension, and appreciation of literary skills.

 Informal Assessment Opportunity

SELECTION OVERVIEW

This selection contains a reference to a vision of an animal spirit, which is an important aspect of many Native American religious or spiritual experiences. After a person has a vision of an animal—which happens when the person is considered "ready" by the elders or by spiritual forces—the animal becomes that person's helper. As a result, the person may call upon the animal in times of crisis and take on the animal's characteristics. For example, if the animal is a bear (as is the case in this selection), the person, with the animal's help, would become strong and powerful during crises. The person might also use this animal as a personal symbol, taking on the animal's attributes during important ceremonies. The vision quest is not defined in a universal way, as the animal vision is very personal. The process of discovering what the animal means to the individual can develop over a lifetime.

In this selection, the author recounts a teenage boy's relationship with his uncle and the knowledge his uncle imparts to him.

ENGAGING THE STUDENTS

Present students with the following hypothetical dilemma: "You have been thrown into a mountain wilderness and must survive for five days with no provisions." Ask students to estimate their chances of survival using the questions below:

- Would you know which berries and plants are safe to eat?
- How would you protect yourself from the elements and from danger?
- Could you find your way out of the wilderness? Can you use the sun and stars to find direction?

When students have completed this activity, explain that in this selection they will read about one young person's voyage of discovery. His journey takes place under conditions similar to those described above.

BRIDGING THE ESL/LEP LANGUAGE GAP

The unusual setting and vocabulary of this selection could be confusing to some students. Help them to read with a purpose by giving them questions to consider as they read:

1. This selection begins with the following statement: "The old man didn't really teach him much; mostly they just lived." As you read the story, think about this statement. Why does the story begin this way? Is this a true statement?

2. Humaweepi realizes that something was happening all the time he was with his uncle. What was it? Have you had a similar experience?

3. What do you think his uncle is trying to teach him? Why?

Ask the class to work in pairs to discuss their answers to these questions when they have finished reading the selection.

☑ PRETEACHING SELECTION VOCABULARY

The author of this selection uses details that appeal to all of the senses, but especially to the sense of sight. (Example: have the students scan the selection to find all the color and texture words—white cotton sack, dark-purple berries, green grass, **succulent,** yellow blossom, pink quartz, **turquoise** beads, black **obsidian** arrowhead, blue images, white sandstone, lichen-covered, white hair, gray boulder, **yucca fiber,** rainbow colors, **coral,** and so on. Explain that the author uses color for contrast between ideas. See if they can figure out his technique as they read the selection. (Colors are used to liven up descriptions; gray and white are used for emphasis.)

The words printed in boldface type are tested on the selection assessment worksheet.

PREREADING

Reading the Author in Cultural Context

Help students read for purpose by asking them to consider the importance that tradition has for the characters in this story. Discuss how the uncle's teaching provides continuity and stability for the community.

Focusing on the Selection

Have students work in pairs. Assign one of the questions in the Student Edition to each member. When they have finished working individually, ask them to collaborate to come up with a single shared response to both questions.

POSTREADING

The following activities parallel the features with the same titles in the Student Edition.

Responses to Critical Thinking Questions

Possible responses are:

1. Growth and change came to Humaweepi in a slow, steady way that he did not acknowledge at first. He learned from watching and living with his uncle, though at first he did not understand exactly what the lesson was.

2. The author most likely learned traditional beliefs from the elder members of her community as well.

3. Though most students will not identify with the lessons in survival that Humaweepi was taught, guide them to understand how the relationship with his uncle also was a rite of passage and a period of growth for the young man.

☑ Guidelines for Writing Your Response to *Humaweepi, the Warrior Priest*

Have students share their journal entries with a partner. Or, as an alternative writing activity, have

students, working in groups of three, each share a favorite scene or idea from the selection. Next, have each group collaboratively answer this question: "Why is this a good story for teenagers of all cultures to read?" Call on each group to read its response aloud.

Guidelines for Going Back Into the Text: Author's Craft

Discuss with students the effective way the author expressed her main theme. Ask: "Was the theme stated directly or indirectly?" Guide students to appreciate the effectiveness of presenting themes that the reader must discover for himself or herself. Possible answers to the questions in the Student Edition are:

1. The general theme is one of change and growth: Humaweepi's growth into a young man and the changes brought on by his new knowledge and ability.

2. The theme in the story is portrayed vividly in the ceremony at the lake in which he sings a prayer to Bear. During the prayer, Humaweepi leaves his adolescence behind to become a warrior priest.

3. Guide students to see the similarities young people might experience today by participating in rituals such as confirmations, graduations, bar mitzvahs, and other cultural ceremonies that mark growth and change.

✔ FOLLOW-UP DISCUSSION

Use the questions that follow to continue your discussion of *Humaweepi, the Warrior Priest*. Possible answers are given in parentheses.

Recalling

1. At first, why does Humaweepi believe he was sent to live with his uncle? (to help the old man do chores)

2. What plant materials are the old man's sandals made of? (yucca fiber)

3. To what animal does Humaweepi sing his prayer? (bear)

Interpreting

4. What kind of peer pressure does Humaweepi encounter while living with his uncle? (His friends are not impressed with what Humaweepi is learning. They do not understand the importance of Humaweepi's lessons. His friends are also still living with their families.)

5. Why doesn't Humaweepi bring food for the trip into the mountains? (His uncle wants to instruct him in survival skills, and perhaps prepare him for the ceremony at the Bear rock.)

6. For what role is Humaweepi's uncle preparing him? Who will Humaweepi one day replace? (Humaweepi is being prepared for the role of warrior priest, when he will take the place of his uncle.)

Applying

7. Why is it necessary for Humaweepi to live apart from the rest of the community? (His role in life is special and requires dedication to learning with no distractions.)

8. What function will Humaweepi's experiences serve for the entire community? (Humaweepi will teach and help his community with his knowledge and pass along what he has learned to future generations of young people.)

ENRICHMENT

To gain an understanding of the approximate region in which this story takes place, students may benefit from watching the videotape *Ancient Indian Cultures of Northern Arizona*. The videotape is available from Facets Multimedia, Inc., 1517 W. Fullerton Avenue, Chicago, IL, 60614.

Cante Ishta—The Eye of the Heart from *Lakota Woman*

by Mary Crow Dog (pages 107–114)

OBJECTIVES

Literary

- to identify the use of personification in the selection
- to understand the elements of an autobiography

Historical

- to understand how cultural losses in the 19th century among Native Americans led to renewed efforts to maintain cultural identity in the 20th century
- to discuss the political goals of the American Indian Movement

Multicultural

- to evaluate the importance of traditional Native Americans' ceremonies and rituals for present-day Native Americans.
- to explain why traditional ceremonies are an important part of cultural heritage

SELECTION RESOURCES

Use Assessment Worksheet 13 to check students' vocabulary recognition, content comprehension, and appreciation of literary skills.

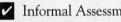 Informal Assessment Opportunity

SELECTION OVERVIEW

During the 1960s, many Native Americans adopted a new militancy in their struggle for civil rights, economic justice, and control over their lands. Some of these struggles continue today. For example, in an ongoing protest that began in the 1960s in the state of Washington, Puyallups and other Native Americans from the Northwest coast deliberately disobey fishing regulations to protest their loss of hunting and fishing rights. For many of the younger protesters the movement that began in the 1960s was coupled with a return to ancient rituals and ceremonies. The American Indian Movement (AIM) was the most militant of the Native American protest groups. Their efforts culminated with the group's seizure (at gunpoint) of the reservation settlement at Wounded Knee in 1973. AIM demanded that Native American lands be returned and reservation programs be managed by tribal governments. After 70 days of confrontation, AIM finally withdrew with none of their demands met.

In this selection the author recounts her exploration of ancestral roots through her marriage to a Native American medicine man and the sweat lodge ritual.

ENGAGING THE STUDENTS

On the board write *sweat.* Call on students to list all the ideas and words that they associate with this term. (Many will associate the word with sports, heat, and so on.) Next list on the board these terms: *purification*, *cleansing*, *uplifting*, *exhilarating*, *spiritual*. Explain to students that in this selection they will read about one person's experience with heat and sweat that is much different from many of the preconceived notions our society holds.

BRIDGING THE ESL/LEP LANGUAGE GAP

The language of this selection is not difficult, but some of the concepts of spirituality are complex. Introduce the selection with the quote by Lame Deer: "You got to look at things with the eye in your heart, not with the eye in your head."

Discuss what the students think this means. Do they agree or disagree with this statement? What kind of values do they think the person who made this statement has?

Mary Crow Dog talks about "spiritual gifts" her husband had even as a child. What are spiritual gifts? Are they important? Why are they important to some people and not to others? As they read, have the students look for things the author considers spiritual gifts.

After they have read the selection, ask the students to list some of the spiritual values the author seems to have. Does the student think they are important? Why or why not?

☑ PRETEACHING SELECTION VOCABULARY

On the board write *ceremony* and *ritual*. Ask students what kinds of experiences they associate with these terms. Next write the words **purification, enraptured, peyote, sham,** and **intellectualism** on the board. Do a webbing exercise to find associations between these terms. Use the questions below to guide you:

• Did ceremonies you attended in the past have any purification rites? (Many world religions have some form of purification ritual.)

• What are some possible associations between *sham* and *intellectualism*? What connotations do these

words have for you? Do you see any connection between these ideas and Native American rituals?

The words printed in boldface type are tested on the selection assessment worksheet.

PREREADING

Reading the Author in Cultural Context

Help students read for purpose by asking them to consider the importance that tradition has for the narrator. Remind them that her family adopted Christian beliefs. Discuss how her interest in her ancestral roots might have caused conflict in her family.

Focusing on the Selection

Have students form pairs, and assign one of the questions in the student text to each member. When they have finished working individually, ask each group to write a shared response to this question: "Would you endure a ritual that entailed some physical pain or discomfort in your search for your ancestral roots?" Discuss the group responses.

POSTREADING

The following activities parallel the features with the same titles in the Student Edition.

Responses to Critical Thinking Questions

Possible responses are:

1. It awakened her talents, knowledge, and awareness of her heritage; she learned traditional beliefs and participated in traditional rituals and ceremonies; she was no longer an "outsider" looking in.

2. Her marriage to a medicine man and the subsequent life adjustments probably were the main experiences that led her to write about the Lakota people and their rituals.

3. Responses will reflect students' own heritages and viewpoints, but guide them to understand

how the discovery of our ancestral roots can provide a deeper understanding of who we are.

☑ Guidelines for Writing Your Response to "Cante Ishta—The Eye of the Heart"

Have students share their journal entries with a partner. Or, as an alternative writing activity, have each student share his or her favorite scene or idea from the selection. Next, ask students to each write a personal evaluation of the sweat ritual, including whether or not he or she would be willing to participate in it. Encourage them to incorporate this writing into their journal entries.

Guidelines for Going Back Into the Text: Author's Craft

Tell students to skim the selection for examples of personification. Discuss what qualities these examples bring to the writing. Ask students to name cases when personification would not be a good type of language to use. (Example: in technical descriptions)

Possible answers to the questions in the Student Edition are:

1. The "heart" is given human qualities—an eye so that deeper meanings can be understood.

2. The use of personification gives the reader different perspectives and ways of viewing events and ideas.

3. In *Humaweepi, the Warrior Priest*, a holy rock becomes an animal which in turn is endowed with human qualities.

☑ FOLLOW-UP DISCUSSION

Use the questions that follow to continue your discussion of "Cante Ishta—The Eye of the Heart."

Possible answers are given in parentheses.

Recalling

1. At what age was Leonard singled out for his role as medicine man? (8 years old)

2. What item did Bessie Good Road use in her rituals? (a buffalo skull)

3. What was the spoken message that signaled too much heat during the sweat lodge ritual? (*Mitakuye oyasin*—All my relatives)

Interpreting

4. Why did the Lakota elder prevent Leonard from attending a white school? (His people believed that a white education would destroy the qualities he needed to be a medicine man.)

5. Was being the wife of a medicine man easy or hard for the narrator? Explain your answer. (Adjusting and learning what role the wife of a medicine man had posed some problems, but these were outweighed by the wisdom and uniqueness he offered.)

6. How did the narrator's view change after her first time in a sweat lodge? (It changed from fear and discomfort to elation.)

Applying

7. What did the sweat lodge ritual teach the narrator about her heritage? (It taught her about purification and rebirth through ritual, and opened her eyes to traditional ceremonies.)

8. Do you think the sweat lodge ritual would appeal to other cultures besides Native Americans? Explain your answer. (Guide students to recognize the benefits of physical and spiritual rejuvenation.)

ENRICHMENT

Students may benefit from viewing a videotape that offers explanations of the Lakota religion. *Eyanopopi: The Heart of the Sioux* is available from Barr (Centre) Films, PO Box 7878, 12801 Schabarum Road, Irwindale, CA, 91706.

Black Hair
by Gary Soto

ALONE/december/night
by Victor Hernandez Cruz (pages 115–118)

OBJECTIVES

Literary

- to identify the poets' attitudes toward their new heritage
- to appreciate the effects of imagery and repetition

Historical

- to recognize the post World War II gains and setbacks of the Mexican American population
- to compare the post World War II civil rights struggles and tactics of African Americans with those of Mexican Americans

Multicultural

- to discuss themes of alienation and isolation among different cultural groups
- to analyze the role of popular heroes for culture groups

SELECTION RESOURCES

Use Assessment Worksheet 14 to check students' vocabulary recognition, content comprehension, and appreciation of literary skills.

✔ Informal Assessment Opportunity

SELECTION OVERVIEW

After World War II many Mexican Americans (and other immigrant groups from Latin America) hoped that their status as "outsiders" in U.S. society was ending. During the war, industries that had been closed to Mexican Americans opened. These hopes were short-lived, however, as returning Anglo veterans squeezed Mexican Americans out of high-paying jobs.

In these selections the authors express the isolation of living in a predominantly white culture. In the first poem, the narrator escapes through a baseball hero. Cultural pride is a theme in both selections.

ENGAGING THE STUDENTS

Ask students to consider the following question that you write on the board: "Do all races have the same economic opportunities in the United States?" Before they answer give them these statistics:

- The average household income for Hispanic families is nearly 30 percent *lower* than it is for Anglo families.
- Twenty-seven percent of all Anglo workers hold blue collar jobs. *Forty-nine* percent of all Hispanic workers hold these lower paying jobs.

Now call on students to answer the question on the board. Tell them to remember their discussion as they read the selections.

BRIDGING THE ESL/LEP LANGUAGE GAP

"Black Hair": Discuss, as a class, people the students admire now or admired as children. Ask the students to look for the role that Hector Moreno played as a hero in this boy's life.

"ALONE/december/night": Have the students write three words that relate to each of the title words of the poem. Ask the students to discuss what they think this poem might be about. What kind of feelings might it express?

☑ PRETEACHING SELECTION VOCABULARY

On the chalkboard write: **ring** of **heat**; shade **rose**; gloves eating baseballs; **altar** of worn baseball cards; and Moreno had hard, **turned** muscles; Hector **lined** balls. Call on students to speculate on the subject matter of the first poem based on the vocabulary in these phrases.

The words printed in boldface type are tested on the selection assessment worksheet.

PREREADING

Responding to Literature: Modeling Active Reading

"Black Hair" and "Alone/december/night" are annotated with the comments of an active reader. These sidenotes, prepared to promote critical reading, emphasize the cultural content of the piece, address author values, call attention to literary skills, invite personal response, and show how the selection is related to the theme. If you have the time, read the entire selection aloud as a dialogue between reader and text. Encourage students to add their own responses and to compare these with the ones printed in the margins. Remind them that they can disagree with the annotated comments. Model these skills by adding your own observations too.

Reading the Author in Cultural Context

Guide students in a comparison of the different ways each author has conveyed his cultural background.

Student responses should address the themes of isolation and ethnic pride.

Focusing on the Selections

Have students work in groups of three. Assign each student one of the questions in the text. When they have finished working individually, challenge them to look for an answer to this question as they read: "How do these poems illustrate change and growth for the characters?" Discuss the responses and tell students to draw on these ideas in their journal writing.

POSTREADING

The following activities parallel the features with the same titles in the Student Edition.

Responses to Critical Thinking Questions

Possible responses are:

1. Both poets express some degree of isolation and hardship. In "Black Hair" the character deals with these feelings by "escaping" into baseball and his fantasies about a baseball hero. The character in "Alone/december/night" deals with these feelings by spending time alone, "listening to the music that is me."

2. The character in "Black Hair" was motivated in part by problems at home and by racism. The character in "Alone/december/night" seems to be motivated by his longing for home and a familiar culture.

3. Sample responses: graduation ceremonies, sporting events (including participatory ones), obtaining a driver's license, summer camps

☑ Guidelines for Writing Your Response to "Black Hair" and "ALONE/december/night"

Have students share their journal entries with a partner. Or, as an alternative writing activity, encourage students to summarize in a few sentences what they believe to be the meaning of each poem and how the poems illustrate growth and change. Have them share their writing in small groups to discuss what they have learned from the responses of the other group members.

Guidelines for Going Back Into the Text: Author's Craft

Tell students to skim the selections and find examples of imagery. Discuss how each author uses imagery differently. Possible answers to the questions in the Student Edition are:

1. In "Black Hair" the author describes the sights, sounds, and tastes of watching his favorite baseball player. In the second poem, the author uses sparse, abstract imagery to convey loneliness.

2. In both poems, the authors use imagery to give the reader a feeling for their own experiences— imagery makes the reader feel, hear, and see what the authors are conveying.

3. Students should select poems or songs that use imagery to create a feeling or tone.

☑ FOLLOW-UP DISCUSSION

Use the questions that follow to continue your discussion of the poems.

Possible answers are given in parentheses.

Recalling

1. Who was Moreno? (a Hispanic baseball player)

Interpreting

2. Why were baseball and Hector Moreno so important to the narrator in "Black Hair"? (Baseball was an escape from home, and Hector Moreno was a source of cultural pride.)

Applying

3. Why is the use of themes an effective way to make the author's point about searching for cultural identity? (By drawing on universal themes of loneliness, isolation, and a lack of love, the author creates a bridge between himself, with his longing for his home, and readers of all cultures who can identify with these universal feelings.)

ENRICHMENT

Students may enjoy watching a somewhat dated but entertaining documentary on Roberto Clemente, who was perhaps the most popular Latino baseball player in major league history. *Roberto Clemente: A*

Touch of Royalty is available from Major League Baseball Productions, 1212 Avenue of the Americas, New York, NY, 10036, (212) 921-8100.

RESPONSES TO REVIEWING THE THEME

1. Sample answer: Mary Crow Dog's theme of growth and change is illustrated in an obvious way—her experiences in the sweat lodge. Here, she discovers a ritual that is part of her heritage and that gives a new perspective on who she is. In "Black Hair," Gary Soto describes personal growth also, but through a more subtle "rite of passage"— a young boy's love for baseball and his infatuation with a Hispanic ball player. Through this experience the author comes to grips with his own heritage and the struggle that can come from growing up in a predominantly white society.

2. Sample answer: In some selections the family is both a source of cultural heritage and conflict. For the character Rosie in "Seventeen Syllables," family represents a traditional past away from which the young girl is moving. For Mary Crow Dog, her new family, is a source of growth and change for her.

3. Some possible ceremonies of change include graduation and religious rites (confirmations, baptisms, bar and bas mitzvahs).

☑ FOCUSING ON GENRE SKILLS

Leslie Marmon Silko effectively conveyed the important themes of change, learning, and human growth in her novel. She conveys the surprise of learning just by living in her novel. Select another novel (such as Jessamyn West's *The Friendly Persuasion* or Kim Ronyoung's *Clay Walls)* and ask the students to analyze the major themes of it..

BIBLIOGRAPHY

Books related to the Theme:

Takaki, Ronald. *Strangers from a Different Shore: A History of Asian Americans.* Boston: Little Brown and Company, 1989

COOPERATIVE/ COLLABORATIVE LEARNING

Individual Objective: to participate in a discussion about the theme of a heritage of traditional stories

Group Objective: to develop a five-minute presentation inspired by viewpoints expressed in the discussion

Setting Up the Activity

Have students work in heterogeneous groups of four. Stipulate that each group member is responsible for researching and expressing an individual viewpoint in the group discussion and for making a contribution to the group presentation. The individual topics are as follows: Student 1 expresses the overall theme of Unit One; Student 2 explains how the selections in Unit One illustrate this theme; Student 3 highlights the different cultures represented by the selections; Student 4 finds five quotations that demonstrate each of the writing styles in Unit One.

When they have finished their discussion, have each group elect a representative to summarize the group's discussion for the entire class. After each group's presentation use the questions below to discuss and evaluate its report.

- What purpose did storytelling serve for each of the communities in Unit One?

- In your group's opinion, is storytelling an effective means of communication? Explain your answer.

Assisting ESL and LEP Students

To give students with limited English proficiency more structure in this closing exercise, ask them to choose two selections from this unit. The first selection should be one that they liked, for any reason; the second choice should be one that they did not like, for any reason. Form groups of three to four students and ask them to compare their choices. Have them answer these questions in their groups: Why did you choose these selections? Did you choose one because you could relate to it, or because you liked the writing style, characters, or plot? Did you select the other one because you don't believe the values in it are valid? Why do you think the author wrote this selection?

Have students discuss whether they agree or disagree with the other members of their group.

Assessment

Before you begin the group activity, remind students that they will be graded both on an individual and a group basis. Without individual contribution, the group presentation will not be complete. Monitor group progress to check that all four students are contributing and that a presenter has been selected.

Time Out to Reflect As students do the end-of-unit activities in the Student Edition, provide time to let them make a personal response to the content of the unit as whole. Invite them to respond to the following questions in their notebooks or journals. Encourage students to draw on these personal responses as they complete the activities.

1. What have I learned about cultural values? Do diverse cultures have different values, or are there similarities?
2. What have I learned about the tradition of storytelling? Was this an effective means of communication? What examples from the selections in Unit 1 had the most significance for me?
3. What have I learned about the importance of understanding cultural heritage? Why is cultural heritage especially important for some communities in the United States? What culture values do I identify with in my own community?
4. What have I learned about growth and change? What characters in Unit 1 taught about change? Were their changes easy or difficult?

WRITING PROCESS: EXPOSITORY ESSAY

Refer students to the model of an expository essay found on pages 397–398 in the Handbook section of the Student Edition. You may want to discuss and analyze the model essay if you are working with less experienced writers.

Assisting ESL and LEP Students

You may want to provide a more limited assignment for these students so that they can complete their first drafts somewhat quickly. Then have them work at length with proficient writers in a peer-revision group to polish their drafts.

Guidelines for Evaluation

On a scale of 1-5, the writer has

- clearly followed all the stages of the writing process.
- made clear and specific references to the chosen selections.
- made clear and specific references to the literary elements and techniques used in the selections.
- provided sufficient supporting details from the selctions to prove his or her main points.
- clearly organized the paper so that the conclusion is easy to follow.
- written on an opening or a closing paragraph that clearly summarizes the main points of the paper.
- made minor errors in grammar, mechanics, and usage.

PROBLEM SOLVING

Encourage students to use the following problem-solving strategies to analyze one of the traditional stories portrayed in Unit One. Afterwards, have students reflect on their use of the strategies and think of ways they could have used the strategies more effectively.

Strategies (optional)

1. Use a semantic map, brainstorming list, or flow chart (works best for Unit 1 quest tales), to identify the origins, traditions, and ceremonies in Unit 1. Have groups of students choose the topic they most want to investigate further and the selection or selections that best shed light on the topic.
2. Use a second graphic organizer to explore in detail the topic they chose. Include information both from the selections and from personal experience.
3. Use the overlapping circles of a Venn diagram to organize information into similarities and differences. Then refine the diagram by putting check marks next to the most significant points.
4. Decide which topics need further research or discussion. Set up a plan that shares responsibility for doing the research.

Allow time for students to work on their presentations.

Guidelines for Evaluation

On a scale of 1-5, the student has

- provided adequate examples, facts, reasons, anecdotes, or personal reflections to support his or her presentation.
- demonstrated appropriate effort.
- clearly organized the presentation so that main ideas and supporting details are logical and consistent.
- demonstrated an understanding of challenges faced by many groups and individuals in the United States.

2 Arrival and Settlement

In Unit II, you will read about several groups of people who helped create what is now the United States. The experiences of these people provide distinct perspectives on arrival and settlement in the United States. They also provide an understanding of the nation's cultural diversity.

UNIT OBJECTIVES

Literary

- to explain how narratives and autobiographies can effectively convey cultural beliefs and values
- to show how authors have used poems, historical documents, and stories to convey the nation's cultural diversity

Historical

- to compare and contrast the settlement experiences of diverse cultural groups
- to understand how the process of arriving and settling has repeated itself throughout U.S. history

Multicultural

- to compare and contrast the struggles groups have faced as newcomers to the United States
- to understand how people share common goals of achieving economic and social equality

UNIT RESOURCES

The following resources appear in the Student Edition of *Tapestry*:

- a full-color **portfolio of theme-related art,** Humanities: Arts of the Tapestry, pages A1–A16, to build background and activate prior knowledge about the unit themes and to generate writing ideas.
- a **unit overview,** pages 120–123, to provide historical background about arrival and settlement in the United States, including Native Americans' perspectives on European immigration and the arrival of enslaved Africans.
- a **time line,** pages 122–123, to help students place the literary works in their historical context. You may wish to have students refer to the time line before they discuss each work.
- the **Focusing the Unit Theme** activities at the end of the unit, page 219, to provide a cooperative/collaborative learning project on the

theme of "Arrival and Settlement," a writing project for a personal essay, and a problem-solving activity in which students analyze one or more of the challenges newcomers faced.

INTRODUCING UNIT 2

Providing Motivation

Before students begin to read the unit overview in the Student Edition, you might want to stimulate their interest in the theme of "Arrival and Settlement" with some of the following activities.

- Read to students this quotation:

  Hold fast, this is most necessary in America. Forget your past, your customs, and your ideals. Select a goal and pursue it with all your might. No matter what happens to you, hold on. You will experience a bad time, but sooner or later you will achieve your goal.

Explain to students that this quotation is from a 1891 manual advising newcomers on how to act in the United States. Ask: "Why do you think the manual advised newcomers to forget their customs and their ideals?" Discuss whether newcomers have a "duty" to assimilate into U.S. culture, as the manual suggests.

• On the board write this question:

Is there anyone in this classroom whose family members were not once classified as newcomers to the United States?

Have students work in pairs. Each student should write a short summary of his or her family or a friend's family roots. When they are finished, have the pairs exchange their work. Partners should look for any history of arrival and settlement.

• Ask: "If you moved to a foreign country, would you change your name to 'blend in' better with that country's society? Would you give up your religion or customs to get along better with your new neighbors?" Before they respond, explain to students that in the past, some newcomers to the United States changed their foreign-sounding names and gave up their ancestors' religions in the face of discrimination and prejudice.

• Use the time line for the following activity. Have students work in groups of three. Assign one student in each group one of the following eras on the time line: 1550–1700; 1700–1850; 1850–1980. Ask students to use the events on the time line and what they might already know about these

time periods to speculate on possible literary topics for each of these eras.

Cross-Discipline Teaching Opportunity: Unit Theme

Collaborate with the social studies teacher to help students develop a broad perspective on the theme of arrival and settlement in the United States. Although there have been a number of different migrations to North America through the years, emphasis has often been placed on the *reasons* for the migrations, such as the search for religious and political freedom. However, another theme is the way newcomers have been treated after their arrival in the United States. Discrimination against newcomers and economic exploitation, ranging from the enslavement of Africans to the more subtle prejudices against some European ethnic groups, are also part of the immigration experience. Ask the social studies teacher to help students examine the economic factors underlying the importation of Africans and the later waves of immigration from Europe, Asia, and Latin America. Students should compare and analyze these factors and see how they might be related to the difficulties the immigrants faced in securing jobs. Sources on these topics are: *First Generation: In the Words of Twentieth Century American Immigrants* (Beacon Press, 1978). A collection of photographs can be found in *Portal to America: The Lower East Side* (Holt, Rinehart and Winston, 1967).

Setting Personal Goals for Reading the Selections in Unit 2

Have students keep a copy of the following chart in their journals or notebooks. Provide class time every few days for students to review and expand their charts. Encourage them to add topics of their own.

ARRIVAL AND SETTLEMENT			
Topic	What I know	What I Want To Learn	What I Have Found Out
Reasons why immigrants came to the Americas			
The value of reading first-hand accounts of past events			
Understanding the causes and effects of discrimination			
The value of cultural diversity			

Theme **4**

Voices of the First Nations

THEME PREVIEW

Voices of the First Nations			
Selections	**Genre/Author's Craft**	**Literary Skills**	**Cultural Values & Focus**
"Wasichus in the Hills," Black Elk, as told to John G. Neihardt, pages 63–65	written oral history	simile	conflict between Native Americans and colonial settlers
"The Council of the Great Peace," from *The Constitution of the Five Nations,* pages 66–68	primary source political document	primary source from oral tradition	sophistication of Native American government
"The Indians' Night Promises to be Dark," Chief Seattle, pages 69–71	written speech	elements of a speech	effects of colonial settlement on Native American culture
MAKING CONNECTIONS: Native American Arts	Linking Literature, Culture, and Theme, page 144		
"They Are Taking the Holy Elements from Mother Earth," Asa Bazhonoodah, pages 73–75	essay	elements of an essay	environmental harm caused by strip mining

Assessment

Assessment opportunities are indicated by a ✓ next to the activity title. For guidelines, see Teacher's Resource Manual, page xxxiii.

CROSS-DISCIPLINE
TEACHING OPPORTUNITIES

Social Studies To help students understand the diversity of North American Native American groups, work with the history teacher to give an overview of the hundreds of nations living in the United States today. Explain to students that each of these groups has a unique history, demography, and cultural heritage. An excellent overview of Native American cultures can be found in Carl Waldmen's *Atlas of the North American Indian* (Facts on File Publications, 1985).

Geography Work with the geography teacher to help students examine the evidence and ongoing theories about where, when, why, and how Native Americans came to the Americas. Use maps to trace the route it is believed that the forebears of Native Americans took on their migratory path from Asia over the Bering Strait land bridge approximately 20,000 to 30,000 years ago. (Note: For some Native American traditionalists, the Bering Strait migration evidence conflicts with their beliefs as expressed in their origin stories. Some of these stories describe how the people rose from sacred land in North America.)

Art Work with the art teacher to give students the opportunity to examine Native American art. Remind students that Native American cultures continue to thrive today and that some artists, particulary Fritz Scholder, T.C. Cannon, and Jolene Rickard, combine Native American traditions with contemporary artistic methods. The Native American Public Broadcasting Consortium offers for rent a six-part video series on contemporary Native American artists. (NAPBC, P.O. Box 83111, Lincoln, NE, 68501, (402) 472–3522.) See also, Margaret Archuletta and Rennard Strickland's *Shared Visions: Native American Painting and Sculpture in the Twentieth Century* (Phoenix: The Heard Museum, 1991).

SUPPLEMENTING
THE CORE CURRICULUM

The Student Edition provides humanities materials related to specific cultures in Making Connections on pages 144–145 and in the Humanities: Arts of the Tapestry on pages A1–A16. You may wish to invite students to pursue the following activities outside of class.

- Among the great artifacts created by Native Americans throughout North America, none are more impressive and arresting than those of the nations of the Pacific Northwest. Their towering totem poles and magnificently crafted boats immediately come to mind. Guide students to recognize in the artifacts on pages A6 and A7 that the superb craftsmanship extended to smaller objects as well. What is often overlooked is the cultural or religious significance of these artworks. The three 19th-century objects all conferred special status on their bearer or wearer. Ask students to read the captions and discuss the possible purpose for each object. Then, guide the students to recognize the comparable objects found in cultures remote from these. (The headdress could be seen as a crown, for example, or the speaker staff, a scepter.) Have students compare and contrast symbols of prestige in other cultures as well. Also ask students if they think the modern pastel drawing, *Jumper,* represents superb craftsmanship. Ask them to justify their opinions.

- A pivotal event in modern Native American history was the seizure of Alcatraz (California) in the early 1970s. For 19 months, Native American groups occupied the island in protest of civil rights violations. One result of the occupation was the unification of many different Native American groups in the United States into a social and political movement, sometimes referred to as Red Power. A good overview of the movement can be found in Herman J. Viola's *After Columbus* (Smithsonian Books, 1990).

- Native American history often deals only with events after Columbus's arrival in 1492. Remind students that Native Americans inhabited North America for about 15,000 years or more *before* 1492. Interested students might like to research pre-Columbian Native American history as a way to compare the impact of European contact on Native American ways of life. An excellent overview of this era can be found in the October 1991 issue of *National Geographic*.

INTRODUCING THEME 4:
VOICES OF THE FIRST NATIONS

Using the Theme Opener

On the board write these terms: settler, wilderness. Call on students to define these terms. As they respond, point out that both terms are often used to describe the immigrant experiences of European "settlers" taming the "wilderness" of North America. Now read to students this quotation:

> Only to white man was nature a "wilderness" and only to him was the land "infested" with "wild" animals and "savage" people. To us it was tame. Not until the hairy man from the east came and with brutal frenzy heaped injustices upon us and the families that we loved was it "wild" for us. When the very animals of the forest began fleeing from his approach, then it was that for us the "Wild West" began.

> Luther Standing Bear

Discuss with students the perspective on settlement expressed in the above quotation. Point out to students that for Native Americans, North America was already "settled." Suggest to students that as they study the selections in Theme 4, they consider how Native Americans perceived the arrival and "settlement" of other cultures on their land.

✔ Developing Concept Vocabulary

Use a webbing exercise to help students understand the nuances and implications of the terms *settler* and *immigrant*. Write both terms on the chalkboard. Call on students to list words that they associate with each term. Link their responses with connecting lines.

To get students started, point out that the term *settler* can have very different interpretations. Native American writers have pointedly remarked that European "settlers" were actually "resettlers," since they were settling on land already inhabited by Native Americans.

ASK:

- What do you think were some of the effects of European settlers on Native American nations?
- What do you think Native Americans thought of European settlers?
- Was migrating always a good experience? Do you think immigrants always found what they were searching for in America?

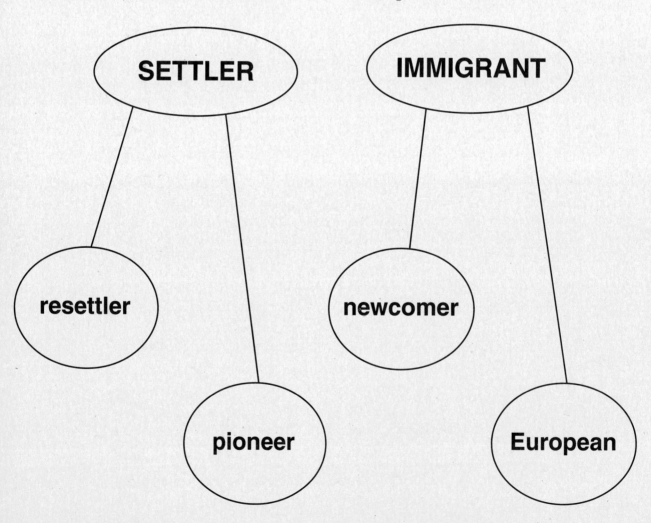

Wasichus in the Hills
from *Black Elk Speaks*
as told to John G. Neihardt (pages 125–130)

OBJECTIVES

Literary

- to identify the use of similes
- to analyze an autobiography for fact and opinion

Historical

- to evaluate how the discovery of gold affected Native Americans living in the Black Hills
- to discuss the U.S. approach to dealing with Native Americans during the 19th century

Multicultural

- to contrast cultural concepts of "wealth" and land ownership
- to explain the cultural repercussions of forced resettlement for Native Americans

SELECTION RESOURCES

Use Assessment Worksheet 15 to check students' vocabulary recognition, content comprehension, and appreciation of literary skills.

 Informal Assessment Opportunity

SELECTION OVERVIEW

The steady movement of U.S. settlers west during the 1800s led to a series of confrontations between the Lakota and U.S. soldiers. In 1865, the Lakota had been asked to sign a treaty that would give passage to settlers and miners from Fort Laramie (located along the Powder River mentioned in the selection) to the gold fields in Montana. Red Cloud, a Lakota chief, refused to sign the treaty and for the next three years he and other Lakota warriors fought the soldiers who tried to build forts along the trail. Finally, in 1868, the government surrendered and ceded the contested land along the Powder River to the Lakota. Red Cloud signed a peace treaty and made the following recommendation to his fellow Lakota: "If you wish to possess the white man's things, you must begin anew and put away the wisdom of your fathers. When your house is built, your store room filled, then look around for a neighbor whom you can take advantage of and seize all he has."

Black Elk Speaks has become an important link to the ancestral past for generations of young Native

Americans. In this selection, Black Elk describes his people's resettlement in the face of advancing soldiers and settlers.

ENGAGING THE STUDENTS

Ask: "Should land stolen from Native Americans 100 years ago be returned to them? What if one of the nation's most famous monuments is situated on this land?" Before students respond, show them a picture of the rock carvings on Mount Rushmore. Tell them that thousands of people visit this historical monument yearly. Explain that the Black Hills, where Mount Rushmore is located, has been sacred land to the Lakota for hundreds of years and, by treaty, is legally theirs. The controversy over who owns the Black Hills continues today. The U.S. government has offered the Oglala more than 100 million dollars as compensation for the loss of the land; the Oglala reject the monetary award.

Ask: "Should we close the monument and give the land back to the people from whom it was taken?"

Explain to students that in this selection they will read a firsthand account of how the U.S. government and settlers took this land from the Lakota.

BRIDGING THE ESL/LEP LANGUAGE GAP

The historical context of the settlement of the United States by Europeans and its effects on Native American civilization may be new to some students; this, combined with the unfamiliar Native American perspective, could make this selection difficult.

Before they read, ask the students what they know about Native Americans. Ask: "What caused the conflicts between U.S. settlers and Native Americans? Do you know of any similar conflicts in other parts of the world today? Have you heard of Custer? Sitting Bull? Crazy Horse?"

As students read the selection, have them look for the differences in values between the Lakota and the soldiers. When they have finished reading, ask them to pay particular attention to the sentences on page 29 (excerpted below):

> . . . He told me that the Grandfather at Washington wanted to lease the Black Hills so that the Wasichus could dig yellow metal, and that the chief of the soldiers had said if we did not do this, the Black Hills would be just like melting snow held in our hands, because the Wasichus would take that country anyway.
>
> It made me sad to hear this. It was such a good place and the people were always happy in that country. Also, I thought of my vision, and of how the spirits took me there to the center of the world.

Ask: "Who is the Grandfather at Washington? Who are the Wasichus? What does this name mean, and what does this tell you about how the Lakota viewed the settlers? What does the image about melting snow mean? How does this passage show the different values of U.S. settlers and Native Americans?" Ask students to work in pairs to find two examples in the text of each of the following: (1) names that the author gives to things, places, and people that show a distinct perspective; (2) images he uses to describe things or events; or (3) the different values of the settlers and the Lakota. Discuss how these differences led to conflict.

✔ PRETEACHING SELECTION VOCABULARY

Do a webbing exercise with students using the names of people and places in this story that contain references to animals or nature. First, tell students to skim the selection and write down four names or locations that contain references to nature or animals. (Examples: Split-Toe Creek, **Crazy Horse**, **Red Cloud**, Horse Creek, **Smoky Earth River**, **Wasichus**, **Plain of Pine Trees**, Sitting Bull.) Use the questions below to find similarities among the students' work:

- Which of your words refer to animals?
- Which of your words refer to nature?
- Why might Native Americans choose to name people and locations with these kinds of references?

The words printed in boldface type are tested on the selection assessment worksheet.

PREREADING

Reading the Author in Cultural Context

Help students read for purpose by asking them to consider how forced resettlement would affect their own lives. Discuss how being forced to leave a neighborhood or community could destroy one's cultural identity. As they read, have them consider how forced resettlement affected Native Americans.

Focusing on the Selection

Give students a quickwrite on the following question: "What value does gold have?" Tell students to consider the practical uses of gold as they write. When they have finished, call on students to read their writing aloud and discuss the different interpretations. Ask them to watch for the opposing viewpoints on the value of gold as they read the selection.

POSTREADING

The following activities parallel the features with the same titles in the Student Edition.

Responses to Critical Thinking Questions

Possible responses are:

1. Students should note that Black Elk and his people have a different sense of values than the settlers. Black Elk notes that his people have always known there was "yellow metal in little chunks" but they never considered it valuable.
2. They would have been captured or killed.
3. Responses will reflect students' prior knowledge. Sample answers: Native Americans did not value gold, Native Americans believed in visions, some Native Americans allied themselves with U.S. soldiers and were ridiculed by other Indians.

☑ Guidelines for Writing Your Response to "Wasichus in the Hills"

Have students share their journal entries with a partner. Or, as an alternative writing activity, have students work in pairs to write two short versions of the events in the selection—one from Black Elk's perspective, and the other from a gold miner's perspective. When they have finished, ask the student pairs to exchange their writing and discuss the differences. Then, have them answer the questions in the Student Edition collaboratively. Tell students to consider their answers as they write their journal entries.

Guidelines for Going Back Into the Text: Author's Craft

Ask students to write a simile that describes their favorite activity. Collect their writing and read each one aloud. Challenge students to "guess" the author and discuss how each one makes the reader understand the idea, event, or feeling compared in the writing. Possible answers to the questions in the Student Edition are:

1. There was much talking, but no action was taken.
2. "The Black Hills would be just like melting snow in our hands." This simile is used to describe what would happen if miners came into the Black Hills to mine gold.
3. Student answers should acknowledge that an effective simile can make points clear and meaningful.

☑ FOLLOW-UP DISCUSSION

Use the questions that follow to continue your discussion of "Wasichus in the Hills." Possible answers are given in parentheses.

Recalling

1. What is the name of the medicine man who made a sacred ornament for Crazy Horse? (Chips)
2. Where do Black Elk and his people spend the winter? (Soldiers' Town)
3. Why are the Wasichus flowing into the Black Hills? (to mine for gold)

Interpreting

4. What implications do you think the phrase "Hangs-Around-The-Fort" carried for Native Americans? (It implied that these people were allied with U.S. soldiers.)
5. What is happening in the Black Hills to make Black Elk and his people uneasy? (White soldiers and gold seekers were moving in.)
6. Why do you think Black Elk feels uneasy shooting squirrels? (Possibly, Black Elk had a vision or premonition of what was about to happen.)

Applying

7. How does a selection such as this provide a direct link to the past? Why is it valuable? (The selection provides a voice that gives first-hand information about ancestral roots and history.)
8. What impact does resettlement have on a community and its culture? (Resettlement can lead to disruption and destruction of a community and its culture, especially when a culture is closely tied to the land and nature.)

ENRICHMENT

For a clearer understanding of the significance of the Black Hills to Native Americans living in South Dakota today, students will benefit from viewing the video *EYANOPOPI: The Heart of the Sioux*. The video is available from Barr (Centre) Films, P.O. Box 7878, 12801 Schabarum Road, Irwindale, CA 91706. For an explanation of the current controversy surrounding the Black Hills, see the two-part video *The Black Hills: Who Owns the Land*, available from NETCHE, Box 83111, Lincoln, NE, 68501, (402) 472-3611.

The Council of the Great Peace

from *The Constitution of the Five Nations* (pages 131–137)

OBJECTIVES

Literary

- to identify the document as a written translation of an oral constitution
- to analyze a primary source

Historical

- to compare the Five Nations Constitution with the United States Constitution
- to evaluate the Five Nations Constitution as a political document

Multicultural

- to compare political beliefs among different cultures
- to summarize the cultural values embodied in the Five Nations Constitution

SELECTION RESOURCES

Use Assessment Worksheet 16 to check students' vocabulary recognition, content comprehension, and appreciation of literary skills.

✔ Informal Assessment Opportunity

SELECTION OVERVIEW

The Iroquois Confederacy was one of the most highly developed political organizations of its time north of Mexico. Each of the five nations that made up the confederacy sent ten members to a council, or governing body. Women, who played an important role in Iroquois life, nominated the representatives of each nation to the council. The council members served life terms, but if a representative turned out to be a bad choice, the matriarch of his clan could have him removed from office. A second level of representatives also was part of the confederacy government. The "Pine Tree chiefs" could obtain their positions through merit. In this respect, the confederacy foreshadowed the present two-house legislature of the U.S. government.

In this selection, the readers will examine a Native American political document that created a Confederacy of Nations in pre-colonial times.

ENGAGING THE STUDENTS

On the board write this question: "Who were the first people in North America to create a representative form of government based on a constitution?" Acknowledge to students that many Americans and historical scholars point to the birth of the United States as an experiment in constitutional and representational (republican form) of government. Suggest to them, however, that after reading the selection they might have a very different response to the question on the board.

BRIDGING THE ESL/LEP LANGUAGE GAP

A number of elements make this passage a challenge for many students: the vocabulary, style, and content are unfamiliar and difficult. Direct a prereading discussion that focuses students on the concepts of uniting many people into one nation. Point out that one of the tasks of the Five Nations was to develop a code of laws. Ask: "What would be difficult about unifying so many people? What is a constitution? How could a constitution play a role in this unification?"

To aid students in their comprehension, give them a list of questions, referencing specific sections from the text for each one. Sample questions are:

Section 2: Who is welcome to take shelter beneath the Tree of Great Long Leaves? What does this mean?

Sections 8–9: How are decisions made?

Sections 17–18: What roles do women have in the Confederate Council?

Section 23: What is wampum? How is it used?

Section 25: Could a Confederate Lord belong to any other council or confederacy? Why or why not?

Arrange the students in mixed groups of five to six English-proficient and LEP students. Ask them to work together to find the answers to these questions as they read. They may want to divide further within their groups and have each subgroup take several questions and the corresponding section of text. Then, each subgroup can share what it learned with the group. As a class, discuss the answers they found. Then, ask each group to look at one section and to discuss how the section illustrates either: (1) the values and ideals of the Five Nations or (2) the strength of the confederacy.

☑ PRETEACHING SELECTION VOCABULARY

Have students scan the selection and create a word bank by writing all the terms from the selection that relate to government. (Examples: Five Nations, council fire, Lords, **candidate,** rules, **allies,** Great Law, **contumacious** War Chiefs, Installation Ceremony, **Confederate** Speaker of the Council, **royaneh,** justice) Call on students to state the words they found, and write these on the board. Use the questions below to help students speculate on what the selection is about:

- Which of the words in the word bank do you think are related to power? organization? control?
- Based on this vocabulary activity, what do you think the selection is about? Explain how you arrived at your answer.

The words printed in boldface type are tested on the selection assessment worksheet.

PREREADING

Reading the Author in Cultural Context

Help students to read for purpose by discussing the relationship between politics and nature in the Five Nations Constitution. Ask: "Why do you think Native Americans used a tree and an eagle to describe the confederacy?" (Point out to students that the eagle figures prominently in U.S. political symbolism as well.) Tell the students to find other nature symbols in the selection as they read.

Focusing on the Selection

Ask students to write a short paragraph summarizing the basic beliefs upon which the United States is founded. When they have finished, call on them to read their writing aloud. On the board, make a list of the values and beliefs they have mentioned. Leave this list on the board and refer to it as the students read and analyze the selection. Challenge students to find similarities between the list on the board and the values and beliefs represented in the selection.

POSTREADING

The following activities parallel the features with the same titles in the Student Edition.

Responses to Critical Thinking Questions

Possible responses are:

1. The Iroquois obviously put great value on cooperation, organization, personal accountability, and honesty; they have a low regard for dishonesty, gossip, and people who do not consider the general welfare of the community.

2. The Tree of Peace was a symbol for the confederacy. It is a fitting symbol in that the confederacy members (roots) all feed into the greater good of the confederacy.

3. Similarities include: representative government, two-house legislature, attention to the general public welfare, honesty, and peace. Some differences include: the selection process, the active involvement of women, and the general lack of universal suffrage.

✔ Guidelines for Writing Your Response to "The Council of the Great Peace"

Have students share their journal entries with a partner. Or, as an alternative writing activity, ask all students to identify and list the qualities Dekanawidah expects the confederation members to have. Next, in pairs, have students discuss and answer the final question in the Student Edition.

Guidelines for Going Back Into the Text: Author's Craft

Discuss with students the value of a primary source. Ask them to evaluate the accuracy of the one they have just read.

Possible answers to the questions in the Student Edition are:

1. a committee of Iroquois Chiefs
2. It was written to chronicle the heritage of the Iroquois.
3. 1900
4. They were related through their heritage.
5. The writer identifies himself as Dekanawidah.
6. Sample answers: historical accounts of this period by other cultural groups, oral histories (in transcript form) of other Native American groups

✔ FOLLOW-UP DISCUSSION

Use the questions that follow to continue your discussion of "The Council of the Great Peace." Possible answers are given in parentheses.

Recalling

1. Where did the affairs of Five Nations take place? (under the Tree of the Great Long Leaves)
2. When did the Council meetings end? (when darkness set)
3. What emblem signified lordship in the confederacy? (deer antlers)

Interpreting

4. What does the Constitution tell you about the role of women in the Five Nations? (Guide students to understand the important role women played in the selection and punishment of representatives. This role takes on added significance when the fact that women colonists had *no* role or rights in the first U.S. government is considered.)
5. Why do you think the Council had specific measures for dealing with members who tried to establish independent jurisdiction? (Any confederacy is only as good as its ability to keep members *interdependent* and to prevent breakaways and secession.)
6. What do you think this final sentence from the Constitution means: "Look and listen for the welfare of the whole people and have always in view not only the present but also the coming generations, even those whose faces are yet beneath the surface of the ground—the unborn of the future Nation"? (This is a warning to consider future generations and to be farsighted in all actions.)

Applying

7. What was the overall purpose of the confederation? What advantage did confederation members receive? (to prevent disputes between confederation members and provide a framework for peaceful coexistence; settling disputes)
8. Do you think that the settlers who came to North America in the 18th and 19th centuries were aware of the political sophistication the Iroquois had achieved? Explain your answer. (Considering the general belief among settlers that Native Americans were "savages," it is unlikely that they recognized any achievements of the native population.)

ENRICHMENT

Students may benefit from viewing *The Faithkeeper*, a video in which Oren Lyons, a present-day chief of the Turtle Clan of the Onondaga Nation (one of the Five Nations), discusses similarities between Native American philosophy and the U.S. government. The video is available from Mystic Fire Video, P.O. Box 1092, Cooper Station, New York, New York, 10276, (800) 727-VIDEO.

The Indians' Night Promises to be Dark
by Chief Seattle (pages 138–143)

OBJECTIVES

Literary

- to analyze the speaker's attitudes toward moving onto the reservation
- to understand the elements of a speech

Historical

- to evaluate the relationship between the U.S. government and Chief Seattle
- to predict the impact of settlement on the Pacific Northwest Native American groups

Multicultural

- to discuss Native American views of the afterlife
- to compare Native American and European concepts of settlement and migration

SELECTION RESOURCES

Use Assessment Worksheet 17 to check students' vocabulary recognition, content comprehension, and appreciation of literary skills.

✔ Informal Assessment Opportunity

SELECTION OVERVIEW

As settlers streamed into the Pacific Northwest in the early 1800s, Native American leaders struggled with how to deal with the encroachment. Seattle, a principal chief of the Suquamish and Duwamish, encouraged friendship and trade with settlers and worked to avoid the costly wars of attrition being fought by other Northwest peoples. In 1855, when the region burst into renewed warfare, Chief Seattle signed the Fort Elliot Treaty, in which he agreed to relocate his people to a reservation. Throughout the Yakima wars of 1856, Seattle remained allied with U.S. forces, and successfully defended his people from an attack by the neighboring Nisqually.

In this selection, Chief Seattle compares two cultures—Native American and European American—to illustrate the differences that separated his people and the settlers. The selection illustrates how Native Americans' views of faith, nature, and the afterlife differed from those of the settlers moving into the Pacific Northwest.

ENGAGING THE STUDENTS

Ask students to consider this hypothetical situation:

You are a community leader. Your people have been asked by invaders to leave their homes and move to a new location where they will be safe. You must now decide whether to stay and fight or to protect lives by moving to the new settlement.

Ask: "Would you move or would you stay and fight?"

BRIDGING THE ESL/LEP LANGUAGE GAP

Great oration relies on language and images that stir strong emotions in people. While some of the vocabulary in this selection is difficult, the images and emotions evoked provide a focus for students. (Some students will need to concentrate on vocabulary before proceeding with this activity. It would be helpful to use the Preteaching Selection

Vocabulary section first with these students.) Write the excerpts below on the chalkboard or copy them and distribute to the class. Ask the students to look for the lines as they read.

> Yonder sky that has wept tears of compassion upon my people for centuries untold, and which to us appears changeless and eternal, may change.
>
> . . . Revenge by young men is considered gain, even at the cost of their own lives, but old men who stay at home in times of war, and mothers who have sons to lose, know better.
>
> Your dead . . . are soon forgotten and never return. Our dead never forget the beautiful world that gave them being.
>
> Your time of decay may be distant, but it will surely come, for even the White Man whose God walked and talked with him as friend with friend, cannot be exempt from the common destiny.
>
> . . . the very dust upon which you now stand responds more lovingly to their footsteps than to yours, because it is rich with the blood of our ancestors and our bare feet are conscious of the sympathetic touch.

For each excerpt, ask the students to discuss, in pairs, the language and images used. Ask: "Are the images memorable or weak? Why?" Ask them to discuss: (1) what they think the speaker was feeling as he said this; and (2) what he wanted his audience to feel.

☑ PRETEACHING SELECTION VOCABULARY

Read this sentence from the selection:

> We are two distinct races with separate origins and separate destinies. There is little in common between us.

Explain that Chief Seattle makes this point throughout the selection by using juxtaposed concepts and words that are opposites. Write the following examples on the board, and ask students to add to the list as they read.

- grass that covers vast prairies / scattering trees of a storm-swept plain
- **wax / ebb**
- day/night
- Red Man / White Man
- deserted / **throng**

- thrill
- happy-hearted / **somber**
- swarm / alone

The words printed in boldface type are tested on the selection assessment worksheet.

PREREADING

Modeling Active Reading

"The Indians' Night Promises to be Dark" is annotated with the comments of an active reader. These sidenotes, prepared to promote critical reading, emphasize the cultural content of the piece, address author values, call attention to literary skills, invite personal response, and show how the selection is related to the theme. If you have time, read the entire selection aloud as a dialogue between reader and text. Encourage students to discuss and add their own responses to the ones printed in the margins. Model these skills by adding your own observations, too.

Reading the Author in Cultural Context

Ask: "Do cultural groups lose their identity when they merge into another culture?" Tell students to consider, as they read the selection, the decision Chief Seattle must make. They should think about the implications this decision will have on the culture of his people, including their ties to their environment.

Focusing on the Selection

Remind students that they have already read the writings of Native American authors who emphasized the roles of land and nature in Native American beliefs. For this selection, ask students to consider the political conflict caused by opposing concepts of land. On one hand, settlers have left distant homes to find new places to live. On the other hand, Native Americans have ties to the land and consider it their home.

POSTREADING

The following activities parallel the features with the same titles in the Student Edition.

Responses to Critical Thinking Questions

Possible responses are:

1. His feelings are a mixture of anger and resignation. He is angry over Washington's insensitivity toward his peoples' connection to the land, yet he acknowledges that the resettlement of his people is inevitable.

2. Differences: views on the afterlife, God, nature, importance of grave sites; religion for settlers is written, but for Indians it is in the dreams and traditions of ancestors. Similarities: Seattle accepts the words of the "Great White Chief"; Seattle believes that the destiny of his people is to disappear as settlers take over their land.

3. Students should recognize the wisdom and sense of justice in Seattle's speech. They should also acknowledge his view of humanity as one.

☑ Guidelines for Writing Your Response to "The Indians' Night Promises to be Dark"

Have students share their journal entries with a partner. Or, as an alternative writing activity, have students list the pros and cons of Seattle's decision to sign the treaty and move to a reservation. On the chalkboard make two headings: *Reasons for Resisting* and *Reasons for Resettling*. Record student suggestions under the appropriate heading. When you have finished, ask students whether they believe Seattle made the right decision.

Guidelines for Going Back Into the Text: Author's Craft

Ask students if Seattle's prediction that his people would disappear has come true. Discuss the effectiveness of Seattle's presentation of ideas including those on the tragic destiny of his people. Possible answers to the questions in the Student Edition are below:

1. His purpose is to maintain relations with settlers, while making a plea for understanding about his people and their culture. He uses this forum to make a condition as well—that his people may visit sacred burial grounds unmolested.

2. Sample answer: The structure of the speech is very effective. By first making reference to the differences between his people and the settlers, and then nonetheless professing his willingness to sign the treaty, Seattle rises above pettiness and gives his request with honor and dignity.

3. Sample answer: I was moved most by the fact that Seattle knew his people were doomed, yet he didn't believe that death would be the end of their existence.

☑ FOLLOW-UP DISCUSSION

Use the questions that follow to discuss "The Indians' Night Promises to be Dark."

Possible answers are given in parentheses.

Recalling

1. What conditions does Seattle make in his acceptance of the treaty? (the right to visit sacred burial grounds)

2. If Seattle accepts the treaty, where will his people live? (on a reservation)

Interpreting

3. What do you think is the meaning of this phrase: "The Indians' night promises to be dark?" (The future bodes ill for Seattle's people.)

4. Whom do you think Seattle is referring to when he uses the term "Great White Chief"? (U.S. President)

Applying

5. What do you think Native Americans today might have to say about Seattle's speech and his decision? (Point out to students that 2 million Native Americans live in the United States today, and many are still fighting the same issues addressed by Seattle almost 150 years ago.)

ENRICHMENT

Students may benefit from viewing *Sacred Ground: The North American Indian's Relationship to the Land,* available from Wood Knapp & Co., Inc., Knapp Press, 5900 Wilshire Boulevard, Los Angeles, CA, 90036, (800) 521-2666.

OBJECTIVES

Overall

- to compare and contrast the traditional values in Native American artwork with the literary themes expressed in this section

Specific

o to compare the themes of humans' relationship to nature and nature's relationship to art in the selections

- to discover how Native American artwork and literature pass traditions from generation to generation
- to appreciate aspects of Native American culture including these aspects' relevance to contemporary society

ENGAGING THE STUDENTS

Allow students to answer this question in writing: "How can you see art in nature?" Call on volunteers to read their responses aloud. Have them look at each of the photographs on pages 144 and 145 to see how the Navajo and the people of San Ildefonso Pueblo incorporate nature into their art. Guide them to see that for Native Americans, the division between nature and art is not as clear cut as it is in Euroamerican culture. Finally, have students look for ways that Chief Seattle evokes the importance of nature for him and his people in his speech, "The Indians' Night Promises to be Dark."

BRIDGING THE LEP/ESL LANGUAGE GAP

Have students work in pairs comprised of language proficient and LEP students. Ask them to read and discuss each of the captions. Be sure students understand the importance of nature in Native American arts. Finally, have the pairs collaborate on a statement about one way nature is represented in a Native American art form.

EXPLORING ART

The place that nature holds in the lives of Native Americans is reflected in their artworks: the overriding subject of Native American art is nature, and the materials from which the artworks are made are natural. The natural objects depicted are usually abstracted, which means they are changed so that they do not look exactly as they appear in nature. For example, the bird in the San Ildefonso pottery does not look like any real bird. Have students work in small groups to read the captions and discuss the artworks. Then ask students to identify some of the natural things abstracted on these pages. Ask: "What natural materials were used to create these artworks? How do the subjects and materials used in these artworks reflect the role that nature plays in Native American culture?"

You might wish to introduce students to art forms from other Native American groups. For example, you might introduce two art forms made from trees—totem poles and False Face masks. Totem poles, made by Northwest Coast Indians, are carved and sometimes painted with symbolic animal images signifying clan lineage. False Face masks, made by the Iroquois of the Northeast, are carved from living trees and believed to be so powerful that they are hidden except when not in use. If possible, have students view the video Art of the Navajo Weaving, which presents a contemporary Navajo family of weavers and gives the origins of this traditional art form.

LINKING LITERATURE, HISTORY, AND THEME

Guidelines for Evaluation

Student answers should note that in Navajo art, beauty is represented by depictions of the natural world. Navajo artisans draw inspiration from nature. In this section, literary selections such as "They Are Taking the Holy Elements from Mother Earth" and "The Council of the Great Peace" also use nature as a metaphor and as an underlying theme. For example, in "The Council of the Great Peace," a large tree with roots is the symbol for a political organization.

Guide students to understand that Native American attitudes about nature reflect their beliefs in humans' oneness with their physical surroundings and the importance of achieving harmony with nature.

Similarities include nature as an object of reverence, nature as an important source of life, and nature as an entity that must be treated with respect.

They Are Taking the Holy Elements from Mother Earth

by Asa Bazhonoodah (pages 146–151)

OBJECTIVES

Literary

- to analyze the author's attitudes toward strip mining
- to evaluate the effectiveness of an essay

Historical

- to evaluate the impact of technology on traditional Native American ways of life
- to compare past and present perspectives on land use and conservation

Multicultural

- to discuss how cultural beliefs are related to concepts of wealth and prosperity
- to list the Native American customs discussed in the selection

SELECTION RESOURCES

Use Assessment Worksheet 18 to check students' vocabulary recognition, content comprehension, and appreciation of literary skills.

✔ Informal Assessment Opportunity

SELECTION OVERVIEW

The process of strip mining—in which huge earth movers several stories high strip away plants and soil to expose coal—has been controversial since its inception. The removal of topsoil has made the process controversial for nearly everyone living near a strip mine, as well as others who are concerned about how strip mining affects the environment. Federal and state laws now require strip mine operators to fill in and replant an area after the coal is removed. As the author of this selection points out, however, even these measures are little solace for a people who consider the land sacred. Furthermore, the author raises several other environmental problems that replanting does not solve, including water contamination and air pollution.

In this selection, the author, a Navajo woman, writes an informative essay on the harmful social and environmental effects of strip mining.

ENGAGING THE STUDENTS

On the chalkboard write the following questions:

- Has technology caused the human race to progress or regress?
- Are Native Americans environmental extremists?

Use these questions as launching points for a discussion of technology, the environment, and Native American concepts of land. Tell students they will have more information to help form their opinions after reading the selection.

BRIDGING THE ESL/LEP LANGUAGE GAP

The vocabulary of this selection is not difficult; it presents a good opportunity for students to focus on style and purpose of writing. After the students have read the selection, ask them to define the following words: *praise, protest,* and *mourn.*

Survey the class to find out which words they think most accurately describe the author's purpose in this selection. Have students work in groups of three or four (mixing proficient and LEP students), according to which descriptions they chose. Ask them to look for details that support their choices.

☑ PRETEACHING SELECTION VOCABULARY

Do a webbing exercise to help students find all the vocabulary terms relating to strip mining in the selection. On the chalkboard, write *strip mining* and draw lines from it. Tell students to skim the selection and search for words or phrases relating to strip mining. (Examples: **particles,** coal dust, **contaminate,** explosions, pollution.) Connect the terms they find to the central theme. Next, work with students to complete a vocabulary web for words relating to Mother Earth. (Examples: **Holy Elements, pollen,** water, air, **herbs.**) When the two webs are finished, use them to discuss how strip mining affects the earth and how the students think the author feels about strip mining.

The words printed in boldface type are tested on the selection assessment worksheet.

PREREADING

Reading the Author in Cultural Context

Help students read for understanding by reminding them that although many of the selections they have read so far focus on Native Americans' past, roughly two million Native Americans live in the United States today. Many of these Native Americans continue to follow traditional ways of life. Ask students to consider the impact of technology and modern land use for Native Americans living today with strong cultural bonds to the earth.

Focusing on the Selection

On the chalkboard write the word *anger.* Ask: "What are some words a writer could use to convey anger?" List the student responses on the board, then ask: "Is anger a good way to convince readers of your position?" Tell students to consider these ideas as they read the selection and "listen" to the voice of the author and her way of conveying anger.

POSTREADING

The following activities parallel the features with the same titles in the Student Edition.

Responses to Critical Thinking Questions

Possible responses are:

1. Sample response: She feels confused about the actions of whites; rather than being accusatory, however, she simply states her case: "I don't know the white man's ways, but to us the Mesa, the air, the water, are Holy Elements."
2. Sample response: Navajo respect for and dependence on the land are evident from the writer's many concrete examples.
3. Student response will reveal their personal attitudes about the environment.

☑ Guidelines for Writing Your Response to "They Are Taking the Holy Elements from Mother Earth"

Have students share their journal entries with a partner. Or, as an alternative writing activity, ask each student to briefly summarize the selection, including the author's viewpoint and message. Next, have students exchange their writing with a partner. Ask them to collaborate on a single summary including the points they believe the author was trying to make.

Guidelines for Going Back Into the Text: Author's Craft

Ask students to make a list of the kinds of topics they believe are best suited for essay form. Challenge students to discover the advantages and disadvantages of this writing form as a method of persuasion. Discuss whether or not fictional accounts can be as persuasive as essays. Possible answers to the questions in the Student Edition are:

1. Topic: the effect of strip mining on the Navajo culture. Purpose: to convince the reader that strip mining has caused irreparable harm to her way of life.
2. The writing is structured and the author's viewpoint is stated with supporting details.
3. Guide students to understand the advantages of essay writing when a topic must be clearly stated and then supported with facts and information.

☑ FOLLOW-UP DISCUSSION

Use the questions that follow to continue your discussion of "They Are Taking the Holy Elements from Mother Earth." Possible answers are given in parentheses.

Recalling
1. What evidence does the author have that the mines are killing livestock? (the death of sheep and cows, and low birth rates among her animals)

Interpreting
2. To what do you think the title of this selection is referring? What are "holy elements"? (The title is a reference to strip mining and the belief by the author that the land and what is in it are sacred.)

Applying
3. Do you think attitudes about land use and conservation have changed since the time this essay was written? Support your answer with facts. (Answers will vary, but students should be able to point to both progress and the lack thereof on the environmental front.)

ENRICHMENT

Students may benefit from viewing a video on the effects of strip mining on Navajo and Hopi reservations: *Black Coal, Red Power*, available from Indiana University Audio-Visual Center, Bloomington, IN, 47405-5901, (812) 335-8087.

RESPONSES TO REVIEWING THE THEME

1. Sample answer: In "The Indians' Night Promises to be Dark," the speaker is from the past and his concerns are for the political issues of the day, including whether or not to sign a reservation treaty. In "They Are Taking the Holy Elements from Mother Earth," the speaker has ties both to the past, through her heritage, and to the present, in her concerns about technology and the environment.
2. Student answers should include new knowledge about Native American perspectives on the afterlife; Native American views on spirituality and religion; concepts of wealth, progress, and happiness; views on technology and the environment.
3. Sample answer: I agree with the viewpoint that the earth is sacred and that humans must achieve a balance with it.

☑ FOCUSING ON GENRE SKILLS

Asa Bazhonoodah makes effective use of the essay, combining elements such as fact, opinion, description, and narration. She clearly states her purpose, to object to the strip mining of her homeland. She lists numerous and specific damages. She appeals to the reader's sense of love and goodness and justice. Select, or have the students select, another essay (such as "The Creative Process" by James Baldwin) Have students analyze the elements used by the writer and evaluate the effectiveness of the essay.

BIBLIOGRAPHY

Books Related to the Theme:

Black Elk, Wallace and William S. Lyon, *Black Elk Speaks Again: The Sacred Powers of a Lakota Shaman*. New York: Harper & Row, 1990.

Curtis, Edward, *Portraits from North American Indian Life*. New York: Promontory Press, 1989.

The Long Road from Slavery

THEME PREVIEW

The Long Road from Slavery			
Selections	Genre/Author's Craft	Literary Skills	Cultural Values & Focus
"The Drinking Gourd" and "Steal Away," African American Spirituals, pages 79–81	African American Spiritual	repetition; including refrains	musical traditions in slave culture
"The Slave Auction" and "The Slave Ship," Frances E.W. Harper and Olaudah Equiano, pages 82–84	poem/autobiography	elements of autobiography	African American stories of the Atlantic slave trade and the slave passage
"The Slave Who Dared To Feel Like a Man." Harriet A. Jacobs, pages 85–87	narrative	elements of personal narrative	literary themes of human dignity in the face of oppression
"The People Could Fly" and "Runagate Runagate," Virginia Hamilton and Robert Hayden, pages 88–90	folktale/poem	universal elements of folktales	themes of freedom and escape in African American literature

ASSESSMENT

Assessment opportunities are indicated by a ✓ next to the activity title. For guidelines, see Teacher's Resource Manual, page xxxiii.

CROSS-DISCIPLINE TEACHING OPPORTUNITIES

Social Studies Collaborate with the social studies teacher in examining Abraham Lincoln's actions with respect to African Americans during the Civil War. Students should be aware that initially he did not regard the war as a crusade to end slavery. He viewed it as an effort to save the Union. However, gradually he became convinced that slavery was a basic cause of the war and had to be abolished. On January 1, 1863, his Emancipation Proclamation went into effect. It liberated the African American slaves in those areas of the South that were still in rebellion; that is, those areas not held by Union forces. This left some 8,000,000 African American slaves still in bondage. Thus, the commonly held view that the Proclamation freed the slaves is only partly true. Nevertheless, the African American leader Frederick Douglass hailed the Proclamation as a "righteous decree," and African Americans, both enslaved and free, celebrated January 1, 1863, as a day of deliverance. Two years later, the 13th Amendment to the Constitution brought slavery to an end in the entire United States.

Geography Use a map to illustrate the trade routes used by English merchants engaged in the slave trade. The routes started in West Africa, where in exchange for rum, guns, and other manufactured goods, the merchants purchased slaves. The captives were jammed onto specially equipped vessels to undergo the most inhumane conditions on the voyage to the Americas. In the early years of the slave trade, the English took most of their captives to the West Indies, where they worked on plantations that raised sugar cane. Quantities of the cane were shipped to New England, where it was converted into rum. The rum, along with tobacco and goods manufactured in the 13 Colonies, was then transported to England. Some of these products were then used by the slave traders in purchasing slaves form West Africa. In time, many slaves were shipped directly to the American mainland, where they worked on plantations that produced cotton, which joined the transatlantic trade.

Music African American spirituals offer invaluable insight into the lives, thoughts, and dreams of Africans and African American slaves. A useful discussion of African music can be found in Dena J. Epstein's *Sinful Tunes and Spirituals: Black Folk Music to the Civil War* (University of Illinois Press, 1977).

The volume discusses the instruments and musical influences of enslaved Africans and African Americans, including the reactions of whites.

SUPPLEMENTING THE CORE CURRICULUM

The Student Edition provides humanities materials related to specific cultures in Making Connections on pages 348–349 and in the art insert on pages A1–A16. You may wish to invite students to pursue the following activities outside of class.

- The three paintings by African Americans on pages A8 and A9 are in primitive style. Explain to students that *primitive* here does not mean "unskillful," but deliberately simple or naive. These artists have chosen to depict scenes from African American life by imitating styles and techniques used by African artists for centuries. Point out the faces like traditional African masks in *Into Bondage* and the severe, hand-carved style of *Forward, 1967* that resembles woodcutting on decorated African shields. Then tell students that two of the most important European artists of the 20th century, Pablo Picasso and Georges Braque, were inspired by African art, particularly masks and sculptures, to create art that revolutionized modern Western art. If possible, show the students some early Cubist paintings by these artists. Then ask students to discuss why artists from such different cultures have turned to African art for inspiration. Ask them how the paintings in *Tapestry* are more effective for being primitive.

- Rebellion and uprising occupied the dreams of many enslaved Africans and African Americans. Some, like Nat Turner, died in their efforts to gain freedom. Ask students to research the Nat Turner uprising in 1831. A very readable and informative source is Stephen Oates' *The Fires of Jubilee* (New American Library, 1975).

- Emancipation was only one step in African Americans' quest for equality. Immediately following the Civil War, African Americans used their newly attained voting rights to gain political representation in Congress. Have students research the Reconstruction Congresses and the first African American members of the House and Senate. Some of these members included: Robert C. DeLarge of South Carolina, Jefferson H. Long of Georgia, Hiram Revels of Mississippi, and Josiah T. Wall of Florida. A good overview of this period can be found in W.E.B. DuBois' *Black Reconstruction* (World, 1962).

- *Uncle Tom's Cabin*, by Harriet Beecher Stowe, was the most popular American novel of the 19th century. The novel was adapted for the theater in 1853, and the play ran into the 20th century. The play, which was probably performed more often than any other play in U.S. history, was so popular that acting companies called "Tommers" performed nothing else. Have interested students read the novel and critique it for its depiction of African Americans and slavery.

INTRODUCING THEME 5: THE LONG ROAD FROM SLAVERY

Using the Theme Opener

Introduce Theme 5, The Long Road From Slavery, with this quotation:

> I have a dream that one day on the red hills of Georgia the sons of former slaves and the sons of former slave owners will be able to sit down together at the table of brotherhood.
>
> Martin Luther King, Jr., 1963

Tell students to consider the implications of this quotation and ask: "How is slavery a part of *all* Americans' heritage?" Acknowledge that slavery is necessarily considered an important part of African American heritage. Point out, however, that as the King quotation suggests, slavery is also a part of other peoples' histories as well.

Allow students three minutes to write a list of words that they associate with the United States. Before they write, ask them to consider the variety of perspectives on what the United States stands for. Point out that many of these different viewpoints are the result of the diverse cultural backgrounds of Americans. For example, for some Americans the United States is a symbol of freedom, but for others it also represents oppression and slavery. After students have finished writing, call on volunteers to read their lists. Discuss the diversity of perspectives in your own classroom. Explain to students that in *Tapestry* they will explore further the many views on the United States that exist in our multicultural society.

☑ Developing Concept Vocabulary

In Theme 5 students will examine closely the concepts of freedom and equality. On the board create two columns labeled *freedom* and *equality*. Lead a guided discussion on these concepts using the questions below. During the discussion, fill in the columns with the words and ideas relating to each heading.

- Do *freedom* and *equality* have the same meaning? How are they different? How are they similar? (Guide students to understand the nuances of each term.)
- Can a person be free and yet not have equality? (Point out to students that freedom does not guarantee that society will treat all people equally.)
- Can a person be physically enslaved and still achieve some degree of freedom? (Guide students to examine the concept of freedom of the spirit and the mind.)

When you have finished the discussion, call on students to draw connecting lines among the concepts and terms that are related to one another. Challenge them to justify their decisions.

FREEDOM	EQUALITY
saying whatever you like	obtaining any job you are qualified for
traveling where you want	receiving equal treatment by the police
choosing the people you associate with	sitting where you like in a restaurant

The Drinking Gourd African American Spiritual
Steal Away African American Spiritual (pages 153–156)

OBJECTIVES

Literary

- to identify the relationship between repetition and rhythm
- to analyze the moods created by songs

Historical

- to describe conditions on plantations with enslaved people
- to evaluate the chances of escape for enslaved African Americans in the early 19th century

Multicultural

- to discuss the synthesis of African culture and the Christian religion in African American spirituals
- to explain the social and political role of music in the culture of enslaved African Americans and to compare it to the role of music in other cultures

SELECTION RESOURCES

Use Assessment Worksheet 19 to check students' vocabulary recognition, content comprehension, and appreciation of literary skills.

 Informal Assessment Opportunity

SELECTION OVERVIEW

African Americans developed a lively culture during the years of their enslavement that was testimony to their strength as a people under the most oppressive circumstances. Their African heritage was particularly evident in their music. Musical instruments were patterned on West African instruments and decorated with African motifs. Drumming and dancing followed African patterns and rhythms. Also echoing the African past were call and response patterns, in which a leader sang out a bit of music and the people sang out an answer.

Reflecting the African Americans' newly acquired Christianity were the religious songs known as spirituals. In the two spirituals included, students will find both religious themes and a glimpse into the lives of the enslaved African Americans.

ENGAGING THE STUDENTS

On the board write:

Ancient Egypt, Ancient Greece, and Ancient Rome

Ask if any students know what these ancient civilizations have in common with the United States of the early 19th century. (slavery) Tell students that slavery has been in practice since ancient times. During Greek and Roman times slavery reached its most advanced stage. The Greeks and Romans used slaves as laborers, artisans, and house servants. Discuss the economic reasons for slavery. Tell students that in the selection they are about to read, they will have the opportunity to examine why slavery existed in our civilization for a period of time.

BRIDGING THE ESL/LEP LANGUAGE GAP

These selections are meant to be *heard,* or at least read aloud, if not sung. Bring in recordings of spirituals (some of the most famous are "Swing Low, Sweet Chariot," "Deep River," and "Go Down Moses"). Listen to some of the songs with students, introducing each with a short description of its theme. Ask: "What lines/ themes stood out as you listened to each song? Are there any themes common to all the songs? What kinds of feelings do these spirituals evoke? Why? Do you know any other spirituals? What are these songs about? What words or lines are repeated in them?"

Have students open their books and follow along as you read the spiritual texts aloud. Ask them to discuss in groups of three to four why they think enslaved peoples would have sung each of the songs. Ask: "What do the repeated lines mean? Why are they important?"

☑ PRETEACHING SELECTION VOCABULARY

Tell students that repetition is used in the selections to provide rhythm and accentuate ideas. Tell students to find words and phrases that are repeated in the selections (for example: **steal away; the trumpet sounds; old man** is **a-waiting; drinking gourd;** I ain't got long to stay here). When students have finished, remind them of the theme: The Long Road from Slavery. Challenge them to explain how the words and phrases they have found are related to the theme. Ask: "Do you think the selections you are about to read offer a hopeful or pessimistic message? Explain how you arrived at your answer."

The words printed in boldface type are tested on the selection assessment worksheet.

PREREADING

Reading the Songs in Cultural Context

Help students read for purpose by asking them to consider the importance that religious themes have in the selections. Discuss how the culture of enslaved African Americans combined African culture with Christian religious beliefs.

Focusing on the Selection

Give students this quickwrite before they read the selection: "Summarize what you have learned about how enslaved African Americans were treated and what efforts were made to abolish slavery." (Tell students to recall anything they might have learned in past classes or in prior readings.) While they write, write the following categories on the board: *Day-to-Day Life* and *Abolition Movement.* Call on students to fill in the categories with details from their writings. After they have read "The Drinking Gourd" and "Steal Away," you might ask them to compare what they already know with what they have read in these selections.

POSTREADING

The following activities parallel the features with the same titles in the Student Edition.

Responses to Critical Thinking Questions

Possible responses are:

1. The road is northward, follows a river, and is lined with dead trees.
2. Students may note: lack of freedom, harsh treatment by owners, separation of families.
3. Responses will reflect students' own emotional responses.

☑ Guidelines for Writing Your Response to "The Drinking Gourd" and "Steal Away"

Have students share their journal entries with a partner. Or, as an alternative writing activity, ask students to consider one of their favorite pieces of music or lyrics and compare it to the selections they have just read. Have students work in pairs to describe how songs can create or invoke feelings in the listener. Call on each pair to explain and discuss its choice.

Guidelines for Going Back Into the Text: Author's Craft

Ask: "What makes a good refrain? What makes a refrain effective? What makes a refrain memorable?" As students respond, remind them that refrains often contain repeating lines at regular intervals. Ask: "How does this limit (or expand) the use of refrains in a song or poem?" Possible answers to the questions in the Student Edition are:

1. Phrases are repeated, creating a rhythmic meter.
2. The repetitions are: *Follow the drinking gourd* and *Steal away.*
3. Phrases are repeated in strategic places to emphasize ideas and meaning.

☑ FOLLOW-UP DISCUSSION

Use the questions that follow to continue your discussion of "The Gourd" and "Steal Away." Possible answers are given in parentheses.

Recalling

1. According to the narrator in "The Drinking Gourd," where does the river end?" (between two hills)
2. According to the narrator in "Steal Away," how does the Lord call him or her? (through the thunder and lightning)
3. According to the narrator in "The Drinking Gourd," who is waiting to carry him or her off to freedom? (an old man)

Interpreting

4. What does the "Drinking Gourd" symbolize? (the Big Dipper constellation/escape)
5. What are two interpretations the title "Steal Away" might have for the listener? (steal away to Jesus, or steal away to the north)
6. Why do you think religious themes were incorporated into African American spirituals? (The religious theme of a better life after death struck a resonant chord in an enslaved people.)

Applying

7. What importance did spirituals such as "Steal Away" and "The Drinking Gourd" have for African Americans in the 19th century? (They were a link to the past and a means of creating a cultural identity.)
8. Why is it important for young Americans today to learn and understand the culture of enslaved African Americans? (Learning about the past helps put into perspective the struggle to achieve harmony in the United States today.)

ENRICHMENT

For an explanation of how West African languages were mixed with English to create Black English—recognized as the heart of spirituals—see *The Story of English: Black on White*. The video is available from Films, Inc., 5547 N. Ravenswood Avenue, Chicago, IL, 60640.

The Slave Auction
by Frances E. W. Harper
and The Slave Ship
by Olaudah Equiano (pages 157–163)

OBJECTIVES

Literary

- to recognize the authors' attitudes about the way African Americans were enslaved
- to understand the elements of an autobiography

Historical

- to explain the trade route taken by ships transporting enslaved Africans and to define the Middle Passage
- to describe the conditions aboard slave ships during the Middle Passage

Multicultural

- to trace the cultural roots of African Americans and compare their experiences to those of other ethnic groups
- to analyze how cultural groups maintain their identities in the face of oppression

SELECTION RESOURCES

Use Assessment Worksheet 21 to check students' vocabulary recognition, content comprehension, and appreciation of literary skills.

 Informal Assessment Opportunity

SELECTION OVERVIEW

For students to get a true picture of how Africans were enslaved, you might have to eliminate some misconceptions. Students often ask, "Why did Africans allow Europeans to make them slaves?" Explain that Europeans did not go into the interior of West Africa and "take" slaves. Until the late 19th century, Europeans had not penetrated the African interior to any degree. Europeans purchased captives at coastal trading stations from other Africans, who had taken captives to sell. Remind students that West Africa was divided into a number of states and kingdoms with different cultures. Many of these groups warred with one another and took prisoners, many of whom were sold as slaves. The author of "Slave Ship" believes that Europeans seemed to have a special talent for cruelty to those Africans they transported to the Americas. An often overlooked feature of slavery in the South was the fact that on one plantation there might be enslaved Africans from many parts of West Africa who spoke many different languages and came from a variety of cultures. The fact that many enslaved Africans could not initially communicate with one another prevented them from forming alliances and organizing uprisings. Only toward the end of the 18th century were most African Americans born in the 13 colonies.

In the selections, the authors describe two aspects of the slave experience: the voyage of enslaved Africans from Africa to North America and the sale of enslaved Africans at a slave auction.

ENGAGING THE STUDENTS

On the board write the following quotation:

> Frederick, is God dead?

Explain to students that the quotation was said by the African American abolitionist Sojourner Truth to Frederick Douglass. Discuss the possible reasons Sojourner Truth might have for wondering whether God was alive or dead. After the discussion read to students this quotation by Frederick Douglass:

> Every tone [of the songs of the slaves] was a testimony against slavery, and a prayer to God for deliverance from chains.

Explain to students that in the following selections, they will have the opportunity to discover why, in the face of their oppression, "God" remained alive for the enslaved African Americans.

BRIDGING THE ESL/LEP LANGUAGE GAP

The vocabulary and style of these selections (especially "The Slave Ship") are very difficult. In preparation for reading "The Slave Auction," explain the plot briefly and then ask students to look for the answer to this question as they read the poem: "What might the young girl, mother, wife/husband, and child be thinking and feeling?"

In preparation for reading "The Slave Ship," ask students what they know of the Middle Passage. Do they know or can they imagine what it was like? Ask: "What kind of physical, mental, and emotional suffering did abducted Africans go through?"

Read the selection aloud together as a class, stopping to discuss each section. Ask: "How did the people suffer? What were their fears? How do you think the slave traders felt as they abducted and later sold the Africans as slaves?"

✔ PRETEACHING SELECTION VOCABULARY

Have students scan the selections and create a word bank by writing all the terms that relate to the categories of *travel* and *slavery*. (Responses should include: slave auction, **slave ship,** Africans, **avarice, fetters, cruelty, pestilential.**) While they

are writing, use these categories to make two columns on the board. Call on students to fill the columns with the words they have found in the text. Use the categories to discuss the questions below:

- Do the words in the word bank relating to travel have a bright tone, or are they somber and dark?
- Which of the word bank terms are related to African heritage?
- Based on this vocabulary activity, what do you think the selections are about? Explain how you arrived at your answer.

The words printed in boldface type are tested on the selection assessment worksheet.

PREREADING

Reading the Author in Cultural Context

Help students read for purpose by asking them to consider the impact that selections such as these might have had on the Abolitionist Movement in the 19th century. Discuss how literature and firsthand accounts of the experiences of enslaved Africans and African Americans could help the abolitionist cause.

Focusing on the Selection

Have students work in groups of three. Assign one member in each group one of the following roles: slave trader, African American, and bystander (These are outlined in the student text). Ask each group to summarize the emotions of the people involved in slave trading from three different perspectives. Call on the groups to read their summaries aloud and discuss the various interpretations.

POSTREADING

The following activities parallel the features with the same titles in the Student Edition.

Responses to Critical Thinking Questions

Possible responses are:

1. The attitudes include hatred, fear, amazement, anger, and disgust.

2. He was influenced by his experiences of being abducted and sold into slavery.

3. Sample answer: Sadly, the readings point out a frightening aspect of human nature—the ability to discriminate on the basis of race and to inflict unimaginable cruelty on fellow humans.

☑ Guidelines for Writing Your Response to "The Slave Auction" and "The Slave Ship"

Have students share their journal entries with a partner. Or, as an alternative writing activity, ask each student to write in a few sentences what he or she believes are the main points of the selections and what the authors were trying to convey. Next, lead a discussion on the slave trade. Ask: "How did white slave owners defend their use of enslaved people?" Call on students to read their interpretations of the story, and ask: "What message do these selections have for teenagers in the 20th century? Has our society put racism and intolerance behind us?" Discuss the student responses with the class.

Guidelines for Going Back Into the Text: Author's Craft

Help students to understand the impact an autobiography can have by asking them to consider what kinds of information and thoughts they would put in their own autobiographies. Ask: "Suppose you kept a diary for your entire life—would reading this be a good way for a person to learn about who you are? Would the information in the diary be factual or fiction?" Guide students to understand how an autobiography may not always be objective, but it can still be a good source of information. Possible answers to the questions in the Student Edition are:

1. Factual information includes: conditions aboard slave ships, how Africans were abducted and transported, how enslaved Africans were sold, how some enslaved Africans escaped.

2. He was initially both fascinated and horrified by his captors. Their cruelty shocked him.

3. One method would be to compare Equiano's description with other personal accounts from this time period.

☑ FOLLOW-UP DISCUSSION

Use the questions that follow to continue your discussion of "The Slave Auction" and "The Slave Ship." Possible answers are given in parentheses.

Recalling

1. What is Equiano's first reaction upon seeing a slave ship? (astonishment)

2. When they first arrive in Barbados, what do Equiano and his people believe is planned for them? (They believe they are to be eaten.)

3. What observation leads Equiano to believe that the slave traders have magical powers? (The slave traders' ability to ride animals.)

Interpreting

4. According to the narrator in "The Slave Auction," what is the cruelest part of a slave auction? (the separation of children from their parents)

5. Why do some of the abducted Africans in Equiano's ship jump overboard? (They felt that suicide was better than continuing the voyage on the slave ship.)

6. Explain the meaning of this phrase: "Ye may not know how desolate/Are bosoms rudely forced to part, And how a dull and heavy weight/Will press the life-drops from the heart." (The implication is that separating families drains the life out of humans.)

Applying

7. Why are firsthand accounts a good way to learn about the historical events surrounding slavery? (Guide students to understand how these accounts provide poignant, accurate insights into the feelings, beliefs, ideas, and values of the people living at this time.)

8. What lessons do these selections have for a multicultural society struggling with ethnic tensions today? (Understanding cultural heritage will hopefully lead to an overall acceptance of and tolerance of different cultures.)

☑ ENRICHMENT

Students may benefit from viewing *Slavery and Slave Resistance*. The video is available from Coronet/MTI Film and Video, 108 Wilmot Road, Deerfield, IL, 60015-9990, (800) 621-2131.

The Slave Who Dared to Feel Like a Man
by Harriet A. Jacobs (pages 164–171)

OBJECTIVES

Literary

- to recognize the expression of various attitudes toward slavery
- to understand the elements of a personal narrative

Historical

- to describe living conditions in slave quarters
- to explain how enslaved Africans and African Americans coped with their captivity and the loss of freedom

Multicultural

- to discuss the impact of slavery on African American families and compare this to other cultures
- to explain why understanding cultural heritage is important for all ethnic groups in the United States

SELECTION RESOURCES

Use Assessment Worksheet 21 to check students' vocabulary recognition, content comprehension, and appreciation of literary skills.

 Informal Assessment Opportunity

SELECTION OVERVIEW

In 1850, a strict Fugitive Slave Act was passed that for the first time gave the federal government a role in capturing African Americans who had escaped from slavery and returning them to their "owners." The law had been passed to stem the rising tide of runaways from slavery, which was testimony both to the discontent of enslaved African Americans and their growing ability to throw off the bonds of their enslavement. Many fled the South with the aid of the Underground Railroad, which provided safe havens for escapees along a route that led from the South to border states such as Kentucky and Maryland and eventually to freedom in the North and Canada. By the 1850s the antislavery movement was making its presence felt in fugitive slave cases. For example, in

1851, a Maryland slaver named Edward Gorsuch was murdered by a crowd of abolitionists in Pennsylvania when he attempted to recapture fugitive slaves.

In this selection, the author describes one African American's attempt to escape from slavery, at the same time describing the conditions of her enslavement.

ENGAGING THE STUDENTS

Ask: "Were enslaved African Americans justified in using violence, including the taking of lives, to escape captivity?" Before they answer, tell students that several plantation owners who were attempting to bring runaways back to their plantations were murdered by mobs. Lead a class discussion on the use of violence and whether or not it was justified.

BRIDGING THE
ESL/LEP LANGUAGE GAP

Have LEP students work in cooperative groups of three to four. Each group should select one or two characters and describe their reactions to the situation. The groups can then exchange ideas. When all are in agreement, a list of specific feelings can be produced. Then have groups discuss what happened to Ben after he was caught, and how Ben's outlook on life changed after he was free.

☑ PRETEACHING
SELECTION VOCABULARY

On the board write: *slavery* and *freedom*. Tell students to skim the selection and find all the words relating to these concepts. Challenge them to be creative in their search. Explain that they can choose words they associate with freedom, even if the asociation is not immediately evident. (Examples: **indecorum, audacity,** sorrow, north, imprisoned, **master, overseer, chattle, yoke**) As they search, call on students to fill in the categories on the chalkboard with their words. Challenge students to justify their choices.

The words printed in boldface type are tested on the selection assessment worksheet.

PREREADING

Reading the Author
in Cultural Context

Help students read for purpose by asking them to consider the historical importance a firsthand account such as this has for readers today. Ask them to compare the kinds of historical information contained in this account to the facts in a history textbook. Inform students that it was against the law to teach African American slaves to read. Those who did learn to read had to do it in secret. In light of this, discuss the importance literacy had for African Americans during their enslavement.

Focusing on the Selection

Ask students to consider the importance family bonds play in this selection, even though the family has been torn apart by slavery. Ask them to compare the

role of the grandmother in the selection with their own ideas about parenting and parental roles. Finally, discuss how slavery changed and modified the roles of family members in African American slave families.

POSTREADING

The following activities parallel the features with the same titles in the Student Edition.

Responses to Critical
Thinking Questions

Possible responses are:

1. Many slaves, despite horrible conditions and continual oppression, continued to struggle for freedom and dignity—even when it meant risking their lives.
2. Slavery caused the forced separation of many African American families. In a perverse way, however, it also brought some families closer together in their struggle to survive and resist oppression.
3. Similarities: willingness to stay together in times of crisis; source of love and friendship. Differences: Most families today do not have to undergo the disruptive conditions faced by enslaved families.

☑ Guidelines for Writing Your
Response to "The Slave Who
Dared to Feel Like a Man"

Have students share their journal entries with a partner. Or, as an alternative writing activity, ask students to list the differences and similarities between the two characters. On the board make two categories: *Harriet's Grandmother* and *Benjamin*. Ask: "Did the two characters agree on the feasibility of escape? Explain why or why not." Tell students to think of the factors that tie families together as they write their responses.

Guidelines for Going Back
Into the Text: Author's Craft

Suggest to students that they too one day may want to write a narrative about their experiences. Ask:

"What writing techniques does Harriet Jacobs use to describe her past?" Have students work in pairs to make a list of the qualities contained in an effective narrative. Call on a student from each pair to read his or her writing aloud and discuss the different interpretations. Possible answers to the questions in the Student Edition are below.

1. Factual information includes: slave conditions and relations with owners; punishments faced by escaped slaves.

2. The selection highlights the author's respect for personal courage, and the ability of humans to feel dignity and pride, even under adverse conditions.

3. A historian would compare it to other accounts of this era, including fugitive slave laws, black codes, and other government documents.

☑ FOLLOW-UP DISCUSSION

Use the questions that follow to continue your discussion of "The Slave Who Dared to Feel like a Man." Possible answers are given in parentheses.

Recalling

1. Why does Mrs. Flint make Harriet run a long errand in bare feet? (She is angry about the noise Harriet makes when walking in new shoes.)

2. What causes Benjamin to try to escape? (He was to be whipped in public for not immediately responding to the summons of Mr. Flint.)

3. Whom does Benjamin meet in New York? (his brother Phil)

Interpreting

4. What is your interpretation of the selection's title? (Guide students to understand how Benjamin's daring to be free gives him dignity and pride.)

5. How does the author's grandmother rationalize their condition of slavery? (Her grandmother tells her children that their enslavement is the will of God.)

6. Why does Benjamin send his chains to Dr. Flint? (It is symbolic that he has not been beaten.)

Applying

7. What was Benjamin's reasoning for continually trying to escape? (Guide students to understand that for Benjamin, death is preferable to slavery—without freedom life has no value for him.)

8. Consider what you have read. What message do you think this selection has for students today? (Humans, even in the face of extreme oppression, can find hope and the will to improve their condition.)

ENRICHMENT

Students may benefit from viewing any of the ten programs in the University of Michigan's *The Black Experience*. The videos are available from Michigan Media, University of Michigan, 400 Fourth Street, Ann Arbor, MI, 48109, (313) 764-8228.

The People Could Fly by Virginia Hamilton
and Runagate Runagate by Robert Hayden

(pages 172–180)

OBJECTIVES

Literary

- to analyze various reactions to oppression
- to identify themes of a folktale and a poem

Historical

- to describe conditions for enslaved African Americans on Southern plantations
- to discuss the relationship of enslaved African Americans to farm and plantation owners in the southern United States

Multicultural

- to discuss how arrival and settlement is a common theme in U.S. culture
- to analyze common themes of freedom in multicultural literature

SELECTION RESOURCES

Use Assessment Worksheet 22 to check students' vocabulary recognition, content comprehension, and appreciation of literary skills.

 Informal Assessment Opportunity

SELECTION OVERVIEW

Life on a plantation for African American slaves was an unending grind of work, over which always hung the threat of physical punishment. Although conditions of slavery varied, a general pattern ran through slaves' work routines. Typically, field workers rose before dawn, ate a breakfast of corn pone or mush and then tended to farm chores. At daybreak workers were summoned to the fields by a blast of a horn. Spring was planting time on a cotton plantation. In summer, slaves worked with hoes to cultivate the ground, meanwhile fighting off the armies of gnats and mosquitoes. In the fall, when the cotton was ready to be picked, workdays became longer and the work more frantic. Cotton bolls had to be harvested before winter frosts. When the crop was in, there were fences to be mended, new fields to be cleared, and countless other chores. In spring, the cycle began again.

In these selections, the authors explore the hopes of enslaved African Americans for freedom and the plans of some for escape.

ENGAGING THE STUDENTS

Read aloud or write on the chalkboard the following quotation:

> Slavery is a good—a positive good. It is absolutely essential for race control, and it cannot be subverted without drenching the country in blood, and destroying one or the other of the races.

Explain to students that the quotation is paraphrased from a speech made by John C. Calhoun, a U.S. Senator and former Vice President in the 1800s. Ask students to respond to the quotation and discuss this defense of slavery.

BRIDGING THE ESL/LEP LANGUAGE GAP

After students have read the story "The People Could Fly," ask them to work in groups to make two separate lists of the realistic and the fantastic elements in the story. Have them discuss why these two elements are mixed in one story. What does the author's statement at the end (excerpted below) mean?

> "The People Could Fly" was first told and retold by those who had only their imaginations to set them free.

In the same groups, ask the students to read the poem "Runagate Runagate" aloud, taking turns with the stanzas. After they have read it once, ask them to read it again, one section at a time. Have them discuss each section. Ask: "Who is speaking? What emotions are conveyed? What do the details tell you about the experience of enslaved African Americans who attempted to escape?"

As a class, discuss the differences between the two selections.

☑ PRETEACHING SELECTION VOCABULARY

Tell students that in the selections they are about to read, there are many references to escape and flight. Ask: "What emotions do you associate with these topics?" Write their responses on the chalkboard. Tell them to skim the selections to make a second list of vocabulary words that the authors have used to describe escape and flight. (Examples from "The People Could Fly": **croon,** eagle, fly, wings, freedom-bound; examples from "Runagate Runagate": reward, dead or alive, runagate, **thicketed, plausible, motif, movering, underground**) When they have finished, tell students to write a short poem or verse on the theme of escape and freedom using the words from the two lists. Call on students to read their verses.

The words printed in boldface type are tested on the selection assessment worksheet.

PREREADING

Reading the Author in Cultural Context

Help students read for a purpose by asking them to consider the importance the themes of freedom and hope play for the characters and narrator, even in the face of oppression. Discuss why understanding the hopes and aspirations of enslaved African Americans is perhaps as important as understanding the repression of their lives.

Focusing on the Selection

Ask: "Can slavery subdue the human spirit? How do humans cope with a life of oppression?" Discuss human beings' ability to undergo and survive extreme hardships. Call on students to explain the characteristics they believe make survival possible during such times. Tell students to consider these topics as they read.

POSTREADING

The following activities parallel the features with the same titles in the Student Edition.

Responses to Critical Thinking Questions

Possible responses are:

1. The selections suggest both physical and mental escapes from slavery. In "The People Could Fly," escape comes via a fantasy—the ability to fly. In "Runagate Runagate," escape refers to running away and being transported via the Underground Railroad.

2. In "Runagate Runagate," Harriet Tubman brings enslaved African Americans to freedom. In "The People Could Fly," enslaved workers are helped to escape by a mythical "Toby" who teaches them to fly.

3. Guide students to understand how the message in both selections contains both pessimism and optimism. In "The People Could Fly," for example, the message could be interpreted both ways: optimistic because some African Americans fly away from their fate; pessimistic because the plot is fantasy.

☑ Guidelines for Writing Your Response to "The People Could Fly" and "Runagate Runagate"

Have students share their journal entries with a partner. Or, as an alternative writing activity, ask each student to summarize briefly the emotional impact of both selections. Next, have students exchange their writing with a partner. Ask them to collaborate on a single list of their ideas and feelings about the selections.

Guidelines for Going Back Into the Text: Author's Craft

Discuss with students the role of African folktales in the culture of enslaved African Americans. Remind them that slaves were cut off from their ancestral roots. Ask: "Why was an oral tradition of storytelling especially important for enslaved African Americans?" Discuss the language and literacy barriers confronting enslaved African Americans. Possible answers to the questions in the Student Edition are:

1. The story is about enslaved African Americans who can flee their captors by gaining the ability to fly. The characters in the story range from courageous African Americans to cruel overseers and plantation owners. One character, Toby, has the magic to make some of the field workers fly.

2. Most folktales teach a lesson, perhaps centering on a struggle.

3. Answers will reflect students' prior knowledge of folktales.

☑ FOLLOW-UP DISCUSSION

Use the questions that follow to continue your discussion of "The People Could Fly" and "Runagate Runagate." Possible answers are given.

Recalling

1. What happened to the people in "The People Could Fly" when they became enslaved? (They shed their wings.)

2. Where are the runaways escaping to in "Runagate Runagate"? ("The mythic North, star shaped yonder Bible city")

Interpreting

3. How do you interpret the character of Toby in "The People Could Fly"? (He is an imaginary magic man with powers stemming from ancient African beliefs.)

4. According to the narrator in "The People Could Fly," what method will the people who are left behind use to escape? (They must wait for a chance to run.)

Applying

5. What social purpose do you think folktales such as these served for enslaved African Americans? (offered hope, dreams of escape)

ENRICHMENT

Students may benefit from viewing *The Freedom Station*, a video that tells the story of one young African American girl's experience on the Underground Railroad. The video is available from PBS Video, 1320 Braddock Place, Alexandria, VA, 22314-1698, (703) 739-5380.

RESPONSES TO REVIEWING THE THEME

1. Sample answers: right of free movement, free thought, free expression, right to happiness and family. Guide students to understand how enslaved African Americans created a cultural identity of their own.

2. Sample answer: knowledge about the hopes enslaved African Americans had for escape, and the effect slavery had on African American families.

3. Sample answer: "The Slave Ship." Because it was a first-hand account, the descriptions of the Africans' life were especially meaningful.

☑ FOCUSING ON GENRE SKILLS

Harriet A. Jacobs makes effective use of several elements of personal narratives. She bases her narrative on factual events but fictionalizes the characters. She shows a breadth of responses to slavery in her characters, the most extreme antislavery feelings being held by her brother Benjamin. Have the students select, another personal narrative or autobiography (such as Maya Angelou's *I Know Why the Caged Bird Sings*) and ask the students to read passages to reveal opinions held by the author.

BIBLIOGRAPHY

Books Related to the Theme:

Andrews, William L., *To Tell a Free Story: The First Century of Afro-American Autobiography, 1760–1865.* Urbana and Chicago: University of Illinois Press, 1986.

Davis, Charles T. and Henry Louis Gates, Jr., *The Slave's Narrative.* New York: Oxford University Press, 1985.

Theme **6**

Stories of Newcomers

THEME PREVIEW

Stories of Newcomers			
Selections	**Genre/Author's Craft**	**Literary Skills**	**Cultural Values & Focus**
"Lali," from *In Nueva York* Nicholasa Mohr, pages 94–96	modern fiction	setting and use of flashback	the isolation and longing for home felt by immigrants
from *Picture Bride*, Yoshiko Uchida, pages 97–99	modern fiction	characterization	cultural traditions and prearranged marriages
"I Leave South Africa," from *Kaffir Boy in America*, Mark Mathabane, pages 100–102	autobiography	irony	life under apartheid contrasted with life in the United States
"Ellis Island," Joseph Bruchac, pages 103–105	poem	figurative language	opposing Native American and European perspectives on immigration

MAKING CONNECTIONS: The Immigrant Experience Linking Literature, Culture, and Theme, p. 106

"Those Who Don't," "No Speak English," and "The Three Sisters," Sandra Cisneros, pages 107–109	modern fiction	tone	confronting racial discrimination and community values

Assessment

Assessment opportunities are indicated by a ✓ next to the activity title. For guidelines, see Teacher's Resource Manual, page xxxiii.

CROSS-DISCIPLINE TEACHING OPPORTUNITIES

Geography Allow students the opportunity to explore the geographical origins of immigrant groups in the late 19th and early 20th centuries. Work with the geography teacher to develop a class project on this topic. For example, students working in groups could do research regarding the countries and regions from which immigrant groups have come and continue to come. Assign each group one of the following countries or regions: Northern Europe, Eastern Europe, Southern Europe, China, Southeast Asia, or Russia. Have each group research its designated place and the people that came from it. Ask the groups to plot, on a large world map, the course of migration the immigrants took. Groups should also explain how the immigrants traveled, and identify the years during which the largest numbers of immigrants from each region came.

Mathematics Work with the math teacher to help students chart the rate of immigration to the United States over several decades. Immigration data can be found easily in encyclopedias, reference books, such as the *Statistical Abstract of the United States* and *Historical Statistics of the United States*, and in many social studies texts. Have students create graphs that illustrate the rates of immigration over time. You might want to begin the graph in the year 1825, when only 10,000 people came to the United States. (By 1845, the number had reached 100,000.) When the graphs are completed, discuss with students some possible causes of fluctuations in the numbers. (For example, immigration to the United States typically has decreased during wars and economic hard times.)

Social Studies Americans continue to wrestle with cultural diversity in U.S. society. The concept of a "melting pot" society is no longer accepted. Traditional concepts of assimilation are being challenged by a new respect for ethnic pluralism, leading one sociologist to suggest that the melting pot metaphor be replaced with a "salad bowl" metaphor. (In a salad, the individual identities are not lost.) This discussion raises important questions about ethnic tolerance in U.S. society. *In Their Place: White America Defines Her Minorities 1850–1950*, Lewis H. Carlson, ed. (John Wiley, 1972), contains a useful collection of primary source readings on the topics of prejudice toward and discrimination against ethnic groups in the United States. The June 1985 National Geographic contains an inclusive photo essay on cultural diversity and exchange along the U.S.-Mexican border. Interested students may read selections from *In Their Place*, or study the photo essay in *National Geographic*. They may then summarize what they have read in an essay or in an oral report.

SUPPLEMENTING THE CORE CURRICULUM

The Student Edition provides humanities materials related to specific cultures in Making Connections on pages 210–211 and in Humanities: Arts of the Tapestry on pages A1–A16. You may wish to assign students the following activities outside of class.

- The artworks on pages A10 and A11 reflect the emotions of people who have grown up under comparatively repressive conditions. Romare Bearden shows us an urban scene as viewed by an African American who grew up in the rural South. Both Hung Liu and Bo Jia grew up in China, where the government exercises strict control over artistic expression. It is often the tendency of artists in these circumstances to depict idealized visions of life under other circumstances. Have the students examine the art on these pages and read the captions. Challenge them to express how any or all the works might show optimism about a new place. Have students interested in an artistic, but more realistic, view of what immigrants to the United States experienced from the 1880s on research the work of Jacob Riis, the Danish American reporter/photographer.

- A global revolution in industrialization, a decreasing infant mortality rate, and an increase in life expectancy were factors that changed the world in the late 19th and early 20th centuries. Food production in many nations did not keep pace with the growing population. In part, immigration to the United States had its roots in these changes. Have interested students research the causes of migration during this time period. Have them look for the causes of the decrease in the standard of living in many parts of Europe and Russia.

- For many immigrants, the worst part of the immigrant experience was the trip itself. Have interested students research the conditions many immigrants endured during their travel to the United States. Ask students to include information about the amount of baggage travelers were allowed to carry, how much passage cost, and conditions aboard the ships. This information can be found in a number of readily available books, including Arthur Mann's *Immigrants in American Life* (Houghton Mifflin, 1974).

• Ask interested students to use their own research skills to compile some firsthand accounts of immigrant experiences. Ask them to interview family members, neighbors, or other people they know who have recently immigrated. Discuss with students the proper interview techniques and approaches to obtaining the interviews. Many books on conducting oral histories are available. Three such books are: *How to Tape Instant Oral Biographies*, by William Zimmerman; *Portraits of Our Mothers*, by Frances Kolb; and *Your Family History: A Handbook for Research and Writing*, by David E. Kyvig and Myron A. Marty. All are available for purchase from the National Women's History Project, 7738 Bell Road, Windsor, CA, 95492, (707) 838-6000.

INTRODUCING THEME 6: STORIES OF NEWCOMERS

Using the Theme Opener

Read the following quotation aloud:

> So at last I was going to America! Really, really going at last! The boundaries burst! The arch of heaven soared! A million suns shone out for every star. The winds rushed in from outer space, roaring in my ears, "America! America!"

Explain that this quote was written by a young Russian teenager in the early 1900s when she learned that she and her family were moving to the United States. Ask: "What do you think the United States symbolized for this teenager?" Tell students that for many immigrants, the United States has symbolized a chance to "start all over again." Allow students five minutes to write on this topic: Imagine that this teenage girl is about to move right now, and you have the opportunity to speak to her before she comes to the United States. What will you tell her about living in this country? What advice will you give to help her adjust to life in the United States?"

When the students have finished writing, call on volunteers to read their writing aloud. Discuss the different topics they have addressed. Ask students to evaluate whether their perspectives on immigration are generally negative or positive. Tell them that in Theme 6, they will explore further the experiences of newcomers to the United States. Ask them to save their writing so that they can determine whether they still feel their advice is appropriate when they have finished studying this theme.

☑ Developing Concept Vocabulary

On the chalkboard write the following:

What Newcomers Expected   What Newcomers Found

Explain to students that in Theme 6, they will read a number of stories and poems about the dreams of newcomers to the United States and the realities they confronted upon arrival. Ask students to group the following vocabulary terms under the appropriate heading. As students place the terms, have them justify and explain their choices. Keep the chart on the chalkboard and refer to it as your class reads Theme 6.

WHAT NEWCOMERS EXPECTED	WHAT NEWCOMERS FOUND
wealth	hard work/ low wages
freedom	prejudice
equality	discrimination
opportunity	despair
	homesickness

Lali from *In Nueva York*
by Nicholasa Mohr (pages 182–189)

OBJECTIVES

Literary

- to identify the element of setting
- to recognize the element of flashback in the selection

Historical

- to explain the economic reasons immigrants came to America
- to describe urban life for immigrants to the United States

Multicultural

- to contrast urban and rural values in diverse immigrant cultures
- to observe common themes of settlement among diverse cultures

SELECTION RESOURCES

Use Assessment Worksheet 23 to check students' vocabulary recognition, content comprehension, and appreciation of literary skills.

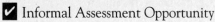 Informal Assessment Opportunity

SELECTION OVERVIEW

Residents of the island of Puerto Rico began coming to the United States in the early 1900s, but did not come in great numbers until after World War II. Nearly two thirds of the 1.5 million Puerto Ricans who came to the United States during the first half of the 20th century found homes in New York City. Most Puerto Rican immigrants came to the United States for economic reasons; they wanted to escape the poverty of their homeland. Upon arrival, they traditionally took the semi-skilled and unskilled occupations that were previously predominantly occupied by "new immigrants."

In this selection, the author describes the experiences of a newly arrived Puerto Rican, and the loneliness she must overcome in her new home.

ENGAGING THE STUDENTS

Ask: "If a referendum were held today, how do you think a majority of U.S. citizens would vote on the issue of banning the entry of newcomers to the United States?" Before students answer, point out that sociologists have detected a clear trend indicating that when economic times are bad, newcomers are often blamed for clogging social service systems, schools, and health care systems. Now call on students to answer the question and to determine the sources of discrimination against newcomers. Point out that in the past, anti-immigration laws have been passed in the United States.

BRIDGING THE ESL/LEP LANGUAGE GAP

This selection, with its vivid descriptive passages, will be very relevant to many LEP students. Help students to expand their vocabulary by focusing on these descriptive passages.

Provide students with a copy of the text that they can write on. Have students mark the text with an *R* if the author is describing a rural home and a *U* if she is describing an urban setting. Have students exchange their work with a partner. Each partner should check the other's work to see if he or she has marked the same passages. Ask students to use their work to write short paragraphs describing either Lali's home in Puerto Rico or her new home in the United States.

☑ PRETEACHING SELECTION VOCABULARY

On the chalkboard make two columns labeled *urban* and *rural*. Tell students to skim the selection to find words that apply to each category. (Responses should include: **tenement, luncheonette, taciturn, countryside,** flower gardens, chicken coop, trucks, cars, honking horns, traffic, **pollution.**) As the students search, fill in the columns with their choices. Discuss the differences in tone and feeling between the words in each column. Explain that in this selection, they will read about one person's struggle to make the transition from a rural village to a large city.

The words printed in boldface type are tested on the selection assessment worksheet.

PREREADING

Reading the Author in Cultural Context

Help students read for purpose by asking them to consider Lali's marriage. Ask: "How do finances and work roles affect Lali's relationship with her husband?" Discuss the economic realities many newcomers must face when they arrive in the United States and the economic factors that probably caused them to make the move in the first place. Remind students that many newcomers give financial assistance to family members who have also come to the United States.

Focusing on the Selection

Tell students to think about the conflict faced by the main character in the selection. Discuss the contrast between her memories of rural life in Puerto Rico and the realities of her new life in New

York City. As they read, students should look for the consequences of Lali's decision to leave her small rural village.

POSTREADING

The following activities parallel the features with the same titles in the Student Edition.

Responses to Critical Thinking Questions

Possible responses are:

1. Sample response: Newcomers must adjust to new ways of life. This adjustment can be especially difficult if the newcomers are from rural areas and must learn to adjust to life in a large city.

2. These aspects might be her remembrances and knowledge about life in a Puerto Rican village.

3. Guide students to compare the conflicts faced by Lali and incidents in their own family's heritage.

☑ Guidelines for Writing Your Response to "Lali"

Have students share their journal entries with a partner. Or, as an alternative writing activity, ask each student to briefly summarize the main points of the selection. Next, have students exchange their summaries with a partner. Ask them to collaborate on a single summary of the selection's main points.

Guidelines for Going Back Into the Text: Author's Craft

Ask students if they have ever watched a movie that used flashbacks to give information about a character. Ask: "How do flashbacks allow the writer or moviemaker to make a point, to give information about a character, or to support the main idea of the piece?" Challenge students to discover the advantages and disadvantages of flashbacks as a writing technique. Possible answers to the questions in the Student Edition are:

1. The author contrasts life in Puerto Rico and life in New York.

2. It creates a mood of serenity, calm, and longing for home.

3. The setting is important because it provides the contrast for Lali's past and her struggle to adjust in her new life.

☑ FOLLOW-UP DISCUSSION

Use the questions that follow to continue your discussion of "Lali." Possible answers are given in parentheses.

Recalling

1. What kind of business do Lali and her husband own and operate? (a restaurant)

2. How much education did Lali have when she left Puerto Rico? (one year of high school)

3. What is Lali learning at night school? (English)

Interpreting

4. Why did Lali marry a man who was much older than she was? (She was shy, he talked to her in a more mature way, and he offered escape from her village.)

5. Why is Chiquitín special to Lali? (He talked to her and understood her loneliness.)

6. Why is Lali disappointed about snow in New York? (It was not as white and beautiful as she had imagined it would be.)

Applying

7. What would you tell Lali about life in a large U.S. city that could help her cope? (Answers will reflect students' opinions of urban life, but students should correlate what they read in the selection with present-day urban life.)

8. Weigh the pros and cons of Lali's new life in New York. Do you think she made a good decision to leave her home? Explain your answer. (Guide students to weigh the economic and social factors that caused her to leave with her misgivings over loss of home, family, and friends.)

ENRICHMENT

Students may benefit from viewing *Puerto Rico: History and Culture*, available from Video Knowledge, Inc., 29 Bramble Lane, Melville, NY, 11747, (516) 367-4250.

from **Picture Bride**
by Yoshiko Uchida (pages 190–198)

OBJECTIVES

Literary
- to identify the element of characterization
- to evaluate the author's development of character

Historical
- to understand some of the reasons why immigrants came to America
- to compare and contrast the different waves of immigration from the 18th century to the present

Multicultural
- to evaluate how U.S. culture has affected diverse immigrant groups
- to compare and contrast Japanese cultural identity with that of other ethnic groups in the United States

SELECTION RESOURCES

Use Assessment Worksheet 24 to check students' vocabulary recognition, content comprehension, and appreciation of literary skills.

✔ Informal Assessment Opportunity

SELECTION OVERVIEW

Chinese, Japanese, and other Asian immigrants in the 19th century entered the United States through an immigration center on Angel Island in San Francisco Bay. Here they were subjected to the same rigorous medical examination by immigration authorities as the far larger numbers of immigrants who entered the United States through New York City. Until the 1870s Chinese immigrants had been welcomed because they supplied much-needed labor on the West Coast. But after the depression of 1873, they were seen as a threat to the livelihood of white workers. This economic fear, together with prejudice against Chinese, led to the passage of the Chinese Exclusion Act of 1882, which barred Chinese immigration for ten years, a prohibition that was made permanent in 1892. Before long, immigration by Japanese was also attacked. The state of California in 1901 called for an end to Japanese immigration, and in 1908, in a

"gentleman's agreement," the Japanese government agreed to prohibit Japanese from leaving for the United States. Local laws by several Western states limited the rights of Japanese living there to hold property or attend public schools.

In this selection, a young woman from Japan travels to the United States to meet her future husband, a man she has never met.

ENGAGING THE STUDENTS

Ask: "What makes more sense—to marry someone with whom you are in love, or to marry someone with whom you have a mutual background and share similar needs and interests?" Before they answer, explain to students that in some countries, marriages are prearranged by family members. Point out that even today, families can exert both subtle and not-so-subtle pressure in regard to their children's choices for marriage. Discuss the question above, then tell students that in this selection they will read about a prearranged marriage and find out why these kinds of betrothals were sometimes part of the immigrant experience.

BRIDGING THE ESL/LEP LANGUAGE GAP

Provide the students with a copy of the three selection excerpts below. Read each excerpt to the students. Ask them to consider these questions about each excerpt: "What does this passage show about Hana's character? Does that characteristic make her similar to or different from other characters in the story? What does this passage tell you about her values?"

For the last excerpt, ask them why Hana would be feeling both excited and terrified at the same time. Ask: "What would make her feel that way?"

> . . . Hana knew she wanted more for herself than her sisters had in their proper, arranged and loveless marriages. She wanted to escape the smothering strictures of life in her village. (p. 193)

> . . . In fact, she would have recoiled from a man who bared his most intimate thoughts to her so soon. (p. 195)

> . . . So it was that Hana had left her family and sailed alone to America with a small hope trembling inside of her. Tomorrow, at last, the ship would dock in San Francisco and she would meet face to face the man she was soon to marry. Hana was overcome with excitement at the thought of being in America and terrified of the meeting about to take place. (p. 195)

☑ PRETEACHING SELECTION VOCABULARY

Tell students that there are two main characters in *Picture Bride*: Hana and Taro Takeda. Explain that the author uses characterization to give us a sense of what these people are like both in personality and appearance. Have students skim the selection to

find characterizations of the two main characters (for example: **kimono, pompadour,** hard-working, conscientious, successful, childlike, terrified, eager, **flustered,** nervous, **derby, sallow**). Use the students' vocabulary words to create a comparison chart of the two characters.

The words printed in boldface type are tested on the selection assessment worksheet.

PREREADING

Reading the Author in Cultural Context

Ask students to consider the pros and cons of an arranged marriage—and the cultural implications of this practice. Ask: "How do prearranged marriages ensure the survival of families? Why was family survival especially important for immigrants?" Tell students to consider these topics as they read.

Focusing on the Selection

Have students work in groups of four. Assign one student the role of Taro Takeda, one the role of Hana's mother, one the role of Hana's brother-in-law, and one the role of Hana's uncle. Have each student write a short character sketch of his or her individual. When they have finished, each group must collaborate on an answer to this question: "Which of the characters had Hana's best interests at heart?" Call on student groups to read their answers aloud, and then discuss the questions in the student text.

POSTREADING

The following activities parallel the features with the same titles in the Student Edition.

Responses to Critical Thinking Questions

Possible responses are:

1. Some of the expectations, such as riches and servants, were unrealistic. Some newcomers also had no idea of the culture they were about to enter.

2. She might have chosen arranged marriages because they offered a strong contrast between Japanese traditional culture and U.S. culture.

3. Sample answer: American marriages are based more on romantic love and are influenced less by parental wishes.

☑ Guidelines for Writing Your Response to *Picture Bride*

Have students share their journal entries with a partner. Or, as an alternative activity, have students work in groups of three to discuss their favorite scenes or ideas from the selection. Then have each group collaboratively answer this question: "What will be Hana's most difficult adjustment to living in the United States?" Call on each group to read its response aloud.

Guidelines for Going Back Into the Text: Author's Craft

Tell students to skim the selection for examples of characterization. Discuss what kinds of characterization the author used and how they help the reader to evaluate the characters in the selection. Possible answers to the questions in the Student Edition are:

1. Guide students to understand that Uchida uses all five techniques.

2. Similarities: With the other characters Hana shares common cultural traits including manners, customs, and ideas about the role of women and marriage. Differences: She exhibits some rebelliousness, has more schooling, and is willing to travel to America and marry.

3. Guide students to understand why Hana chooses to marry, and compare her reasons with their own ideas on marriage.

4. Guide students to compare their culture and personality to those of Hana. Their answers should include their ideas on prearranged marriages.

☑ FOLLOW-UP DISCUSSION

Use the questions that follow to continue your discussion of *Picture Bride*. Possible answers are given in parentheses.

Recalling

1. Who arranges Hana's marriage? (her uncle)

2. How much schooling does Hana have? (She has graduated from high school.)

3. Where does Hana first arrive in America? (Angel Island)

Interpreting

4. What qualities make Taro Takeda an attractive husband in the eyes of Hana's family? (He is a successful businessman.)

5. Why does Hana's brother-in-law support the marriage? (Hana's knowledge and "rebellious" character threaten and irritate him.)

Applying

6. What cultural benefits do prearranged marriages provide? (Prearranged marriages allow single people to find suitable mates and help families to survive.)

7. How does Hana's choice to marry Taro illustrate her discontent with her place of origin? (Hana does not want to marry a farmer from her village and end up working in the fields.)

ENRICHMENT

Students may benefit from viewing *Angel Island*, available from Chinese for Affirmative Action, 17 Walter Lum Place, San Francisco, CA, 94108, (415) 982-0801.

I Leave South Africa
from *Kaffir Boy in America*
by Mark Mathabane (pages 199–206)

OBJECTIVES

Literary

- to observe how the author expresses his attitude about his old and new homelands
- to recognize the use of irony in an autobiography

Historical

- to describe the facets of apartheid in South Africa
- to explain the rise of the Black Muslim movement in the United States

Multicultural

- to contrast cultural views on assimilation
- to understand the impact of segregation on cultural groups

SELECTION RESOURCES

Use Assessment Worksheet 25 to check students' vocabulary recognition, content comprehension, and appreciation of literary skills.

☑ Informal Assessment Opportunity

SELECTION OVERVIEW

In the 1960s, most of Africa experienced sweeping changes that ended colonial rule. South Africa was an exception. The white ruling government continued to rule the nation, using a system of forced segregation know as apartheid. Under apartheid, races are completely separated. Laws ban all social contacts between whites, blacks, Asians, and coloreds. Blacks are excluded from most high level jobs and the best universities. In 1959, South Africa extended the segregation policy by creating *bantustan*—separate homelands for South African blacks. The injustice of the program was immediately evident—only 13 percent of the country's land was set aside for 87 percent of its population. In 1989, the government of South Africa moved to abolish apartheid, but the country continues to struggle with its history of segregation.

In this selection, a young black South African travels to the United States and meets an African American who challenges his preconceptions about the United States.

ENGAGING THE STUDENTS

Ask: "Should African Americans build and attend their own schools with African American teachers and students?"

Discuss with students the stance taken by some African Americans to resist integration into U.S. society. Tell them that in this selection, they will confront a character who disdains integration and mocks African Americans who "mix with them [whites]." Challenge students to form their own opinions on integration.

BRIDGING THE
ESL/LEP LANGUAGE GAP

Ask students to consider the following questions as they read:

1. Compare and contrast Mark and Nkwame. What do they have in common? How are they different? How does each feel about his heritage? How can you tell?

2. When Nkwame says he has always wanted to know where his homeland is, Mark writes: "I found this statement baffling for I thought that as an American his homeland was America." Is Mark's statement true or false? Which is Nkwame's true homeland? Why do you think so?

3. Mark uses the following words to describe his reactions to Nkwame: *intimidated, startled, baffled,* and *shaken.* Why does he react to Nkwame this way? What are his expectations of America and of African Americans? How do Nkwame's words meet or change his expectations?

☑ PRETEACHING
SELECTION VOCABULARY

Explain to students that the selection they are about to read contrasts two different views on integration. Tell students to write their own definitions of *integration.* Ask them to read their definitions aloud, and then ask: "In your opinion, is integration good or bad?" After your discussion, tell students to skim the selection for any vocabulary words that give them clues about how the characters in the story feel about integration (for example: good, **truth, apartheid, bondage, liberated, praised,** railed, segregation, worst, fraud, dependent, bitter, nonsense, freedom, opportunity) Tell students to keep their lists and compare their ideas about integration with those of the characters in the story.

The words printed boldface type are tested on the selection assessment worksheet.

PREREADING

Reading the Author
in Cultural Context

Help students read for purpose by asking them to consider the conflict the author faced. Remind

them that his parents disagreed on what path he should take. Discuss his father's belief that a commitment to Mark's African heritage was more important than education. Tell students to consider this conflict as they read about Mark's confrontation with an African American who also believes in the importance of his African heritage.

Focusing on the Selection

On the chalkboard make two columns: *Mark's Expectations of America* and *Mark's Realizations about America.* As they read the selection, call on students to complete the columns and to answer the questions in the Student Edition.

POSTREADING

The following activities parallel the features with the same titles in the Student Edition.

Responses to Critical
Thinking Questions

Possible responses are:

1. Mark expects a land free of all prejudices and segregation. His expectations are challenged by Nkwame's exhortations not to believe in integration. His expectations are validated by his initial sensations and observations upon arriving in America.

2. Besides the descriptions of apartheid, the author tells us that South African blacks imitated and revered African Americans.

3. Sample response: This narrative gave me insight into opposing views of integration and the search for cultural identity among African Americans.

☑ Guidelines for Writing Your
Response to "I Leave South Africa"

Have students share their journal entries with a partner. Or, as an alternative writing activity, have students work in pairs so that each partner writes the answers to different questions in the Student Edition. When they have finished working individually, ask each pair to write a shared response to this question: "Do you think Mark understood Nkwame's comments? Why or why not?" Discuss the group responses.

Guidelines for Going Back Into the Text: Author's Craft

Discuss with students the irony of Mark's conversation with Nkwame. Ask: "Which character do you think was closer to his African heritage?" Challenge them to explain how the encounter between Mark and Nkwame has irony for both characters and their search for roots. Possible answers are:

1. Mark is trying not to attend a "black college" which for him signifies segregation.
2. Sample answer: Nkwame's insistence on using Mark's African name has a certain sense of irony, since Mark is really closer to his African heritage and roots than Nkwame is, yet Mark has not paid much attention to the trappings of his past.
3. Guide students to find irony in everyday events. Suggest that irony often springs from the contrast between expectations about an experience and the reality of the experience.

☑ FOLLOW-UP DISCUSSION

Use the questions that follow to continue your discussion of "I Leave South Africa." Possible answers are given in parentheses.

Recalling

1. Why is Mark going to the United States? (He has a scholarship to Limestone College.)
2. What does Mark's name mean in African? (wise one)
3. Why is Mark almost sent back to South Africa? (He only has a one-way ticket.)

Interpreting

4. Why does Nkwame tell Mark he would be better off at a black college? (Nkwame does not believe in integration.)
5. Why does Mark nearly faint when he hears there are black schools in America? (For Mark, black schools imply apartheid.)
6. Why isn't Nkwame shocked by Mark's statement about blacks being murdered every day in South Africa? (Nkwame responds that these same kinds of conditions exist in the United States as well; note his comment about the Ku Klux Klan.)

Applying

7. What implications do you think the Black Muslim movement has for U.S. society as a whole? (Guide students to discuss the roots of the Black Muslim movement and its efforts to forge ancestral ties for African Americans.)
8. Do you agree with Nkwame's assessment of race relations in the United States? Cite some recent events, positive or negative, to support your answer. (Though Nkwame's analysis might be viewed as simplistic there are certainly negative events [Los Angeles riot] and positive events [growing African American presence in politics and other high-level positions] that prove race relations in the United States are still in a state of flux.)

ENRICHMENT

For a historical look at the problems that have developed because of South Africa's system of apartheid, students may view *South Africa: The Solution*. The video is available from Journal Films, Inc., 930 Pitner Avenue, Evanston, IL, 60202, (800) 323-5448.

Ellis Island

by Joseph Bruchac (pages 207–209)

OBJECTIVES

Literary
- to identify the use of poetry to express inner conflict
- to understand the difference between literal expressions and figurative speech

Historical
- to evaluate the Native American claims to U.S. territory
- to describe the function of Ellis Island in the early 20th century for incoming immigrants

Multicultural
- to understand the significance Ellis Island has had for different immigrant groups
- to compare and contrast Native American and European views on immigration

SELECTION RESOURCES

Use Assessment Worksheet 26 to check students' vocabulary recognition, content comprehension, and appreciation of literary skills.

☑ Informal Assessment Opportunity

SELECTION OVERVIEW

Between 1900 and 1924, Ellis Island was the primary reception center for immigrants coming to the United States from Europe. In 1907, its busiest year, 900,000 people came through Ellis Island. In some years, up to 5,000 people a day streamed through its gates. First, the immigrants endured a medical examination. Those marked with chalk were directed to an examination room for closer scrutiny. The next hurdle was a legal inspection. Immigrants who did not answer the inspector's questions properly were sent to special inquiry rooms. Twenty percent of incoming immigrants were singled out for special inspections, and 2 percent were sent home.

In this selection, the author, who is of Native American and European descent, describes the conflict Ellis Island symbolizes for him.

ENGAGING THE STUDENTS

On the chalkboard write:

> 6,000,000 and 900,000

Explain that the first figure represents the number of legal immigrants who came to the United States in the 1980s. The second figure is the highest number of immigrants to enter the United States in one year (1907). Ask: "Is a high rate of immigration a good trend or should it make Americans wary?" Acknowledge that the answer to this question depends on one's perspective. Tell students that in this selection, they will read opposing perspectives on immigration—from one person.

BRIDGING THE ESL/LEP LANGUAGE GAP

Read the poem aloud and have students listen without looking at their texts. Ask them what images, words, or phrases stood out. Write the students' ideas on the chalkboard, then read the poem again.

Have students work in groups of three to four. Ask them to discuss what the images mean and what emotions they express. Ask: "What other words or images can you think of that might describe the same emotions?"

Ask: "What do the lines 'Yet only one part of my blood loves that memory./Another voice speaks/of native lands/within this nation' mean?" Have them tell how they can relate to this type of conflict.

✔ PRETEACHING SELECTION VOCABULARY

Have students create a chart of words based on the geographical descriptions in the selection. Tell them that one column of the chart should contain the geographical landmarks and places in the poem (**Ellis Island,** Europe, forests, meadows) and the second column should list the adjectives used to describe the landmarks (**red brick, old empires, dreams,** invaded). When they have finished the charts, ask the questions below to lead a discussion on the importance of place in the selection.

- What kinds of words does the author use to describe the places in the poem?
- Does the author describe landmarks in purely physical terms, or do the descriptions go beyond physical characteristics?
- Based on this vocabulary activity, what do you think Ellis Island symbolized for the author? Explain your answer.

The words printed in boldface type are tested on the selection assessment worksheet.

PREREADING

Reading the Author in Cultural Context

Help students read for purpose by asking them to consider the different values land ownership has among diverse cultures. Explain that land ownership in many European cultures is synonymous with wealth and success. For Native Americans, however, land cannot be "owned" by people. Land is a resource with which humans must maintain a balance.

Focusing on the Selection

Help students to read for purpose by asking them to consider the unique dilemma facing the author. On one hand, Ellis Island represents an important positive aspect of his heritage. On the other hand, Ellis Island is a puzzling symbol in his Native American past. Challenge students to determine how the author deals with this conflict in the poem.

POSTREADING

The following activities parallel the features with the same titles in the Student Edition.

Responses to Critical Thinking Questions

Possible responses are:

1. He points out that immigrants passed through Ellis Island and waited in long quarantine lines.
2. Native Americans view the immigrants as invaders.
3. Guide students to understand the inner conflict faced by the author who must reconcile two opposing views of immigration into North America.

✔ Guidelines for Writing Your Response to "Ellis Island"

Have students share their journal entries with a partner. Or, as an alternative writing activity, ask each student to write in a few sentences what he or she believes are the opposing viewpoints contained in the poem. Next, have students exchange their writing with a partner. Ask them to collaborate on a single summary of the conflict contained in the selection.

Guidelines for Going Back Into the Text: Author's Craft

Discuss with students the difference between a literal expression and figurative speech. Ask: "Why do you think figurative speech is commonly found in poems?" Guide students to understand how figurative expressions add color to writing—often in a few short phrases. Possible answers to the questions in the Student Edition are:

1. "island of the tall woman, green as dreams of forests and meadows"
2. He is comparing the Statue of Liberty to a forest of dreams.
3. The simile conveys to the reader the hopes and dreams of immigrants.

☑ FOLLOW-UP DISCUSSION

Use the questions that follow to continue your discussion of "Ellis Island." Possible answers are given in parentheses.

Recalling

1. Why did some immigrants have to be detained? (They were quarantined.)
2. How long did immigrants come through Ellis Island? (nine decades)
3. According to the poem, from where did the immigrants come? (old empires of Europe)

Interpreting

4. Who is the "tall woman, green" mentioned in the poem? (Statue of Liberty)

5. What did the immigrants symbolize for Native Americans? (invaders)
6. What does the poem say about land ownership? (Land was not "owned" until Europeans and their concept of private property came to the United States.)

Applying

7. What reasons does Bruchac give to justify European immigration into the United States? In your opinion, are these justifications valid? Support your opinion. Encourage students to try and put themselves in the positions of both cultures. (Bruchac suggests that land ownership and dreams for a new life were the reasons Europeans came to the United States. Student opinions will vary on whether these reasons justify the taking of Native American land.)
8. List two different ways that Ellis Island is a symbol in this poem. (It is a symbol to European immigrants of the hopes and opportunity in the United States. To Native Americans it is a symbol of European encroachment.)

ENRICHMENT

Students may benefit from viewing *Ellis Island*, a video that shows how a choreographer depicts the immigration experience through an interpretative dance. The video is available from Send Video Arts, 650 Missouri Street, San Francisco, CA, 94107, (415) 863-8434.

MAKING CONNECTIONS THE IMMIGRANT EXPERIENCE

OBJECTIVES

Overall

• to discover the connections between American history, culture, and the theme of immigration

Specific

• to compare and contrast a poem on the immigrant experience with two historical photographs and a painting
• to identify universal aspects of the immigrant experience
• to appreciate the need to commemorate aspects of the immigrant experience

ENGAGING THE STUDENT

Write the words immigrant and immigration on the chalkboard. Ask students to brainstorm a list or words they associate with the two terms. Then have students work in small groups to explore ways of classifying those words. To help them get started, ask students to identify the words on their list that immigrants might use to describe themselves. Then ask them to look for words that seem to reflect the feelings and attitudes of people already living in the United States. Or have them decide which words suggest such emotions as hope or fear. Which suggest attitudes like hostility, curiosity, or welcome? Encourage groups to share their classifications. Then ask which words on the list are reflected in the photographs and the painting on pages 210–211. Which are reflected in Bruchac's poem?

BRIDGING THE ESL/LEP LANGUAGE GAP

Use the photographs and the painting on page 210–211 as well as other illustrations in Theme 6, to help students explore various aspects of the immigrant experience. Have them study each illustration, and explain orally what is suggests about an aspect of the process of immigration. If students are recent arrivals or the children of immigrants, encourage them to compare what they observe in the illustrations to their own experiences.

EXPLORING ART

Early photographers tried to take pictures that looked like paintings. By the turn of the century, however, many photographers were trying to show the world as it really looked. They wanted to use their cameras to document the lives or ordinary people. A number of painters, including George Luks, tried to do the same. Their work focused on commonplace subjects that many people considered "vulgar" or "raw." Have students work in small groups to read the captions and discuss the illustrations. What do the photographs and the painting suggest about the arrival of the newcomers to the United States? About life in immigrant neighborhoods?

Students may also wish to compare and contrast Luks' painting of Hester Street with photographs of the same street taken by such photographers as Jacob Riis and Lewis Hines. A movie called Hester Street is available in many video stores. Made in the 1980s, it presents yet another view of the street and the immigrant experience.

LINKING LITERATURE, HISTORY, AND THEME

Guidelines for Evaluation

In answering the question, students should contrast the dreams of freedom, open spaces, and land ownership expressed in Bruchac's poem with the reality of poverty and crowding suggested by the illustrations.

Students are likely to suggest that newcomers struggled to adjust to differences in language and culture and to overcome prejudice and hostility. Common goals might include taking advantage of the economic and educational opportunities the immigrants found in the United States or simply making the most of the chance for a fresh start in a new land.

Students should note that the two landmarks remind Americans of the magnitude and importance of immigration to the development of the United States. People today visit both sites to learn about and to honor the millions of immigrants who helped build the United States, often at great cost to themselves.

Those Who Don't, No Speak English and The Three Sisters,

by Sandra Cisneros (pages 212–218)

OBJECTIVES

Literary

- to identify the author's attitudes about a culture
- to evaluate the tone in three selections

Historical

- to explain how events in Latin America have affected Hispanic migration patterns to the United States
- to analyze the effects of illegal immigrations on the Hispanic community in the United States

Multicultural

- to compare the experiences of Hispanic immigrants with those of other immigrant groups
- to discuss how diverse cultural groups confront discrimination in U.S. society

SELECTION RESOURCES

Use Assessment Worksheet 27 to check students' vocabulary recognition, content comprehension, and appreciation of literary skills.

 Informal Assessment Opportunity

SELECTION OVERVIEW

More than 18 million people from Latin American countries live in the United States. Mexican Americans constitute about 60 percent of this group. In the mid-1980s, 42 percent of all legal immigrants were Hispanic, making them the fastest growing ethnic group.

Increased competition for jobs and housing in large cities has created a number of social problems. There is disagreement over how to deal with these issues. For example, bilingual education is seen by some as a way to bring Hispanics into the community and give them greater access to economic opportunities. For others, however, bilingual education promotes cultural pluralism.

In these selections, the author describes life in the Hispanic community from several viewpoints.

ENGAGING THE STUDENTS

Ask: "Are there neighborhoods in your city that you have not visited?" Discuss with students what makes neighborhoods seem different and what causes people either to visit them or stay away from them. Tell students that in this selection they will read about one author's perspective on how culture affects our perceptions of ethnic neighborhoods.

BRIDGING THE ESL/LEP LANGUAGE GAP

After the students have read the selections in this theme, ask what they think the author is trying to communicate. Ask: "Why do you think she wrote these pieces?" Then copy the excerpts below on the chalkboard or duplicate them and distribute them to

107

the class. After they read the excerpts, ask: "How do the excerpts communicate ideas that are important to the author? Why are these ideas important to her? Are there other passages that are more significant?"

> All brown all around, we are safe. But watch us drive into a neighborhood of another color and our knees go shakity-shake. (p. 213)
>
> No speak English, she says to the child who is singing in the language that sounds like tin. No speak English, no speak English, and bubbles into tears. No, no, no as if she can't believe her ears. (p. 215)
>
> . . . You can't erase what you know. You can't forget who you are. (p. 217)

☑ PRETEACHING SELECTION VOCABULARY

Have students create a word bank by writing all the terms in the selection that are in Spanish. (Examples: **mamacita, cuando, Esperanza, las comadres**) Using the footnotes on pages 214–216, go over these terms and discuss their meanings. Use the questions below to lead a guided discussion of the author's use of these terms:

• Why do you think the author chose to use Spanish words instead of English ones?
• Based on this vocabulary activity, what do you think the major themes of these selections are?

The words printed in boldface type are tested on the selection assessment worksheet.

PREREADING

Reading the Author in Cultural Context

Help students read for purpose by asking them to consider how discrimination and bias can affect a person's perceptions of his or her own neighborhood. Discuss how popular images of crime and poverty often contain racial stereotypes and generalizations. Ask students to compare the author's view of her neighborhood with their own views.

Focusing on the Selection

Ask: "Have Americans lost their faith in

neighborhoods?" Discuss the importance of community and neighborhoods for immigrants in a strange land and how recent increases in crime have affected people's feelings about the notion of community. Tell students to consider their own views about their communities and to compare them with what they read about in the selections.

POSTREADING

The following activities parallel the features with the same titles in the Student Edition.

Responses to Critical Thinking Questions

Possible responses are:

1. Guide students to examine the issues of discrimination, bias, and isolation caused by the language barrier.
2. They have strong community values and loyalty to their place of origin. There are also religious references in one selection.
3. Guide students to see how cultural stereotypes can lead to discriminatory attitudes and bias toward an ethnic group.

☑ Guidelines for Writing Your Response to "Those Who Don't," "No Speak English," and "The Three Sisters"

Have students share their journal entries with a partner. Or, as an alternative writing activity, ask each student to write in a few sentences what lesson they believe the author is trying to teach about being Mexican American. Call on students to read their interpretations of the author's message, then ask: "What message does this story have for teenagers in the 20th century?" Guide them to compare the author's message with contemporary urban ills.

Guidelines for Going Back Into the Text: Author's Craft

Ask students to skim the selections to find words that set the writer's tone. Call on them to list the words they have chosen on the chalkboard. Discuss how these words can help a reader to identify the tone of a literary work. Possible answers to the questions in the Student Edition are:

1. The tone of the selections ranges from angry ("Those Who Don't"), to humorous ("No Speak English"), to sad ("The Three Sisters").

2. Sample answers: "scared" ("Those Who Don't"), "Holy Smokes" ("No Speak English"), and "Esperanza" ("The Three Sisters").

3. It is unlikely students will find that the tone interferes with the author's message. Guide them to see how the tone in each selection helps to convey the author's message.

☑ FOLLOW-UP DISCUSSION

Use the questions that follow to continue your discussion of "Those Who Don't," "No Speak English," and "The Three Sisters."

Possible answers are given in parentheses.

Recalling

1. In "Those Who Don't," when does the author not feel safe? (when she drives into a neighborhood of "another color")

Interpreting

2. What do you think the three sisters mean by this phrase: "You will always be Mango Street"? (One cannot ignore or leave the past and community of origin behind.)

3. What do you think the sisters "know" about the author? (Sample answer: They know she will leave the neighborhood one day.)

Applying

4. Do neighborhoods continue to have the same importance today as they did to the author in these selections? Explain your answer. (Guide students to consider the ongoing importance of neighborhoods, especially in urban areas with immigrant populations.)

ENRICHMENT

Students may benefit from viewing a video that examines political, social, and economic problems of Hispanic communities in southern New Jersey. *South Jersey: A Dream for Hispanics?* is available from New Jersey Network, 1573 Parkside Avenue, Trenton, NJ, 08625, (609) 292-5252.

RESPONSES TO REVIEWING THE THEME

1. In "I Leave South Africa," Mark has high expectations as a newcomer to the United States. These expectations are challenged by an African American, who seems to have more regard for Mark's African past than Mark himself does. In *Picture Bride*, Hana's high expectations stem more from her prearranged marriage, though economic expectations are a factor for her as well. Both characters had to overcome obstacles to make their trips; Mark had to escape South Africa, and Hana had to escape her family.

2. An author's heritage can give his or her writing poignancy and be informative for the reader in terms of cultural details. Perhaps the most complex heritage in this section is that of Joseph Bruchac, whose European and Native American heritages create a conflict between two opposing views on European immigration into the United States.

3. Guide students to compare their experiences with those of the characters in the selections.

FOCUSING ON GENRE SKILLS

Mark Mathabane makes effective use of *autobiography,* which combines the elements of *fact and opinion.* He recounts the factual details of his arrival in America intertwined with his opinions of his new surroundings. Select, or have the students select, another autobiographical text (such as *Night Country* by Lauren Eisely or Richard Wright's *Black Boy*), and ask the students to read passages to distinguish between facts and opinions.

BIBLIOGRAPHY

Books Related to the Theme:

Daniels, Roger, *Coming to America: A History of Immigration and Ethnicity in American Life.* New York: Harper Collins, 1990.

Santoli, Al, *New Americans: An Oral History.* New York: Viking, 1988.

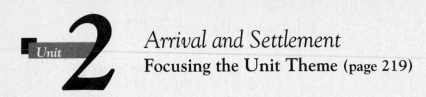

COOPERATIVE/ COLLABORATIVE LEARNING

Individual Objective: to participate in a discussion about the theme of arrival and settlement

Group Objective: to develop a five-minute presentation inspired by viewpoints expressed in the discussion

Setting Up the Activity

Have the class work in heterogeneous groups of four. Stipulate that each group member is responsible for researching and expressing an individual viewpoint in the group discussion and for making a contribution to the group presentation. The individual topics are as follows: Student 1 expresses the overall theme of Unit II; Student 2 explains how the selections in Unit II illustrate this theme; Student 3 highlights the different cultures represented by the selections; Student 4 finds five quotations that demonstrate each of the writing styles in Unit II.

When the groups have finished their discussion, have them elect a representative to summarize the group's discussion for the entire class. After each group's presentation, use the questions below to discuss and evaluate the report:

• What different perspectives on the experiences of newcomers are expressed in Unit 2?

• In your group's opinion, how is arrival and settlement an important part of America's heritage? Explain your answer.

Assisting ESL and LEP Students

To give students more opportunity to participate in a closing discussion of the unit, give them time to prepare with some questions to think about. Ask them to think of the conflicts and/or challenges they have read about in this unit. Ask each of them to choose one or two that seemed particularly difficult, painful, or inspiring to them. Then, for the character and/or setting they chose, ask them to consider these questions:

1. Is this a conflict in a relationship between peoples or within an individual?
2. What caused the conflict(s)/challenge(s)?
3. How did the character deal with the conflict/challenge?
4. What can you see about the character's values from the way he or she dealt with the conflict/challenge?

Assessment

Before you begin the group activity, remind students that they will be graded both on an individual and group basis. Without contributions from each individual, the group presentation will not be complete. Monitor group progress to check that all four students are contributing and that a presenter has been selected.

Time Out to Reflect As students do the end-of-unit activities in the Student Edition, provide time to let them make a personal response to the content of the unit as a whole. Invite them to respond to the following questions in their notebooks or journals. Encourage students to draw on these personal responses as they complete the activities.

1. What have I learned about the experiences of different ethnic groups coming to the United States?
2. What have I learned about autobiographies and personal narratives? Why is reading these genres an effective way to learn about the themes in Unit II?
3. What have I learned about the struggles newcomers face in the United States? How does cultural discrimination affect this struggle? What examples of this struggle can I find in my own community?
4. What have I learned about cultural identity in a diverse society? How do the members of cultures preserve their identities? How does this affect their acceptance by mainstream society?

WRITING PROCESS: EXPOSITION ESSAY

Refer students to the model of an exposition essay found on pages 397–398 in the Handbook section of the Student Edition. You may want to discuss and analyze the model essay if you are working with less experienced writers.

Guidelines for Evaluation

On a scale of 1-5, the writer has

- clearly followed all the stages of the writing process.
- made clear and specific references to the chosen selections.
- made clear and specific references to the literary elements and techniques used in the selections.
- provided sufficient supporting details from the selections to support the main idea.
- written an opening or a closing paragraph that clearly summarizes the main points of the paper.
- made minor errors in grammar, mechanics, and usage.

Assisting ESL and LEP Students

You may want to provide a more limited assignment for these students so that they can complete their first drafts somewhat quickly. Then have them work at length with proficient writers in a peer-revision group to polish their drafts.

PROBLEM SOLVING

Encourage students to use the following problem-solving strategies to analyze one of the stories portrayed in Unit II. Afterwards, have students reflect on their use of the strategies and think of ways they could have used the strategies more effectively.

Strategies (optional)

1. Use a semantic map, brainstorming list, or flow chart to analyze the different cultural perspectives on arrival and settlement in the United States. Have groups of students choose one of the cultural groups they most want to investigate further and the selection or selections that best shed light on their group's experiences.

2. Use a second graphic organizer to explore in detail the topics they chose. Include information both from the selections and from personal experience.

3. Use the overlapping circles of a Venn diagram to organize information into similarities and differences. Then refine the diagram by putting check marks next to the most significant points.

4. Decide which topics need further research or discussion. Set up a plan that shares responsibility for doing the research. Allow time for students to work on their presentations.

Guidelines for Evaluation

On a scale of 1-5, the student has

- provided adequate examples, facts, reasons, anecdotes, or personal reflections to support their presentations
- demonstrated appropriate effort
- clearly organized the presentation so that main ideas and supporting details are logical and consistent
- demonstrated an understanding of challenges faced by many groups and individuals in the United States.

3 Struggle and Recognition

"Sometimes the struggle for equality occurs on a very personal level. Individuals discover the need to break down walls of misunderstanding and mistrust and to appreciate people for who they are, not from where they come. At other times, the vision of equality has greater scope. This larger goal includes a nation in which groups of people overcome social barriers and share common goals of building a multicultural society of diverse traditions."

UNIT OBJECTIVES

Literary

- to discover how authors, through their craft, have used the traditional genres of literature to express the conflicts involved in meshing cultures
- to understand the ways in which literature serves to extend the human understanding needed to produce social progress

Historical

- to explain how cultural groups in the United States have worked to achieve equality, justice, and freedom
- to describe the barriers to cultural pluralism in U.S. society

Multicultural

- to explain the advantages of living in a multicultural society
- to describe how different cultures in the United States have dealt with discrimination and prejudice

UNIT RESOURCES

The following resources appear in the Student Edition of Tapestry:

- a full-color **portfolio of theme-related art**, Arts of the Tapestry, pages A1–A16, to build background and activate prior knowledge about the unit themes and to generate writing ideas.
- a **unit overview,** pages 220–221. The overview for Unit 1 provides historical background about the struggles of various cultural groups in the United States to achieve equality, sustain their special cultural heritages, and at the same time, develop community with all groups.
- a **time line,** pages 222–223, to help students situate the literary works in their historical context. You may wish to have students refer to the time line before they discuss each work.

- the **Focusing the Unit Theme** activities at the end of the unit, page 369, to provide a cooperative/collaborative learning project on the theme of "Struggle and Recognition," a writing project to develop an essay of comparison and contrast, and a problem-solving activity in which students analyze one of the struggles or histories in the unit.

INTRODUCING UNIT 3

Providing Motivation

Before students begin to read the unit overview in the Student Edition, you might want to stimulate their interest in the theme of "Struggle and Recognition" with some of the following activities.

- Ask students if they believe teenagers are more or less likely than adults to accept cultural differences. Discuss whether teenagers are accepting of others who have different cultures and different ways of dressing and speaking. Ask:

• Do teenagers place a great emphasis on "fitting in"? If so, does this make it more or less difficult to recognize and accept cultural differences?

• Do teenagers' concepts of accepted styles and behavior include all cultural groups?

• In general, do you think teenage social groups symbolize equality and acceptance or are they exclusionary and foster separatism? Support your answer with examples.

When you have finished the discussion, tell students that several research studies have shown that teenagers are as likely to adopt prejudiced attitudes as adults. In one study—*Prejudice and Discrimination*, by Frederick Holmes—the researcher found cultural behavior prevalent among high school seniors, although many of the students denied being prejudiced.

Ask: "Do private clubs and organizations that discriminate against cultural groups have a right to exist in U.S. society?" Tell students that in fact such organizations do exist. Explain that in the past, there have been groups in U.S. society with the sole purpose of keeping cultural groups out of the United States. Groups such as the Ku Klux Klan have, through the years, issued threats against not only African Americans, but also Jews, Catholics, and immigrants. Ask: "Why do you think some people are afraid of cultural differences? What can be done to make society more accepting of diverse cultures?" Tell students they will be better able to answer these questions when they have finished studying Unit 3.

Setting Personal Goals for Reading the Selections in Unit 3

Have students each keep a copy of the following chart in their journals or notebooks. Provide class time every few days for students to review and expand their charts. Encourage them to add topics of their own.

• Use the Unit Time Line for the following activity. Have students work in pairs. Assign one student the role of "optimist" and the other of "pessimist." Have each student survey the Unit Time Line and the events listed on it. Tell them that their goals are to assess U.S. society's ability in achieving cultural equality over the past 150 years. Have each student make his or her assessment according to his or her assigned viewpoint. When they have finished, each pair must decide which viewpoint is strongest. Poll the student pairs to obtain a class consensus on how well U.S. society has managed in achieving equality over the past 150 years.

Cross-Discipline Teaching Opportunity: Unit Theme

Remind students that though cultural groups in the United States have been victims of prejudice over the years, these same groups have devised a number of strategies to achieve equality. Work with the social studies teacher to examine the different approaches cultural groups used in the past to achieve recognition and equality in the United States. Some of the methods you might like to explore include:

• boycotts, protests, and marches (African American civil rights activities in the 1950s–1960s)

• labor strikes (Mexican American migrant farm workers, UFW in the 1960s to the present)

• activist organizations (A.I.M., Black Panthers, Chicano movement)

Discuss with students the successes and failures of these movements. Challenge students to determine what the best methods for achieving equality will be in the future.

Struggle and Recognition			
Topic	What I Know	What I Want to Learn	What I Have Found Out
The struggle among cultural groups to achieve equality in U.S. society.			
How U.S. society deals with and accepts cultural differences.			
Barriers to achieving a harmonious multicultural society.			

The Quest for Equality

THEME PREVIEW

The Quest for Equality			
Selections	Genre/Author's Craft	Literary Skills	Cultural Values & Focus
Martin Luther King, Jr., "I See the Promised Land," pages 117–119.	speech	allusion	hopes for ending discrimination and prejudice
Malcolm X, "See for yourself, listen for yourself, think for yourself," pages 120–122.	speech	restatement, repetition, parallelism	use of violence in a violent society
Maxine Hong Kingston, *China Men*, and "Immigration Blues," pages 123–125.	novel/poem	direct characterization, indirect characterization	experiences of Chinese laborers and immigrants in the United States
Ana Castillo, "Napa, California" and Francisco Jiménez, "The Circuit," pages 126–128.	poem/novel	setting	life among migrant farm workers in California
MAKING CONNECTIONS: Mexican American Heritage Linking Literature, Cultures and Theme, page 129			
Sojourner Truth, "Speech of Sojourner Truth," pages 130–132.	speech	rhetorical question	equality for women in U.S. society
Rodolfo Gonzales, "I Am Joaquin," pages 133–135.	poem	free verse	maintaining a cultural identity in U.S. society
Jeanne Wakatsuki Houston and James D. Houston, "Free to Go," pages 136–138.	nonfiction/ autobiography	irony	internment and release of Japanese Americans from detention camps during World War II
Constantine Panunzio, "In the American Storm," pages 139–141.	nonfiction/ autobiography	tone	immigrants struggling to find work in the United States

Assessment

Assessment opportunities are indicated by a ✓ next to the activity title. For guidelines see Teacher's Resoruce Manual, page xxxiii.

CROSS-DISCIPLINE TEACHING OPPORTUNITIES

Social Studies With the social studies teacher, provide students with an overview of the significant civil rights legislation and other laws affecting equality among cultural groups in the United States. A good starting point might be the passage of the 13th, 14th, and 15th Amendments ratified in the 1860s and the 19th Amendment ratified in 1920. Other significant laws include: the Chinese Exclusion Act of 1882 (and its extensions in 1892 and 1902), the *bracero* program of 1942, and the Civil Rights Act of 1964.

Geography The internment of Japanese Americans was related in many ways to geography—many Japanese Americans had settled along the western seaboard of the United States. During the height of anti-Japanese hysteria, some people claimed that these citizens would help Japanese invaders by signaling offshore ships and sabotaging U.S. coastal military installations. Have students use maps to locate and mark the important military centers along the West Coast and the internment camps in the California interior. (Examples: Manzanar in the Owens Valley; Tule Lake near the Oregon border.) Use the maps to lead a discussion on the internment of the Japanese Americans and the impact the relocation had on their lives.

Music In the 1960s, the Student Nonviolent Coordinating Committee (SNCC) was at the forefront of many civil rights battles. Its most important goal was to mobilize communities for protest and for voter registration drives. In 1962, the SNCC established the Freedom Singers, a chorus that traveled throughout the country giving programs to raise money for the Civil Rights Movement. Many of the songs performed by the Freedom Singers were traditional spirituals, but with slightly different words. Work with the music teacher to provide students with examples of protest music, especially the tradition of singing freedom songs, such as "We Shall Overcome," during the Civil Rights Movement of the 1950s and 1960s.

SUPPLEMENTING THE CORE CURRICULUM

- On facing pages, page A12 and A13, are two works that deal directly with the theme, the Quest for Equality. Have students read the captions and examine the two paintings. Then explain to them that the painting by Jane Evershed, *If You Believe In Woman*, is regarded as "fine art". That is, it is the work of a well-known artist whose paintings are exhibited in galleries and may command substantial prices from private collectors and museums. The wall painting in Harrisburg, Pennsylvania is a striking example of "street art." It was probably painted by one or more unknown artists and is free for all to see. Discuss the problem of art conservation. How can street art be preserved? Ask students if it *should* be preserved. Challenge students to explain why one painting is more highly valued than the other. Ask them if one painting is more expressive of the theme than the other or if they think one is better executed than the other. Guide the students to see that the same emotions and concerns underlie artistic expression at all levels.

- Interested students might like to research the so-called "zoot suit riots" of 1943. During this period, some young Mexican Americans wore the "zoot suit," a men's suit of exaggerated style consisting of a thigh-length jacket with extended shoulder padding and very full pants with narrow cuffs. They were originally worn by jitterbug dancers, but were adapted by young Mexican Americans to show cultural solidarity. According to historians Walton Bean and James J. Rawls, "Much of the public believed that not only all zoot-suiters but virtually all young Mexican Americans were *pachucos*, or juvenile hoodlums." On June 3, 1943, mobs of white soldiers and civilians, provoked by such behavior as catcalls and wild automobile driving, began indiscriminately to attack zoot suiters in Los Angeles. The riots continued for six days until the military police stepped in. Perhaps the best treatment of this episode can be found in the play *Zoot Suit*, by Luis Valdez.

- From 1954 to 1965, the Civil Rights Movement for African Americans in the United States reached a peak. Ask students to research the pivotal events of this period, including the well-known and not so well-known figures from this movement. An outstanding history of the movement during this time can be found in *Eyes on the Prize*, by Juan Williams, (Viking, 1987).

INTRODUCING THEME 7:
THE QUEST FOR EQUALITY

On the board write:

Is Equality in the United States a Myth or a Reality?

Before students answer, explain that most Americans believe strongly that the United States is a nation founded and based on the principle of equality. Ask:

• What yardstick do you use to measure equality? Financial success? Political power? Social status?

• Why do you think so many people in the United States hold the concept of equality so dearly?

Point out to students that the long-held belief that people can "get ahead" if they work hard enough is one factor underlying Americans' belief that the United States is a nation based on equality. Discuss this assumption and ask students whether or not they believe it is true today, or was ever true in the past.

Ask students to write a short answer to the question you have written on the board. When they have finished, call on volunteers to read their responses. Discuss the concepts of equality in the students' writings and whether they believe this ideal is attainable. Tell students to keep ideas from their writings in mind as they read the titles in Theme 7.

☑ Developing Concept Vocabulary

Tell students to complete an outline of vocabulary terms to help them explore the ideas and concepts in Theme 7. Have them skim the selections to find words for each outline category. When the outlines are completed have each student write a sentence on what he or she believes is the main idea of Theme 7. Call on each student to read the sentences aloud and discuss how each of them arrived at his or her conclusions. On the board provide the following outline framework:

I. Cultural Groups in Theme 7

a.

b.

c.

d.

II. Problems, Struggles, and Conflicts In Theme 7

a.

b.

c.

d.

III. Cultural Customs and Beliefs in Theme 7

a.

b.

c.

d.

I See the Promised Land
from *A Testament of Hope*
by Martin Luther King, Jr. (pages 225–231)

OBJECTIVES

Literary

- to evaluate the application of the speaker's proposals
- to identify allusions in the selection

Historical

- to discuss the impact of Martin Luther King, Jr., on the Civil Rights Movement
- to evaluate nonviolent versus violent protest methods for effecting change

Multicultural

- to compare the African American Civil Rights Movement to the struggle for equality undertaken by other cultural groups in the United States
- to evaluate the impact of the African American Civil Rights Movement on other cultural groups

SELECTION RESOURCES

Use Assessment Worksheet 28 to check students' vocabulary recognition, content comprehension, and appreciation of literary skills.

☑ Informal Assessment Opportunity

SELECTION OVERVIEW

While Martin Luther King, Jr., was earning his divinity degree at Crozer Theological Seminary, he became intensely interested in the life and teachings of Mahatma Gandhi, the Indian leader who brought independence from colonial rule for India through passive resistance. Ultimately, Gandhi's influence would figure prominently in King's own belief in nonviolent resistence. During the time of the Montgomery bus boycott, even though King was in great personal danger, he maintained the following principles:

> "We will not resort to violence. We will not degrade ourselves with hatred. Love will be returned for hate."

In this selection, King describes his vision of a United States free of prejudice and discrimination, and his belief that African Americans could achieve these goals through nonviolent protest and the use of economic leverage.

ENGAGING THE STUDENTS

On the board write:

IS VIOLENCE EVER JUSTIFIED?

Lead a class discussion of this issue, using the questions below:

- Has violence ever solved a problem? Give examples to support your answer.

- Is violence counterproductive, or is it effective in achieving goals?

Explain to students that in this selection, they will read about King's alternatives to violence.

BRIDGING THE ESL/LEP LANGUAGE GAP

This speech may be difficult for students who are unfamiliar with the history of the Civil Rights Movement. This strategy will help you familiarize the students with King's philosophy. Provide the students with a copy of the text that they can write on. Ask them to scan it and highlight each of the following sentences and phrases, which should be read aloud, one at a time.

1. The masses of people are rising up.
2. . . . nonviolence or nonexistence.
3. We are saying that we are God's children.
4. That's always the problem with a little violence.
5. . . . but we just went before the dogs singing, "Ain't gonna let nobody turn me round."
6. . . . the power of economic withdrawal.
7. "If I do not stop to help the sanitation workers, what will happen to them?" That's the question.
8. I've seen the promised land.

Ask students to read the selection and to use the context to find out how each of these ideas demonstrates King's philosophy and/or his method for causing change. Discuss their answers as a class.

✔ PRETEACHING SELECTION VOCABULARY

Do a webbing exercise based on the references made to freedom in this selection. First, tell students to skim the selection and write four to five words or phrases in the text that they associate with the concepts "freedom" and "equality." (Student choices should include: **Promised Land, human rights,** unity, slavery, **injustice, First Amendment privileges.**) On the board write *freedom* and draw connecting lines from it. Write the student word choices on these lines. When they have finished, challenge students to define freedom further with these questions:

- What is the difference between freedom and equality?
- Can someone be free without achieving equality?

The words printed in boldface type are tested on the selection assessment worksheet.

PREREADING

Reading the Author in Cultural Context

Help students read for purpose by asking them to consider how King's message combined African American history, Christian beliefs, and American patriotism. Discuss whether or not King had faith in the United States and its system of government. Ask: "How does King's message of nonviolence affirm his faith in the United States?" Tell students that when they have finished reading the selection they will be better able to answer this question.

Focusing on the Selection

Ask students to consider King's approach to causing social change. Ask them to consider the violence that surrounded King (and the violence that eventually took his life) and compare it to his philosophy of nonviolent, passive resistance. Invite them to decide for themselves the merit of King's philosophy.

POSTREADING

The following activities parallel the features with the same titles in the Student Edition.

Responses to Critical Thinking Questions

Possible responses are:

1. The tone of King's speech suggests that the time has finally come for African Americans to act, but with economic leverage, not violence. King's focus is on economic coercion through the power of a united force.
2. King's experiences are similar to those of many African Americans living in the United States in 1954—segregation and discrimination.
3. The rights that King fought for are basic human

rights of freedom and equality—rights that all people are entitled to in this country.

☑ Guidelines for Writing Your Response to "I See the Promised Land"

Have students share their journal entries with a partner. Or, as an alternative writing activity, have students work in pairs to write two opinions of King's speech—one from the perspective of an African American living in the 1950s and one from the perspective of an African American today. Tell students to consider the changes U.S. society has undergone in the past 40 years, including the escalation of violence. When they have finished, ask the students to exchange their writing and discuss the similarities and differences. Then, have them collaboratively answer the questions in the student text. Tell students to consider their answers and their writing as they complete their journal entries.

Guidelines for Going Back into the Text: Author's Craft

Discuss the effectiveness of biblical allusions, especially in regard to how King used them in his speech. Ask: "How did these allusions broaden King's appeal?" Possible answers to the questions in the Student Edition are:

1. I See the Promised Land, God's children, Pharaoh in Egypt, streets flowing with milk and honey, new Jerusalem, I've been to the mountain top.
2. The Pharaoh kept his slaves under control by causing them to fight among themselves. King makes the point that unity is essential for African Americans to achieve civil rights.
3. Guide students to find allusions in other selections.

☑ FOLLOW-UP DISCUSSION

Use the questions that follow to continue your discussion of "I See the Promised Land." Possible answers are given in parentheses.

Recalling

1. Whom is King addressing in his speech? (striking workers in Memphis)
2. Who is Bull Connor? (the police chief in Birmingham, Alabama, who sent dogs after protestors)
3. What song did protestors in Birmingham sing while being taken to jail? ("We Shall Overcome")

Interpreting

4. What do you think King means by the phrase "promised land"? (an era when equality and freedom exist for African Americans)
5. Why is King disappointed in the press coverage of civil rights events? (The press only reports negative, violent acts.)
6. Explain King's references to China and Russia on page 00. (King points out that he can understand the denial of basic human rights in totalitarian regimes, but he does not understand why this denial occurs in a nation founded on the protection of basic human rights.)

Applying

7. How does King's understanding of economics inform his strategy for change? (King believes that economic might, gained via merchant boycotts, is more effective than violence.)
8. Why is group cohesion important for any movement advocating change? (Guide students to analyze how fragmentation depletes the strength of any group action, and that this was an important factor in the Civil Rights Movement.)

ENRICHMENT

Many videos on the life of Martin Luther King, Jr., are readily available. Two such videos, *Martin Luther King, Jr.* and *Martin the Emancipator* can be purchased from Britannica Video, 310 S. Michigan Avenue, Chicago, IL, (800) 554–9862.

See for yourself, listen for yourself, think for yourself

by Malcolm X (pages 232–237)

OBJECTIVES

Literary

- to analyze the effectiveness of a persuasive speech
- to evaluate the elements of repetition, restatement, and parallelism in a speech

Historical

- to explain the political events that shaped the philosophy of Malcolm X
- to contrast the philosophies of Malcolm X and Martin Luther King, Jr.

Multicultural

- to compare Muslim religious beliefs with those of other world religions
- to contrast cultural views on violence and methods of protest

SELECTION RESOURCES

Use Assessment Worksheet 29 to check students' vocabulary recognition, content comprehension, and appreciation of literary skills.

☑ Informal Assessment Opportunity

SELECTION OVERVIEW

The Black Muslim sect that Malcolm X embraced rejected Christianity and called for African Americans to arm themselves in self-defense. Malcolm X spent most of his efforts in this cause giving speeches across the United States. His message had a special appeal for teenage African Americans in northern urban areas and for African American civil rights workers who had been victims of violence by police. Eventually Malcolm X's emphasis on African American unity and his willingness to cooperate with sympathetic whites led to conflict within the Nation of Islam. In 1965, Black Muslim leader Elija Muhammad expelled Malcolm X from the organization. Ironically, just as Malcolm X's message was becoming less violent, he was shot and killed.

In this selection, the author describes to a group of students his reasons for rejecting a nonviolent approach to exacting change in the United States.

ENGAGING THE STUDENTS

Explain to students that the excerpt below is from Malcolm X's autobiography. This is the response that Malcolm X's junior high school teacher gave when Malcolm X told him about his wish to become a lawyer.

Mr. Ostrowski looked surprised, I remember, and leaned back in his chair and clasped his hand behind his head. He kind of half smiled and said: "Malcolm, one of life's first needs is for us to be realistic. Don't misunderstand me now. But you've got to be realistic about being a nigger. You need to think about something you can be. You're good with your hands— making things. Why don't you plan on carpentry?

Allow students to respond to the quotation, then tell them that in this selection, they will read Malcolm X's response to racism in the United States.

BRIDGING THE ESL/LEP LANGUAGE GAP

Ask students to read the selection and then have them discuss the following questions in groups of three to four:

1. How does the author's story about the woman on the plane relate to the title of his speech?

2. What does the author mean by the statement below?

    . . . If you form the habit of taking what someone else says about a thing without checking it out for yourself, you'll find other people will have you hating your own friends and loving your enemies. . . . I think our people in this country are the best example of that.

3. What are some of the images that Malcolm X used in this speech? After using several images, he says, "I don't go along with any kind of nonviolence unless everyone is going to be nonviolent." Do you agree with his statement? Are his images persuasive? Why or why not?

4. What do you think the statement below shows about what Malcolm X believed was necessary for changes to take place in society? Do you agree?

    If the leaders of the nonviolence movement can go into the white community and teach nonviolence, good, I'd go along with that.

✔ PRETEACHING SELECTION VOCABULARY

Write the headings **nonviolent** and **violent** on the board. Ask students to skim the last half of the selection (beginning with "I think our people in this country. . .") for words that would fit under each heading. (nonviolent: loving, patient, **forgiving;** violent: **brutality, rumble,** cut up, shot up, busted

up) Ask them which group of words is more appealing and why. Explain that they will learn the views of Malcolm X on this subject as they read the selection.

The words printed in boldface type are tested on the selection assessment worksheet.

PREREADING

Reading the Author in Cultural Context

Tell students that the debate continues over the merits of separatism for African Americans. Ask them if they believe Malcolm X would support or oppose programs such as affirmative action and school busing to achieve integration. As they read, have them compare the integration issues facing the United States today with Malcolm X's background and beliefs.

Focusing on the Selection

Ask students to write a short paragraph summarizing the basic rights they believe all human beings are entitled to. When they have finished, call on them to read their writing aloud. Make a list on the board of the values and beliefs they mention. Leave this list on the board and refer to it as your class reads and analyzes the selection. Challenge students to find similarities between their list and Malcolm X's ideas.

POSTREADING

The following activities parallel the features with the same titles in the Student Edition.

Responses to Critical Thinking Questions

Possible responses are:

1. Malcolm X believes that nonviolence is only practical when all groups are nonviolent—until then African Americans must defend themselves.

2. He may have had experiences with the

organizations he mentions, such as the Ku Klux Klan and White Citizens Council, both of whom use violence against African Americans.

3. Sample answer: After reading this selection, I understand more fully the division between the beliefs of Martin Luther King, Jr., and the other activist African American groups.

☑ Guidelines for Writing Your Response to "See for yourself, listen for yourself, think for yourself"

Have students share their journal entries with a partner. Or, as an alternative writing activity, ask all students to identify and list the reasons Malcolm X uses to justify violence. Next, in pairs, have students discuss and answer the questions in the student text. When they have finished, lead a class discussion on Malcolm X's justifications for violence and whether or not students agree with his ideas. Tell students to draw upon their writing to support their discussion points.

Guidelines for Going Back into the Text: Author's Craft

Help students understand how repetition and restatement can help a speaker make his or her point. Ask them to think of any powerful and motivating speakers they may have heard. Discuss what made these speakers effective. Ask them to look for ways Malcolm X used repetition and restatement. Possible answers to the questions in the Student Edition are:

1. People who talk about nonviolence are often violent, I have nothing against people who can be nonviolent.

2. So I myself would go for nonviolence if it was consistent, if it was intelligent, if everybody was going to be nonviolent, and if we were going to be nonviolent all the time.

3. Guide students to find other examples of parallelism in Units 1 and 2.

☑ FOLLOW-UP DISCUSSION

Use the questions that follow to continue your discussion of "See for yourself, listen for yourself,

think for yourself." Possible answers are given in parentheses.

Recalling

1. Who asked Malcolm X to give the speech? (Mrs. Walker)

2. What organizations does Malcolm X mention that use violence against African Americans? (Ku Klux Klan, White Citizens Council)

3. According to Malcolm X, who were the victims in the Harlem hospital? (African American victims of violence from other African Americans)

Interpreting

4. Why do you think the woman on the plane was surprised to learn the person sitting next to her was Malcolm X? (Answers will vary, but suggest to students that Malcolm X's appearance did not match the woman's preconceptions about him.)

5. What groups does Malcolm X criticize for being violent? (Malcolm X points to violence among African Americans and violence of whites against African Americans.)

6. Why does Malcolm X believe young people should "see, listen, and think" for themselves? (so that they can make their own decisions about race and equality, without prejudice)

Applying

7. How does the title of the speech relate to Malcolm X's message? (He suggests that if people are aware of what goes on around them, they will realize that it is not feasible to be nonviolent in a world that uses violence against them.)

8. How do you think new media technology has influenced our ability to "see, look and listen" for ourselves? Support your answer with examples. (Answers will vary, but guide students to weigh the advantages and disadvantages of society's increasing reliance on television for information.)

ENRICHMENT

Students may benefit from viewing *A Tribute to Malcolm X*, available from Indiana University Audio-Visual Center, Bloomington, IN, 47405–5901.

from *China Men*
by Maxine Hong Kingston
Immigration Blues
from *The Gold Mountain Poems* (pages 238–246)

OBJECTIVES

Literary

- to recognize the writers' means of conveying their attitudes toward their cultural heritage
- to observe character development in a story

Historical

- to explain the economic reasons Chinese immigrants came to the United States
- to discuss the impact of Chinese labor on the construction and success of the transcontinental railroad

Multicultural

- to compare the experiences of Chinese immigrants to those of other immigrant groups in the United States
- to discuss the different effects of assimilation among diverse cultural groups who immigrated to the United States

SELECTION RESOURCES

Use Assessment Worksheet 30 to check students' vocabulary recognition, content comprehension, and appreciation of literary skills.

✔ Informal Assessment Opportunity

SELECTION OVERVIEW

While many Chinese immigrants worked for the railroads during the 19th century, others engaged in other occupations. Chinese occupational patterns for this period can be divided into four phases. From 1850 to 1865, Chinese laborers worked mainly as miners and traders. From 1865 to the late 1870s, they branched out into agriculture, light manufacturing, and common labor. From the late 1870s to the late 1880s, Chinese immigrants competed successfully in all economic fields. From the late 1880s to the turn of the century, however, discriminatory laws and attitudes forced them to abandon many occupations. The 1882 Chinese Exclusion Act ended Chinese immigration and sent a negative message to the Chinese already living in California.

In these selections, the experiences of Chinese laborers and immigrants are described. The authors confront the prejudice and harsh treatment these groups faced.

ENGAGING THE STUDENTS

On the board write the following two lists:

Our People	Their People
Good debaters	Quarrelsome
Patriots	Warmongers

123

Our People	Their People
Cultivated	Snobbish
Eccentric	Odd
True Believers	Heathens
Robust	Fat
Athletes	Jocks
Restrained	Heartless

Use the lists to lead a class discussion on ethnocentrism and how it can lead to discriminatory attitudes toward cultural groups that are different from one's own. Tell students that in this selection they will read about one cultural group's experience with ethnocentrism.

BRIDGING THE ESL/LEP LANGUAGE GAP

China Men: The author of this selection uses familiar words to create a vivid description of the lives of many Chinese immigrant workers. This piece provides an opportunity for students to study the author's use of language and to expand the way they use language themselves.

Ask the students, as they read, to put themselves in Ah Goong's place, focusing on the hardships of the immigrants' lives. How many of those hardships can they find? What images, phrases, or passages stand out to them?

After they have read the selection, have students work in groups of four to five to list some of the hardships that seemed particularly difficult and the phrases that were most effective in communicating these hardships. Ask: "Is the vocabulary difficult? What makes the writing effective?"

Ask them to discuss why the other workers thought Ah Goong was crazy. Ask: "Do you think he was crazy? Why or why not? What do you think about his philosophizing about the mountain and time and life?"

"Immigration Blues": Read the poems out loud and, as a class, discuss the images that the poets use. Ask: "Why did the speakers leave home? What did each find here? What is the last poet trying to communicate with the last question?"

✔ PRETEACHING SELECTION VOCABULARY

Explain to students that the selection from *China Men* describes the experiences of Chinese railroad laborers. Ask students to imagine how grueling the work must have been, based on just the following words from the selection: Tools used: **pickax, shovel, sledgehammer;** Actions of workers: swung, hammering, struck, **jarred,** slam, beating, pounding. Tell them to add to these lists as they read the selection, completing their understanding of the rigors of such labor.

The words printed in boldface type are tested on the selection assessment worksheet.

PREREADING

Reading the Author in Cultural Context

Help students to read for purpose by discussing the relationship between the Chinese laborers and the "demons" in the selection. As they read, have students consider the following questions: "Why did laborers continue to work despite these conditions? How were their occupational choices limited?"

Focusing on the Selection

Ask: "In general, is the mainstream society aware of the contributions made by immigrant laborers?" Discuss with students what they know already about immigrant laborers. For example, discuss what kinds of jobs immigrants were likely to obtain upon arrival in the United States in 1850, in 1890, or in 1920. What barriers might have initially prevented them from obtaining high-paying jobs?

POSTREADING

The following activities parallel the features with the same titles in the Student Edition.

Responses to Critical Thinking Questions

Possible responses are:

1. Chinese immigrants had to endure prejudice and discrimination. These conditions forced them into positions requiring hard manual labor.
2. By having the reader share the thoughts of the Chinese laborers and immigrants, the authors give the reader a sense of their pride and compassion for the experiences of their ancestors.
3. Guide students to understand how reading these selections should help them to appreciate their own freedom.

Guidelines for Writing Your Response to *China Men* and "Immigration Blues"

Have students share their journal entries with a partner. Or, as an alternative writing activity, ask each student to briefly summarize the author's viewpoint and message. Next, have students exchange their writing with a partner. Ask them to collaborate on a single summary including what points they believe the author was trying to make about the Chinese immigrants and their quest for equality. Call on them to read their summaries and to determine if the authors' ideas apply to all immigrant groups.

Guidelines for Going Back into the Text: Author's Craft

Tell students to write a short characterization of themselves. Call on them to read their writing and ask: "Did you use direct or indirect characterization?" Use their responses to define these terms. Possible answers to the questions in the Student Edition are:

1. She used indirect characterization. Example: After tunneling into granite for about three years, Ah Goong understood the immovability of the earth. Men change, men die, weather changes, but a mountain is the same as permanence and time.
2. These techniques allow us to gain knowledge about the character through observation, versus being told directly what the character is like.
3. Example: In "The Slave Ship" by Olaudah Equiano, we learn about the main character through his own words and descriptions of his life.

☑ FOLLOW-UP DISCUSSION

Use the questions that follow to continue your discussion of *China Men* and "Immigration Blues." Possible answers are given in parentheses.

Recalling

1. What was the hardest part of tunneling for Ah Goong? (tunneling through granite)
2. Who were the "demons"? (the railroad bosses)
3. Why was snowfall dangerous for the laborers? (risk of avalanches)

Interpreting

4. What do you think this statement from the selection means: "I saw what's real, I saw time, and it doesn't move." (These were Ah Goong's feelings after tunneling in granite for weeks. It describes the workers' feelings of hopelessness and despair.)
5. Why was the main character in the first poem from "Immigration Blues" deported? (The character had a "lapse of memory" which could signify inability or unwillingness to answer the questions of immigrant officials.)
6. Why is the main character of the third poem from "Immigration Blues" angry at the United States? (He points out that, in a country that prides itself on liberty and democracy, he has been treated with restrictions and autocracy.)

Applying

7. Do you think the experiences of Ah Goong are relevant for immigrants today? Give an example to support your answer. (Immigrants to the United States continue to take low-paying and dangerous jobs to survive.)
8. How might the economic and cultural development of the United States be different without the contributions of immigrant groups? (Answers will vary, of course, but students should acknowledge how immigrant contributions have led to a more diverse American culture.)

ENRICHMENT

Students may benefit from viewing *The Chinese-American: The Early Immigrants* which explores the immigrants' role on the transcontinental railroad. The video is available from Handel Film Corp, 8730 Sunset Boulevard, West Hollywood, CA, 90069.

Napa California

by Ana Castillo

from The Circuit

by Francisco Jiménez (pages 247–257)

OBJECTIVES

Literary

- to identify the function of the writers' values
- to evaluate the function of the setting in the selections

Historical

- to explain the causes and consequences of the *bracero* program
- to understand the impact of the Mexican American labor force on California's development

Multicultural

- to compare the themes of protest among writers from diverse cultural groups
- to contrast the different cultural interpretations of the "American Dream"

SELECTION RESOURCES

Use Assessment Worksheet 31 to check students' vocabulary recognition, content comprehension, and appreciation of literary skills.

☑ Informal Assessment Opportunity

SELECTION OVERVIEW

The *bracero* (strong ones) program was created as a reaction to the labor shortage caused by World War II. In 1942, Congress—with the agreement of the Mexican government—authorized the United States Department of Agriculture to recruit, contract, transport, house, and feed temporary immigrant farm workers from Mexico. Before the war, California growers had been able to persuade the government not to include Mexico in the quota restriction systems being enacted at the time. In 1957, 192,438 *braceros* were brought to California. As the program continued, it created an underclass of impoverished workers. César Chávez became a leading advocate for the farm workers. Chávez, who had been a migrant worker as a child, (he attended 40 different schools in eight years) began the United Farm Workers (UFW) Union. Their efforts culminated in the 1968 boycott of all California table grapes. By 1970, most of the large California grape growers had signed contracts with the UFW.

In these selections, the authors portray the lives of Mexican farm laborers.

ENGAGING THE STUDENTS

On the board write:

Viva la huelga!

Before you explain to students the meaning of this phrase, ask them these two questions:

- What is the single largest industry in California? (agriculture)
- Which cultural group comprises the single largest work force in this industry? (Mexican Americans)

Explain to students that despite these facts Mexican American farm workers have historically been mistreated and underpaid. Explain that the slogan on the chalkboard means "long live the strike," and was the slogan of the United Farm Workers Union. Tell students that in these selections they will read accounts of what life was like for these Americans and why unions were so important for them.

BRIDGING THE ESL/LEP LANGUAGE GAP

Before the students read, ask them to be looking for specific ways their lives are different from those of the people in these selections. After they have read, discuss the selections as a class. Some migrant students might have shared the experiences described in these selections, or know people who have. Ask them to contribute to class discussion. Ask the students:

1. What are some of the things you take for granted that the people described in these selections do not have?
2. How do you think the kind of lifestyle described in the selections might affect your outlook on life?
3. If your life were like Panchito's or Roberto's, what do you think would be the hardest thing for you to accept?

✔ PRETEACHING SELECTION VOCABULARY

Read the following list of words from the selections to the class. Tell students that these words help reveal the location or setting of the selections.

harvest	one hundred degrees
sweat	buzzing insects
sun-beaten	sandy
fields	dirt
dust	**dusk**

Ask students to imagine what working long hours in such a place would be like. Encourage them to find other descriptive words as they read the selections to complete their perceptions of setting.

The words printed in boldface type are tested on the selection assessment worksheet.

PREREADING

Reading the Author in Cultural Context

Help students read for purpose by asking them to consider what impact constant movement from one town to another would have on their own lives. Discuss how being forced to leave a neighborhood or change schools would affect their education and social life. Ask them to consider the impact this movement had on Mexican American laborers as they read.

Focusing on the Selection

Ask: "What gives people hope, even when a situation seem hopeless?" Challenge students to determine what is valued by their own culture and compare this to the lives and beliefs they will read about in the selections.

POSTREADING

The following activities parallel the features with the same titles in the Student Edition.

Responses to Critical Thinking Questions

Possible responses are:

1. Inequality and injustice are part of their daily lives. They endure these conditions to provide food and shelter for their people.
2. They've lived in a society that has discriminated against Mexican Americans and kept some of them in the lowest economic brackets.
3. Guide students to compare the food and shelter they take for granted to the life of the main character in "The Circuit."

☑ Guidelines for Writing Your Response to "Napa, California" and "The Circuit"

Have students share their journal entries with a partner. Or, as an alternative writing activity, ask students to consider their favorite childhood memories and compare them to the memories of the authors they have just read. Ask students to work in pairs. Have them discuss what life was like for the main characters in the selection and compare it to their own way of life. Have each pair collaborate on a paragraph describing what message they believe the selections have for American society today.

Guidelines for Going Back into the Text: Author's Craft

Tell students to write a short description of where they live. Call on them to read their writing aloud and discuss why setting would be important when describing a memory or writing a story about their own past. Possible answers to the questions in the Student Edition are:

1. The setting emphasizes the harsh conditions Mexican American farm workers experienced.

2. The setting invokes both the harsh conditions faced by the family and the constant movement of their lives.

3. The setting is crucial because it contains the elements that dictate the conditions the characters are in—farm, crops, heat, and dust.

☑ FOLLOW-UP DISCUSSION

Use the questions that follow to continue your discussion of "Napa, California" and "The Circuit."

Possible answers are given in parentheses.

Recalling

1. What crops were picked by the workers in the selections? (strawberries, grapes, cotton)

2. Why did the family in "The Circuit" have to move so often? (They had to follow the seasonal rotation of crops.)

3. Why did Panchito and his brother hide from the school bus? (They were afraid they would be picked up for not attending school.)

Interpreting

4. Why was Panchito embarrassed in front of Roberto on his first day of school? (His brother was older and had to work longer— Panchito was able to attend school sooner.)

5. Was Panchito accepted at school? Support your answer with examples from the selection. (Answers will vary. Though Panchito sat in the back of the bus alone, he was accepted and helped by Mr. Lema.)

Applying

6. How do you think being deprived of an education affected Mexican Americans' ability to improve their social and economic standing? (Guide students to evaluate the impact a lack of education had on Mexican Americans during this era.)

7. Why do you think "Napa, California" is dedicated to a union organizer? (Unions offered a chance for better conditions, a degree of power, and better wages.)

ENRICHMENT

Students may benefit from viewing *Migrant Farmworkers*, available from Downtown Community TV, 87 Lafayette Street, New York, NY, 10013, (212) 966–4510.

MAKING CONNECTIONS THE QUEST FOR EQUALITY

OBJECTIVE

Overall

• to discover the connections between Mexican American heritage reflected in art and architecture and the themes in "The Circuit" and "Napa, California"

Specific

• to compare and contrast views on work and labor expressed in "Napa, California" and "The Circuit" with those expressed in the mural by Emigdio Vasquez

• to identify Mexican American heritage and traditions in architecture in the southwestern United States

• to appreciate the values and beliefs expressed in Mexican American literature and art and how they relate to the Mexican American community in the United States today

ENGAGING THE STUDENTS

On the board write the following question:

What is the most striking visual feature of our school community?

Allow the students a few moments to consider many different kinds of visual aspects. List their ideas on the board. Next, ask: "What does this aspect say about our school community? How does it represent our school?" Have students study each of the photographs on pages 258 and 259 and discuss how they represent some aspect of Mexican American heritage. Finally, ask: "If you could choose another visual feature to represent the attitudes and values of our school community, what would that feature be?"

BRIDGING THE ESL/LEP LANGUAGE GAP

Use the photographs on pages 258 and 259 to help students explore various aspects of Mexican American heritage. Have them study each illustration and explain orally what it suggests about Mexican heritage. If students are recent arrivals or the children of immigrants, encourage them to compare what they observe in the illustrations to their own heritages.

EXPLORING ART

The paintings of many Mexican American artists, as with artworks from Mexico itself, combine grandeur and social awareness to make vivid statements. This artistic tradition gained momentum early in the 20th century with the revival of mural-painting by Diego Rivera in Latin America and the United States. Have students work in small groups to examine the artworks on pages 258 and 259 and read the captions. Ask: "What social statements are made by the paintings shown here? What can you say about Mexican American values as reflected in these artworks? In your opinion, why are these values shown on such a grand scale and with such vivid colors?"

Invite students to compare and contrast the paintings on these pages with murals by Diego Rivera, such as Detroit Industry, a mural which celebrates labor, and Agrarian Leader Zapata, a fresco of the revolutionary leader painted in 1931.

LINKING LITERATURE, HISTORY, AND THEME

Guidelines for Evaluation

Sample answers: Hispanic heritage, Christianity, dedication to hard work, contributions of Mexican American laborers

Panchito's family would be most likely to identify with the mural by Emigdio Vasquez depicting Mexican American laborers, since his family also made important contributions to building the United States.

Castillo's poem describes the labor of Mexican American farm workers. The mural by Emigdio Vasquez also celebrates Mexican American workers.

Speech of Sojourner Truth

by Sojourner Truth (pages 260–263)

OBJECTIVES

Literary

- to understand the speaker's attitudes about women's rights
- to understand persuasive techniques in the speech

Historical

- to evaluate the impact of the abolitionist movement on women's rights
- to explain how Sojourner Truth combined the causes of abolitionism and women's rights

Multicultural

- to contrast the role and status of women in diverse cultures
- to determine whether the goals of Sojourner Truth remain goals among women in diverse cultures today

SELECTION RESOURCES

Use Assessment Worksheet 32 to check students' vocabulary recognition, content comprehension, and appreciation of literary skills.

☑ Informal Assessment Opportunity.

SELECTION OVERVIEW

As the abolitionist movement forged ahead in its struggle to liberate enslaved African Americans, the same men who made eloquent pleas for the equality of African American men told women "it was not their time" to fight for equality. Sojourner Truth attacked this double standard. Explain to students that at this time women could not own property, vote, or hold office. Women were openly condemned for speaking in public and rebuked for entering the halls of Congress with their petitions. This discrimination especially enraged women reformers of this period who were trying to deal with problems such as alcoholism and prostitution. One result of women's frustrations with the male-dominated abolitionist movement was the addition of women's rights to the reformers' list.

In this selection, the author admonishes men for not giving women equality and questions some common notions of the day concerning women's abilities.

ENGAGING THE STUDENTS

On the board write these statistics:

- In 1985, women working full time earned 65 percent of what men earned.
- In 1981, 1 percent of senior executive positions were held by women.
- In 1992, the number of women in executive positions had only risen to 3 percent.

Challenge students to explain the causes of these statistics and the disproportionate number of white males in upper-level management positions. Ask: "How long do you think this kind of disparity has existed in the United States?" What do you think are the causes of the disparity between the wages men and women earn? As they respond, tell students that in the next selection they will read a speech given by a woman in 1851. Tell students to compare the concerns of Sojourner Truth to the concerns of people today.

BRIDGING THE
ESL/LEP LANGUAGE GAP

Be certain that students understand the prereading section before they read this selection. Discuss as a class the issue of women's rights. Ask: "How have women been discriminated against in the past in the United States? Are they still discriminated against today? How is this discrimination similar to the way people of various cultural groups have been discriminated against? What is the solution?"

Ask students to read the speech and to find the arguments that Sojourner Truth used as she called for women's rights. Discuss this as a class.

☑ PRETEACHING
SELECTION VOCABULARY

Tell students that there are many biblical references in the selection they are about to read. Have them skim the selection and make a list of vocabulary words and phrases with religious connotations. (Examples: **Eve,** sin, **Jesus, Lazarus, Mary,** faith, love, **God,** blessed) When they have finished, ask students to speculate on why such words might be in a speech about women's rights. Use the questions below to guide your discussion:

- What do religious references tell you about the audience the speaker is addressing?
- Why would a speaker trying to make a point make references to the Bible and to God? In what cases would this approach be to the speaker's advantage?

The words printed in boldface type are tested on the selection assessment worksheet.

PREREADING

Reading the Author
in Cultural Context

Help students read for purpose by asking them to consider the similarities and differences between the women's rights movement and the civil rights movement. Ask: "How would you define basic human rights? Are these rights the same for men and women?" Guide students to consider the goals of the abolitionist movement and their universal appeal as they read.

Focusing on the Selection

Challenge students to identify the "tone" Sojourner Truth takes in the selection. Ask students to determine if the author is humorous or serious, angry or cajoling, confrontational or conciliatory. Tell students to read the selection from the perspective of a 19th-century citizen living in a society where women have no legal rights at all.

POSTREADING

The following activities parallel the features with the same titles in the Student Edition.

Responses to Critical
Thinking Questions

Possible responses are:

1. Truth makes a common-sense case for women's rights on the basis of her own experiences—doing anything a man can do.
2. Her experiences as an enslaved African American and facing sex discrimination while fighting for the abolition of slavery may have influenced her decision.
3. Her main point, that equality should not be denied on the basis of sex, applies to groups struggling for equality denied on the basis of race.

☑ Guidelines for Writing Your
Response to "Speech of Sojourner
Truth"

Have students share their journal entries with a partner. Or, as an alternative writing activity, ask each student to write in a few sentences what he or she believes are the main points of Sojourner Truth's speech. Next, lead a discussion on the women's rights movement today. Ask: "Have women achieved equality in the workplace? in politics? in families?" Call on students to read their interpretations of the speech and ask, "What message does the speech have for teenagers in the 20th century?" Discuss the student responses.

Guidelines for Going Back into the Text: Author's Craft

Ask students to read the definition of *rhetorical question* in the student text. Next, have students write their own rhetorical question on any subject they choose. Call on them to read their questions, and guide them to understand how this technique can be used to make a point in a persuasive speech. Possible answers to the questions in the Student Edition are:

1. She suggests that if women upset the world when Eve caused Adam to sin, the world should be given back to women so they have a chance to set it right again.

2. Truth asks a number of rhetorical questions concerning her abilities to do the same physical tasks as a man.

3. Guide students to find other examples of rhetorical questions in the text.

☑ FOLLOW-UP DISCUSSION

Use the questions that follow to continue your discussion of "Speech of Sojourner Truth." Possible answers are given in parentheses.

Recalling

1. In what areas does Sojourner Truth compare herself to men? (work, eating, and thinking)

2. According to the Bible, who caused man to sin? (Eve)

3. What was one skill Sojourner Truth admitted she lacked? (She could not read.)

Interpreting

4. Explain this quotation from the selection: "You need not be afraid to give us our rights for fear we will take too much, for we can't take more than our pint'll hold." (Women only want equal rights—not to usurp men's rights.)

5. According to Sojourner Truth, why will men feel better if they take her advice? (Sojourner Truth asserts that men will benefit emotionally if they allow women equality.)

6. Whom do you think Sojourner Truth is referring to in her final reference to the "hawk and a buzzard?" (abolitionists and fighters for women's rights)

Applying

7. Why do you think being discriminated against in the abolitionist movement was especially painful for female abolitionists? (Most women probably believed that men who were fighting for freedom and equality for enslaved African Americans would extend the same rights to women.)

8. What does this selection tell you about the struggle for women's rights in the 19th century? (One could conclude that women's rights would be even more difficult to obtain than equality for the races.)

ENRICHMENT

Students may benefit from viewing *American Women: Portraits of Courage*; Sojourner Truth is one of the women featured. The video is available from CRM/McGraw Hill Films, 674 Via De la Valle, P.O. Box 641, Del Mar, CA, 92014, (619) 453–5000.

from *I Am Joaquín*

by Rodolfo Gonzales (pages 264–268)

OBJECTIVES

Literary

- to analyze the poet's attitudes about his culture and identity
- to evaluate the use of repetition and parallelism

Historical

- to list the major events and political figures of the Chicano movement
- to describe the growing political power of Mexican Americans in the United States

Multicultural

- to compare the ways Mexican Americans and other cultural groups have adjusted to U.S. society
- to explain the importance of heritage and literature in retaining one's cultural identity

SELECTION RESOURCES

Use Assessment Worksheet 33 to check students' vocabulary recognition, content comprehension, and appreciation of literary skills.

☑ Informal Assessment Opportunity

SELECTION OVERVIEW

It is important for students to understand that by the 1960s the Mexican American population in the United States no longer consisted mainly of seasonal farm workers. By the 1960s, 85 percent of all Mexican Americans lived in cities, and less than 15 percent were employed as agricultural laborers. From these urban ranks of young Mexican Americans there arose an active search for ethnic identity. During this time, many Mexican Americans adopted the appellation "La Raza," which means "the race" or "the people." Another designation was Chicano, a term used for all Mexican Americans and for some who were from places other than Mexico.

In this selection, the author expresses rage over the assimilation of his people into white American culture, and decries the choices many must make between maintaining cultural identity and achieving economic success.

ENGAGING THE STUDENTS

Ask students:

"Do all Americans have an equal chance for success?"

Before they respond, discuss with students what the images of success are in the United States. Discuss with them role models in our society. Tell students that a number of Mexican American leaders have decried the lack of role models for the Mexican American community. Finally, allow students to respond to the question on the board. Tell them to consider their responses as they read the selection.

BRIDGING THE ESL/LEP LANGUAGE GAP

Group the students in pairs, matching English-proficient and LEP students wherever possible. Give each student a copy of the text to write on. Ask the students to read the poem and to highlight any verbs that stand out. What feelings do these verbs communicate? How do the kinds of verbs used change from the beginning to the end of the poem?

Combine the pairs into groups of four to six and ask each group to consider one section of the poem, to discuss what it means, and then to choose a spokesperson to present the group's ideas to the rest of the class.

The poem could be divided at these lines: (1) I am Joaquín; (2) And now!; (3) I shed the tears of anguish; (4) Part of the blood; (5) I have endured; (6) And now the trumpet; (7) And in all the fertile farmlands; and (8) I am the masses of my people.

☑ PRETEACHING SELECTION VOCABULARY

On the board write the following words: La Raza, **Mejicano, Español, Latino, Hispano, Chicano.** Direct students to the footnotes on page 267 to go over these terms and discuss their meanings. Use the questions below to lead a guided discussion of the author's use of these terms.

- How are these terms related? How are they different?
- Based on this vocabulary activity, what do you think is the author's viewpoint on being Mexican American?

The words printed in boldface type are tested on the selection assessment worksheet.

PREREADING

Reading the Author in Cultural Context

Help students read for purpose by asking them to consider how cultural identities can be blurred or even eliminated by mainstream society. Discuss how popular images in the United States often contain both generalizations about the mainstream society and ethnic stereotypes. Ask students to compare the author's view of Mexican Americans with the images and beliefs that Americans receive through media and advertising.

Focusing on the Selection

Tell students to consider the author's struggle to maintain a strong cultural identity in a society in which the economic system is geared toward domination by whites. Challenge them to find the historical roots of the author's heritage as they read and to discover why these are important to him.

POSTREADING

The following activities parallel the features with the same titles in the Student Edition.

Responses to Critical Thinking Questions

Possible responses are:

1. Gonzales confronts the dilemma he has faced in trying to be economically successful in the United States while also maintaining his cultural identity.
2. Gonzales is aware that people are often judged on ethnicity and, unless they are willing to assimilate, can be held back and discriminated against.
3. Guide students to compare their own awareness of cultural heritage with Gonzales' ideas about ethnic pride.

☑ Guidelines for Writing Your Response to "I Am Joaquín"

Have students share their journal entries with a partner. Or, as an alternative writing activity, ask each student to briefly summarize the emotions he or she felt while reading the selection. Next, have students exchange their writing with a partner. Ask them to collaborate on a single response to the author's charges that American society has tried to "absorb" and "sterilize" his cultural identity.

Guidelines for Going Back into the Text: Author's Craft

Ask: "What qualities do you like best in a poem?" As students respond, remind them that free verse often contains repeating phrases with similar structures. Ask: "Does parallelism limit or expand a poet's freedom in free verse?" Possible answers to the questions in the Student Edition are:

1. "I have endured in the rugged mountains of our country. I have survived the toils and slavery of the fields. I have existed in the barrios of the city, in the suburbs of bigotry, in the mines of social snobbery, in the prison of dejections, in the muck of exploitation."

2. Guide students to understand the effective way that Gonzales uses free verse.

3. Point students to the use of parallel structure in the African folk tales they have read.

✔ FOLLOW-UP DISCUSSION

Use the questions that follow to continue your discussion of "I am Joaquín." Possible answers are given in parentheses.

Recalling

1. What past cultures are referred to in the poem? (Moorish, Spanish, European, Aztec)

2. What is a barrio? (a Mexican American neighborhood)

3. What is a gringo? (a foreigner)

Interpreting

4. What references does Gonzales make to Mexican American culture? (Examples: mariachi strains, smell of chile verde)

5. Why do you think the author chose to name the poem, "I Am Joaquín?" (It is an assertion of pride and self esteem.)

6. Why is asserting cultural identity a "paradox" for the author? (He must choose between a victory of the spirit [maintaining cultural identity] and being able to eat [survive in American society].)

Applying

7. How would a poem such as this instill pride in one's culture? (It speaks of the past struggles and ongoing effort to maintain cultural identity in a society that in many ways is trying to squelch this spirit.)

8. How does this selection emphasize the need for societies to accept cultural diversity? (Guide students to understand the value of cultural diversity and the importance of maintaining a distinct cultural identity for American ethnic groups.)

ENRICHMENT

Students may benefit from viewing *Chicanos in Transition* available from Centre Productions, 1800 30th Street, Suite 207, Boulder, CO, 80301, (800) 824–1116.

Free to Go from *Farewell to Manzanar*
by Jeanne Wakatsuki Houston
and James D. Houston (pages 269–275)

OBJECTIVES

Literary

- to observe the author's attitudes about the paradox of internment camps
- to identify the ways in which irony is used in the selection

Historical

- to explain the causes that led to the internment of Japanese Americans during World War II
- to discuss the constitutionality of the Japanese internment camps

Multicultural

- to explain why Japanese Americans were interned during World War II, whereas German Americans and Italian Americans were not
- to compare the ways different cultural groups have come together during times of discrimination

SELECTION RESOURCES

Use Assessment Worksheet 34 to check students' vocabulary recognition, content comprehension, and appreciation of literary skills.

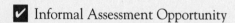 Informal Assessment Opportunity

SELECTION OVERVIEW

The Japanese air attack on Pearl Harbor in 1941 and the incessant rumors of sabotage by Japanese sympathizers in the United States resulted in strong anti-Japanese sentiments. These feelings were aided by the sensationalist publicity that surrounded the Pearl Harbor attack. Fears of a Japanese attack on the mainland led the United States to evacuate from the western coast all persons of Japanese ancestry. Much anti-Japanese feeling resulted from the economic success that this group had achieved in this country in comparison with other cultural groups. Most of these people were imprisoned in "relocation" camps in interior desert areas. Citizens and aliens alike were sent to the camps. Lt. Gen. John L. DeWitt, head of the Western Defense Command, had originally opposed the internment of Japanese Americans, but public pressure finally forced him into lobbying for Executive Order 9066, authorizing the secretary of war "to prescribe military areas . . . from which any or all persons may be excluded." Most of the Japanese American detainees were released by the end of 1944. In the 1980s, the nation's embarrassment over these internments led Congress to pay reparations to many Japanese Americans who lost property, or whose lives were otherwise compromised, as a result of the internments.

In this selection, the author describes her family's reaction to the news that the Japanese internment camps are being closed.

ENGAGING THE STUDENTS

On the board write:

German Americans   Italian Americans
Japanese Americans

Remind students that during World War II the Allies fought against Germany, Italy, and Japan. Ask:

• How were German Americans treated during the war?
• How were Italian Americans treated during the war?

Now tell students that in this selection they will read how Japanese Americans were interned in prison camps. As they read, challenge them to formulate their own opinions about why the Japanese Americans were singled out.

BRIDGING THE ESL/LEP LANGUAGE GAP

The opening of this selection is difficult because of the unfamiliar language and, to some students, unfamiliar concepts. Ask the students to read the opening paragraphs on page 270 about the three court cases, and to discuss with a partner these questions for each case: (1) What happened? (2) What was the Japanese citizen's reason for his actions? (3) What did the Supreme Court decide? (4) What do these cases tell us about the attitudes of white society toward the Japanese at this time?

Read the first sentence of the narrative on page 271 ("In our family . . ."). Ask students why they think this was true. As they read, ask them to look for five reasons that the girl's family was not happy to be freed from the camp. Discuss this as a class after they have read the selection.

Ask the students to consider this excerpt:

> The physical violence didn't trouble me . . . It was the humiliation. . . . Call it the foretaste of being hated. I knew ahead of time that if someone looked at me with hate, I would have to allow it, to swallow it, because something in me, something about me deserved it.

Discuss as a class how feelings of humiliation might limit a person's freedom.

✔ PRETEACHING SELECTION VOCABULARY

Discuss with students the vocabulary terms that are associated with prejudice and discrimination in the selection. On the board write **curfew, evacuation, order, internment camps,** wartime **propaganda,** racist headlines, atrocity movies, hate slogans, fright-mask posters, refusal to **assimilate, nightriders.** (You may want to define some terms for students.) Discuss with students what each of these terms means and ask them to describe possible examples of each term. Ask: "Why do you think Japanese Americans were the victims of these tactics after the attack on Pearl Harbor?" Tell students to look, as they read the selection, for other ways that prejudice and discrimination were present during this era.

The words printed in boldface type are tested on the selection assessment worksheet.

PREREADING

Modeling Active Reading

"Free to Go" is annotated with the comments of an active reader. These sidenotes, prepared to promote critical reading, emphasize the cultural content of the piece, address author values, call attention to literary skills, invite personal response, and show how the selection is related to the theme. If you have the time, read the entire selection aloud as a dialogue between reader and text. Encourage students to add their own responses and to compare these with the ones printed in the margins. Model these skills by adding your own observations too.

Reading the Author in Cultural Context

Help students read for purpose by asking them to consider the impact of prejudice on the family in this story. Discuss why, despite the likelihood of returning to a community prejudiced against them, the father prefers to return home instead of finding a new place to live.

Focusing on the Selection

Have students work in groups of three. Assign one of the three questions in the student text to each member. When they have finished working individually, ask them to collaborate on a single shared response to the final question in the student text.

POSTREADING

The following activities parallel the features with the same titles in the Student Edition.

Responses to Critical Thinking Questions

Possible responses are:

1. She is proud of her ancestry and has a strong sense of cultural identity.
2. Though she considers herself an American, the history of internment camps has given her cause to question the ability and willingness of white Americans to accept other races.
3. Guide students to examine prejudicial stereotypes in their own communities.

☑ Guidelines for Writing Your Response to "Free to Go"

Have students share their journal entries with a partner. Or, as an alternative writing activity, ask students if they were surprised to learn that Japanese Americans were interned during World War II. In groups of three, have students discuss whether this could ever happen again. Challenge each group to devise safeguards for preventing this reoccurrence. Next, have each group collaboratively answer this question: "Why is this a good story for teenagers of all cultures to read?" Call on each group to read their response aloud.

Guidelines for Going Back Into the Text: Author's Craft

To be sure that students understand the term *irony*, have each student write a sentence that demonstrates this concept. Call for volunteers to read their sentences aloud and discuss the different ways they have used this technique.

Possible answers to the questions in the Student Edition are:

1. It is sadly ironic that in a country that prides itself on equality and freedom for all, some citizens were imprisoned unjustly and with no trial.
2. It suggests that she does not fathom entirely her situation and is not aware of the grievous events that have befallen her family.

3. Though they are about to be freed, the Houston family feels trepidation over the likelihood of prejudice and discrimination in their old community.

☑ FOLLOW-UP DISCUSSION

Use the questions that follow to continue your discussion of "Free to Go." Possible answers are given in parentheses.

Recalling

1. Why was the Wakatsuki family being allowed to leave the camp? (The Supreme Court ruled that the camps were unconstitutional.)
2. What was the name of the camp? (Manzanar)
3. To what state were the older brothers and sister moving? (New Jersey)

Interpreting

4. Why was the Wakatsuki family ambivalent about returning home? (Anti-Japanese feelings were running high in California.)
5. Why were the young people more willing to find new places in which to live? (They were not as attached to their homes or as established in businesses as the older family members were.)
6. Explain the author's reference to slaves in the final paragraph of the selection. (The author draws a parallel between her parents and enslaved African Americans who, when finally offered freedom, were unsure how to react.)

Applying

7. In your opinion, should societies try to "forget" episodes like the one you have read about, or should they be remembered and talked about? (Guide students to understand the importance of acknowledging the occurrence of events such as this so that future generations can learn from and prevent them.)

ENRICHMENT

Students may benefit from viewing a film that was made in 1942 that presents the U.S. Army's policy toward Japanese Americans during World War II. The video is available from the University of Washington Instructional Media, Kane Hall, DG-10, Seattle, WA, 98195, (206) 543–9909.

In the American Storm
from *The Soul of an Immigrant*
by Constantine M. Panunzio (pages 276–282)

OBJECTIVES

Literary

- to understand the author's attitude toward finding employment
- to identify tone in an autobiography

Historical

- to describe the working conditions European immigrants experienced in the early 20th century
- to observe the economic impact of immigrant labor in the industrial development of 20th-century United States

Multicultural

- to contrast the work experiences of immigrant groups that came to the United States in the early 1900s
- to compare the contrasting ways different immigrant groups were accepted into U.S. society

SELECTION RESOURCES

Use Assessment Worksheet 35 to check students' vocabulary recognition, content comprehension, and appreciation of literary skills.

☑ Informal Assessment Opportunity

SELECTION OVERVIEW

European immigration followed two distinct patterns. Until 1888, most immigrants came to the United States from northern and western Europe ("old immigrants") After that date, they generally came from southern and eastern Europe ("new immigrants"). Most of the immigrants from southern Europe were from Italy. Between 1891 and 1920, more than 3.8 million Italians came to the United States. Although Italian-American agricultural communities existed in places like southern New Jersey and California, most immigrants (like the characters in this selection) sought cash wages in cities.

In this selection the main character, an Italian, encounters his first taste of ethnic discrimination as he attempts to adjust to his new life in the United States.

ENGAGING THE STUDENTS

On the board write:

Give me your tired, your poor
Your huddled masses yearning to breathe free,
The wretched refuse of your teeming shore.
Send these, the homeless, the tempest-tossed
    to me.
I lift my lamp beside the golden door!

Tell students that these lines by the Jewish American poet Emma Lazarus are engraved on the Statue of Liberty. Next, tell students that between the years 1918 and 1924, anti-immigration laws were passed in the United States that nearly ended immigration. Discuss this paradox, and tell students that in this selection they will read more about the nation's ambivalence toward newcomers and about how the immigrants' quest for employment often became a quest for equality.

BRIDGING THE ESL/LEP LANGUAGE GAP

The unusual syntax and vocabulary of this selection will be difficult for some students. Arrange the students in groups of four to five. Ask them to read the selection one paragraph at a time, focusing on what they *do* understand, and sharing their knowledge. Ask them to come up with questions about the text for sections they do not understand.

When everyone has read the selection, have the class work together to reconstruct the story orally, encouraging all students to share what they understood, as the class "talks through" the story.

☑ PRETEACHING SELECTION VOCABULARY

Have students scan the selection and then create a chart of words based on the descriptions of labor in the selection. Tell them that one column of the chart should contain the kinds of work mentioned in the selection, (sailor, **excavation,** farming, manufacturing), the second column should list tools used to work (**mattock,** ax, plow, **pick and shovel, wheelbarrow**), and the third column should list the words that describe or define the work (**padrone,** sore arms and backs, suffocating, "dagoes," unskilled, hard, **humiliating**). When they have finished the charts, use the questions below to lead a discussion on the meaning of the selection.

- What is the "tone" of the words used by the author to describe work in the selection?
- Why do you think labor and work is such an important subject in a selection about immigrants?
- Based on the vocabulary, do you think the main character in the selection had a difficult or easy transition into U.S. society?

The words printed in boldface type are tested on the selection assessment worksheet.

PREREADING

Reading the Author in Cultural Context

Help students read for purpose by discussing the view, held by many immigrants, that the United States was a land of opportunity. Ask: "Was this view realistic?" Point out to students that for many immigrants, even the hardships they faced in becoming established in a new land were considered a small price to pay when compared with the benefits of life in America. Discuss what these benefits might have been.

Focusing on the Selection

Tell students to consider the thoughts of the main character as they read—are his thoughts different from those of anyone seeking employment? Ask them to consider how the main character's experiences are recognizable in all people, even those who have always lived in the United States.

POSTREADING

The following activities parallel the features with the same titles in the Student Edition.

Responses to Critical Thinking Questions

Possible responses are:

1. For the first time in his life, Panunzio experiences discrimination and the harsh struggle newcomers often face when seeking employment.
2. It is likely his European background shaped his attitudes toward work and culture.
3. Guide students to compare Panunzio's experiences with what they know of their own family's experiences.

☑ Guidelines for Writing Your Response to "In the American Storm"

Have students share their journal entries with a partner. Or, as an alternative writing activity, ask each student to consider the power of an employer in a market full of eager workers. Discuss the discrimination Panunzio encountered while looking for work and what groups discriminated against him. Next, have students work in pairs. Have them collaborate on advice they would have given Panunzio if they had met him on his first day in the United States. Call on each pair to read their writing aloud, and discuss the different ideas.

Guidelines for Going Back Into the Text: Author's Craft

Suggest to students that they may want to write about an experience that affected their lives and attitudes about people. Ask: "What tone would you use to tell your story?" Discuss student responses and the different ways each has interpreted the meaning of "tone."

Possible answers to the questions in the Student Edition are:

1. Sample answers: curious, excited, humorous, and sad.
2. The tone is revealed through the thoughts of the main characters and their descriptions of events.
3. The tone effectively conveys the experiences of the main character. The tone is appropriate because it accurately portrays the feelings and thoughts of Panunzio.

☑ FOLLOW-UP DISCUSSION

Use the questions that follow to continue your discussion of "In the American Storm." Possible answers are given in parentheses.

Recalling

1. Why did the factory superintendent make Louis and Panunzio go? (There were more Russian workers who would be displeased if the Italians stayed.)

Interpreting

2. What bonded Louis and Panunzio to each other? (They were both sailors and had traveled to similar places.)

Applying

3. What does this selection tell you about relations between diverse immigrant groups in the early 1900s? (Guide students to analyze how competition for jobs created animosity between some immigrant groups.)

ENRICHMENT

Students may benefit from viewing *The Immigrant: Journey in America*, available from New York State Education Department, Center for Learning Technologies, Media Distribution Network, Room C–7, Concourse Level, Albany, NY, 12230, (518) 474–1265.

RESPONSES TO REVIEWING THE THEME

1. In "Free to Go," the authors challenge concepts about equality and freedom in the United States, by exposing the injustices faced by Japanese Americans during World War II. In "The Circuit," the author offers a glimpse into the harsh life of seasonal farm workers in California, and how a combination of economics and discrimination led to a life of inequality.

2. Though Martin Luther King, Jr., and Malcolm X were both African Americans and were both concerned with civil rights, their methods are different. In the selections, King espouses nonviolent, passive resistance. Malcolm X asserts that nonviolence is impractical against violent enemies.

3. Student answers should include how the selection applies to their own life. For example, some students might understand why authors such as Malcolm X and Rodolfo Gonzales feel anger toward injustices in U.S. society.

☑ FOCUSING ON GENRE SKILLS

Rodolfo Gonzales uses several elements of poetry effectively in "I Am Joaquín." The poem is written in free verse and he uses repetition and parallelism to give structure to his heartfelt song of strength, confusion, anger, and misery. Select, or have the students select, another free verse poem (such as *Leaves of Grass*, by Walt Whitman, "A Poem for Black Hearts," by Amiri Baraka [LeRoi Jones], "Love Calls Us to the Things of This World," by Richard Wilbur, or "Full Consciousness," by Juan Ramon Jimenez [translated by Robert Bly]) and ask students to read the poem and identify the use of repetition or parallelism. Have students evaluate the poet's use of these elements.

BIBLIOGRAPHY

Books Related to the Theme:

Daniels, Roger. *Coming to America: A History of Immigration and Ethnicity in American Life*. New York: HarperCollins, 1990.

Mindel, Charles H., Robert W. Habenstein, and Roosevelt Wright, Jr. (eds.) *Ethnic Families in America: Patterns and Variations*. New York: Elsevier, 1988.

Recognizing Differences

THEME PREVIEW

Recognizing Differences

Selections	Genre/Author's Craft	Literary Skills	Cultural Values & Focus
Zora Neale Hurston, *Their Eyes Were Watching God*, pages 145–147.	story	dialect	understanding one's cultural identity
Lorna Dee Cervantes, "Refugee Ship," Francisco X. Alarcón, "Letter to America," Alurista, "address," pages 148–150.	poems	mood	dealing with cultural discrimination and prejudice
Richard Wright, *Black Boy*, pages 151–153.	nonfiction/ autobiography	first person narrative	learning to accept and appreciate one's culture
Amy Tan, from "Four Directions" from *The Joy Luck Club*, Diana Chang "Saying Yes," pages 151–156.	novel/poem	plot	recognizing cultural differences in your world
Harry Mark Petrakis, *Stelmark: A Family Recollection*, pages 157–159.	nonfiction/ autobiography	indirect or stated main idea	coping with cultural diversity
August Wilson, *Fences*, pages 160–162	drama	dialogue	breaking down prejudices

Assessment

Assessment opportunities are indicated by a ✓ next to the activity title. For guidelines see Teacher's Resource Manual, page xxxiii.

CROSS-DISCIPLINE TEACHING OPPORTUNITIES

Social Studies Accepting and rejecting cultural differences are recurring themes in U.S. history. One area where many of these battles have been fought is in education. With the social studies teacher, give students the opportunity to study how schools in the United States have changed from institutions where students were "Americanized" to their present role at the forefront of maintaining and teaching cultural pluralism. Topics for discussion include: bilingual education, multicultural studies, integration of cultural groups, and global studies.

Art Work with the art teacher to show students reproductions of the work of some African American artists who have examined cultural differences in their work. A good starting point might be the work of Archibald Motley, who in the 1920s used grotesque caricatures to satirize the insulting stereotypical images of African Americans. Another artist, William Henry Johnson, turned in the 1930s to a deliberately naive style that took ideas from the evolved folk culture of African Americans that celebrates the uniqueness of this culture. Another artist, Romare Bearden, a master of collage, developed prismatic shapes that reflect African American experiences and extend African American associations to Africa through the use of mask-type imagery.

Music Perhaps more than any other medium, popular music has recognized and embraced a variety of cultures. With the music teacher, encourage students to explore popular music that is based on a variety of influences. You might like to begin with Latin American music, which draws upon elements from diverse regional cultures and has become a major musical influence throughout the world—from Argentina's tango to the Caribbean rumba and calypso, and most recently in the form known as reggae. Students might also be acquainted with, or be interested in, the music of Los Lobos, a musical group that combines Mexican and American folk and rock influences. Discuss with students how music can bridge cultural differences and help foster the acceptance of diverse cultures in a multicultural society.

SUPPLEMENTING THE CORE CURRICULUM

- The typical approach to art in the 20th century has been to view it on its aesthetic merits alone. The critical questions are: "Is it well done?" "Does it resemble another artist's work?" "What's new about it?" The three works on pages A14 and A15 were chosen for other reasons. They illustrate the different values and experiences of their artists. Have students look at the three paintings and read the captions, then ask them to describe the significant differences among them. What about each work is special to its artist and his culture? Then challenge the students by asking them if they find the works appealing or interesting for purely aesthetic reasons as well. (For example, ask them if Joe Ben, Jr., might have done a better painting had he chosen oil paint and canvas rather than sand.)

- Ask students to research adolescent attitudes toward accepting cultural differences. Challenge them to determine if attitudes about cultural differences are already formed by middle school and high school. The book *Adolescent Prejudice*, by Charles Glock, Metta Spencer, and Jane Piliavin, or the article "Adolescent Bigotry," by Margie Norman in the November 1976 issue of *Human Behavior*, would be good places for students to begin their research. When students have finished, ask them to study their own community and school to evaluate how cultural differences are recognized and treated.

- Recognizing cultural differences has been both a strong point and a shortcoming in the U.S. military system. Ironically, in the past African American soldiers have gone overseas to fight for "freedom" and "democracy" while living as second-class citizens in their own country. Other ethnic groups have experienced discrimination and inequality in all branches of the U.S. military as well. Have interested students research the experiences of African American soldiers in World War I and World War II. An interesting treatment of this subject can be found in *Unknown Soldiers: Black American Troops In World War One*, by A.E. Barbeau and Florette Henri (Temple University Press, 1974).

- Cultural groups do not always agree on how cultural differences should be recognized. Perhaps nowhere are these differences more evident than in the debate over affirmative action programs. In the 1980s and 1990s, some African Americans began to question whether it is possible to create any "quota based" programs that are not inherently stigmatizing. Have interested students read the debate between Amire Baraka [LeRoi Jones] and Shelby Steele, "A Race Divided," in the February 1991 issue of *Emerge* magazine. Ask them to report to the class on the two men's different viewpoints concerning affirmative action, relations between African Americans and mainstream society, and student racism.

Introducing Theme 8: Recognizing Differences

On the board write:

> Mainstream U. S. Society

Allow students three minutes to write a short paragraph on what they believe constitutes "mainstream U.S. society." When they have finished writing, lead a guided discussion using the questions below. Tell students to draw upon their writing to support their responses.

- Is it difficult to live outside of mainstream society? Explain why or why not.

- Is mainstream society easy or difficult to belong to? Who belongs? Who does not belong?

- Has mainstream society changed in the past decades? If so, explain how. Do you think ideas in mainstream society itself will change during your lifetime? Explain how.

Conclude your discussion by acknowledging to students that "mainstream U.S. society" can be anything they wish it to be, including a society that recognizes and accepts all cultural groups.

☑ Developing Concept Vocabulary

On the board do a webbing exercise based on the term *prejudice*. First tell students to skim the titles of the selections in Theme 8. Ask: "How do you think the theme 'Recognizing Differences' is related to the concept of prejudice?" Guide students to speculate on this connection using the selection titles as reference points. (For example, the titles "Refugee Ship" and *Black Boy* should give students some ideas on the topics of this theme.) Next, ask students to find other words in Theme 8 that they associate with prejudice. Draw connecting lines from the student word choices to the word *prejudice* on the board. Leave the webbing diagram on the board during your discussion of the selections in Theme 8.

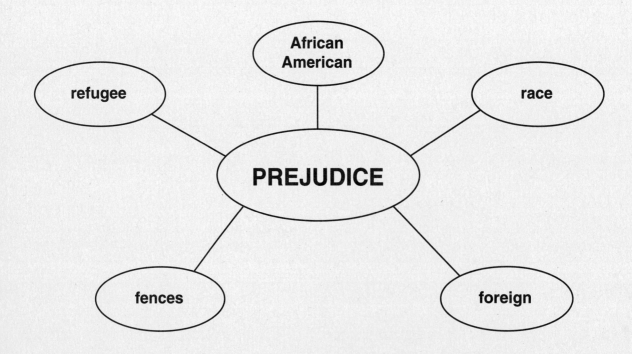

from *Their Eyes Were Watching Watching God*
by Zora Neale Hurston (pages 285–288)

OBJECTIVES

Literary

- to identify how the author portrays a particular region
- to identify the use of dialect to portray character

Historical

- to discuss race relations between African Americans and whites in the early 1900s
- to explain how the sharecropping system replaced a system based on slave labor

Multicultural

- to compare how language and dialects are a component of cultural identities
- to contrast northern and southern U. S. culture in the early 1900s

SELECTION RESOURCES

Use Assessment Worksheet 36 to check students' vocabulary recognition, content comprehension, and appreciation of literary skills.

 Informal Assessment Opportunity

SELECTION OVERVIEW

The abolishment of slavery did not lead to the dismantling of whites' dominance in the Southern United States. In the rural South, enslavement was replaced by the sharecropping system as the primary method of agriculture. This system was reinforced by stereotypes—generated in part by Northern media—that fostered a view in which all African Americans in the South were illiterate, poor sharecroppers. At the time Zora Neale Hurston wrote, many white Americans believed these stereotypes about Southern African Americans. Hurston worked to counteract this perception, and often pointed to seldom acknowledged African American professionals who, like herself, preferred the South to the North, where there was "segregation and discrimination up there too, with none of the human touches of the South."

In this selection, a young African American girl, who has been raised with white children, realizes for the first time that she and her playmates are from different groups. Her realization is triggered by an external source: a photograph.

ENGAGING THE STUDENTS

On the board write:

> "It is a great shock at the age of five or six to find that in a world of Gary Coopers you are the Indian."
>
> James Baldwin

Explain to students that James Baldwin was an African American writer; Gary Cooper was a white actor who often played heroic cowboy roles in western movies. In these movies, Native Americans were usually portrayed as the "villains." Discuss with students what they think the quotation means. Ask why it would be shocking for a young boy to suddenly find himself the villain instead of the hero. Tell them that in this selection, they will read about a similar type of realization that a young African American girl has about herself.

BRIDGING THE ESL/LEP LANGUAGE GAP

The use of dialect in this selection will be difficult for some students. To help them in their comprehension, read the selection aloud once or twice as they follow along.

Present the following questions and ask students to listen for the answers as you read the selection:

1. Where were Janie's mother and father?
2. How would you describe Janie's relationship with the white children?
3. What incident made Janie realize that she was African American?

Define *dialect* as a form or variety of a spoken language special to a region or group of people and discuss with students the use of dialect in the selection. Ask: "Have you heard other dialects (in any language)? Why do you think the author choose to write in dialect? How does it add to the story?"

☑ PRETEACHING SELECTION VOCABULARY

Tell students to skim the dialogue and the footnotes in the selection. Ask: "What is different about the words?" Call on students to find examples of word usage that is unfamiliar to them. (Examples from the dialogue: Ah, **speck, useter, chile, chillun,** mah, dat's, **youngun**) Use the questions below to lead a guided discussion of the author's use of these words.

• How does this language convey tone and feeling?
• How does this language give readers insights into the author's cultural background?

The words printed in boldface type are tested on the selection assessment worksheet.

PREREADING

Reading the Author in Cultural Context

Guide students to analyze how the author has conveyed her cultural background through the use of dialect and tone of language. Help students read for purpose by asking them to consider the importance language and dialect have for the author, as evidenced in the story. Ask students what words and word usages they would use to convey their own cultural background.

Focusing on the Selection

Tell students that this selection has a message for any person who has discovered something about himself or herself that others have always known. As they read, encourage students to compare their own childhood experiences with those of Janie.

POSTREADING

The following activities parallel the features with the same titles in the Student Edition.

Responses to Critical Thinking Questions

Possible responses are:

1. As a young child these differences were insignificant to her.
2. It is likely that growing up in the South during this era gave Hurston many chances to observe and experience first-hand, relations between Southern African Americans and whites.
3. Responses will reflect students' own feelings and experiences, but guide them to consider how their perceptions of race were acquired and at what age.

☑ Guidelines for Writing Your Response to *Their Eyes Were Watching God*

Have students share their journal entries with a partner. Or, as an alternative writing activity, have

students work in pairs, and assign one student the role of Janie and the other student the role of one of the white children. Have each student write a short summary of the story from his or her character's perspective. Challenge students to determine how each character viewed the other.

Guidelines for Going Back Into the Text: Author's Craft

Help students understand how the author used dialogue to create a character that is round and well-developed. Discuss how the reader's perception of the character would be different if Janie did not speak in dialect. Possible answers to the questions in the Student Edition are:

1. The entire selection is dialogue.
2. Dialect distinguishes the characters by providing clues through accents, word usage, and speaking mannerisms.
3. Some examples from *Tapestry* include: *Things Fall Apart*, "In the Land of Small Dragon," and "The Man Who Had No Story."

✔ FOLLOW-UP DISCUSSION

Use the questions that follow to continue your discussion of *Their Eyes Were Watching God*. Possible answers are given in parentheses.

Recalling

1. Who raised Janie? (her grandmother)
2. Who were the Washburns? (the white people who owned the property where Janie lived)
3. Where did Janie grow up? (in west Florida)

Interpreting

4. Why was Janie sometimes called Alphabet? (because she had been given so many different names)
5. How did Janie finally discover she was African American? (She saw herself in a photograph.)
6. Why did Mrs. Washburn punish the children when the picture arrived? (They had allowed a man to take their picture without her permission.)

Applying

7. What does this selection tell you about how cultural stereotypes are acquired? (It suggests that they are acquired later in life and that children do not recognize these stereotypes.)
8. Does this selection suggest that whites and African Americans were treated equally during this era? Explain your answer. (No— note that the African Americans lived in a house behind the white people's house. It does suggest [and this was a common theme for Hurston] that African Americans who followed "correct behavior," according to dominant society's rules, were allowed a degree of equality.)

ENRICHMENT

Students may enjoy viewing *Black Girl*, a story of about a young African American girl's attempt to become a ballet dancer. The video is available from UC Berkeley Extension Media Center, 9 Dwinelle Hall, Berkeley, CA, 94720, (415) 642–2535.

"Refugee Ship"
by Lorna Dee Cervantes
"address"
by Francisco X. Alarcón
"Letter to America"
by Alurista   (pages 289–293)

OBJECTIVES

Literary

- to observe the poets' unique perspectives of cultural differences
- to recognize the mood in the poems

Historical

- to discuss the goals of the Chicano activists
- to explain the economic hardships Mexican Americans have endured in the 20th century

Multicultural

- to explain why themes of isolation are found in the literature of different cultural groups
- to compare and contrast efforts to maintain cultural identity among cultural groups

SELECTION RESOURCES

Use Assessment Worksheet 37 to check students' vocabulary recognition, content comprehension, and appreciation of literary skills.

✔ Informal Assessment Opportunity

SELECTION OVERVIEW

The Chicano political activism that emerged in the 1960s was a result of the prejudice and economic discrimination felt by the Mexican American population for more than a century. By the 1960s, a cycle of undereducation and underemployment had kept Mexican Americans in the lowest economic levels of U.S. society. Hispanics were, and still are, underrepresented in politics. This occurred nationwide, despite the large numbers of Hispanics living in states such as California—a state in which according to the 1990 census, one of every four people is of Hispanic origin. In 1992, only 11 members of the California legislature were Hispanic Americans. Besides striving for political goals, the Chicano movement also spurred the development of the emerging Mexican American arts. In 1965, Luis Valdez founded *El Teatro Campessino*. The theater group toured California's agricultural labor camps with plays and sketches that combined Mexican legends, bilingual dialogues, and political issues.

In the following poems, the poets describe their feelings and thoughts on being Hispanic in America.

ENGAGING THE STUDENTS

Present students with the following hypothetical situation:

Two people are applying for the same job. One applicant, Mr. Smith, is white. The second applicant, Mr. Valdez, is Mexican American.

Ask: "Which of the applicants do you believe has a better chance of becoming employed?" As students respond, use the questions below to lead a guided discussion on hiring practices as related to discrimination.

- Did you assume that the employer was white? male?
- Why are many groups in the United States underemployed? How do you think stereotypes affect employment?

Tell students that several research studies using test cases have shown that white employers are more likely to hire applicants who are white.

BRIDGING THE ESL/LEP LANGUAGE GAP

These poems should not be difficult for most students. Ask students, in groups of three to four (mixing English proficient and LEP students), to read the poems together and then discuss how the poems are similar and how they are different. Ask: "What feelings are the speakers in the poems trying to communicate?"

Next, ask them to discuss the following:

1. What do you think this excerpt from "Refugee Ship" means? Why are these lines important in this poem?

   Mama raised me without language/I'm orphaned from my Spanish name/. . ./I feel I am captive aboard the refugee ship.

2. What do you think these lines from "Letter to America" mean?

   pardon/the delay/in writing you/we were left/ with few/letters

3. What mood is created by the contrast of Spanish and English in "address"? How are the languages used differently in this poem?

☑ PRETEACHING SELECTION VOCABULARY

Tell students that the selections they are about to read include many references to cultural heritage. Tell them to skim the selections and make a list of vocabulary words and phrases that reveal cultural

differences. (Examples: Spanish name, **lag, perdone, foreign, bronzed** skin, black hair, our faces, **race;** Students should also include the Spanish phrases.) When they have finished, tell students to write a short poem or verse using the words they have selected. Call on students to read their verses. Tell them to save their writing and compare their poems to the selections after they have read them.

The words printed in boldface type are tested on the selection assessment worksheet.

PREREADING

Reading the Author in Cultural Context

Help students read for purpose by asking them to consider how injustice is portrayed in the selections. Discuss the Mexican American perspective. Point out that in many regions of the United States, Spanish-speaking people were the first inhabitants of European origin and their culture was the dominant one.

Focusing on the Selection

Ask students to consider their own neighborhood or community. Then have them privately consider the following questions:

- Would you categorize your community as integrated? Are there ethnic neighborhoods in your community?
- Are there clear economic levels in your community? Which groups are at the highest economic levels? Which are at the lowest?
- Do you think that all ethnic groups in your community feel equally "at home"? Support your answer.

Discuss student responses and ask them to consider these issues as they read about the Mexican American community in the selections.

POSTREADING

The following activities parallel the features with the same titles in the Student Edition.

Responses to Critical Thinking Questions

Possible responses are:

1. All three authors are proud of their cultural identities. They are committed to preserving it, even in a society that does not always value cultural diversity.

2. It is likely that the authors have experienced or have knowledge of discrimination and are aware of the economic deprivation many members of their community must face.

3. Guide students to compare their experiences with the ideas expressed in the selections.

✔ Guidelines for Writing Your Response to "Refugee Ship," "address," and "Letter to America"

Have students share their journal entries with a partner. Or, as an alternative writing activity, ask each student to write a short summary of the image he or she found most memorable in the selections. Tell them to include in their summaries the criteria they used to select the image. Ask: "Did the image invoke a feeling or a memory in you, or was it a new image for you?" Next, have students exchange their writings with a partner. Ask them to collaborate on a list of reasons why these poems have impact on the reader.

Guidelines for Going Back into the Text: Author's Craft

Discuss with students how the first lines of each poem set the mood. Ask: "How does the phrase 'Like wet cornstarch, I slide' set a mood immediately?" Call on students to find other phrases that capture the selections' moods. Possible answers to the questions in the Student Edition are:

1. anger, sadness, isolation, pride
2. All three poets create mood with word choice and diction. In "address," for example, the poet chooses to use direct, succinct, cold words to describe a job interview and to convey sadness and frustration.
3. Answers will reflect students' own literature choices.

✔ FOLLOW-UP DISCUSSION

Use the questions that follow to continue your discussion of "Refugee Ship," "address," and "Letter to America." Possible answers are given in parentheses.

Recalling

1. What language is difficult to speak for the main character in "Refugee Ship?" (Spanish)
2. To what does the main character in "Letter to America" compare himself? (furniture)
3. What procedure is being described in "address"? (a job application or interview)

Interpreting

4. What do you think Cervantes hopes to convey by her reference to a "refugee ship"? (her feelings of isolation and of being outside the mainstream of American society)
5. What do you think Alarcón means by the phrase "our faces reflect your future"? (It is possible that the author's intent is to suggest that today's recent immigrants are tomorrow's societal mainstream.)
6. Why does the main character in "address" respond in Spanish? (He or she does not understand the questions.)

Applying

7. Consider the exchange in "address." Why might these be difficult questions to answer for some job-seekers in the United States? (Students should note that language barriers and migrant status would make these difficult questions to answer for some.)
8. According to these poets, is the mainstream society willing to recognize and accept differences? Use examples from the poems to support your answer. (Guide students to analyze the poets' implications that in many cases the mainstream society does not recognize cultural differences. For example, the employer in "address" is oblivious to the culture of the interviewee.)

ENRICHMENT

Students may benefit from viewing a film about Mexican American miners who struggled for labor reforms for almost 50 years. *The American Experience: Los Mineros* is available from PBS Video, 1320 Braddock Place, Alexandria, VA, 22314–1698, (703) 739–5380.

from *Black Boy*
by Richard Wright (pages 294–299)

OBJECTIVES

Literary

- to observe the author's attitudes toward discrimination
- to understand the elements of an autobiography

Historical

- to summarize the effects of segregation
- to discuss civil rights movements before the 1950s

Multicultural

- to explain the impact of segregation on cultural groups in America
- to discuss the diversity of national origins in United States culture

SELECTION RESOURCES

Use Assessment Worksheet 38 to check students' vocabulary recognition, content comprehension, and appreciation of literary skills.

☑ Informal Assessment Opportunity

SELECTION OVERVIEW

When Richard Wright's first books were published between 1938 and 1945, the United States was still a nation that segregated its people on the basis of race. The progressive reforms that had swept the United States in the early 1900s had made little difference to the lives of African Americans. This does not mean that African Americans were silent bystanders. During this era, a number of activists such as W.E.B. Du Bois, Mary McLeod Bethune, and A. Philip Randolph led a frustrating battle against discrimination and racial prejudice. Since 1910, the NAACP had waged and won some significant court battles. Nonetheless, parks, buses, theaters, trains, restaurants, and schools were still legally segregated in the South.

In this selection, the author describes his memory of a childhood experience that left him knowing for the first time, what it meant to be an African American in white society.

ENGAGING THE STUDENTS

On the board write this question:

> Are Americans still segregated on the basis of race?

Acknowledge to students that while segregation is now *illegal*, some sociologists believe U.S. society is still segregated. Challenge students to examine their own schools and community to determine if desegregation has actually been achieved. Point out that busing is still mandated in some cities to achieve a "racial balance" in schools.

BRIDGING THE ESL/LEP LANGUAGE GAP

After students have read the story, use the two questions below to generate a vocabulary list.

- What are some words that give clues to the attitude of Wright's mother in their conversation?
- What words would you use to describe the beginning of Wright's life in Elaine?

Discuss the following excerpt:

> I did not object to being called colored, but I knew there was something my mother was holding back. She was not concealing facts, but feelings, attitudes, convictions which she did not want me to know.

Ask: "What was his mother concealing from him? Why? Do you think she should conceal this from him? Explain your answer."

☑ PRETEACHING SELECTION VOCABULARY

List on the board the following words from the selection:

irritated	angrily
countered	**peevishly**
evaded	**grudgingly**
taunting	mockingly

Explain that these words are all used to describe Wright's mother's attitude when she is questioned about issues of race. Ask students what these words reveal—both about Wright's mother and the topic discussed. Encourage them to add to the list other appropriate words they find as they read the selection.

The words printed in boldface type are tested on the selection assessment worksheet.

PREREADING

Reading the Author in Cultural Context

Help students read for purpose by asking them to consider the events they are about to read from the perspective of an eight-year-old child. Ask them to consider the impact of discrimination on a young child and how these experiences could influence him or her for the rest of his or her life.

Focusing on the Selection

Tell students to write a remembrance of when they became aware of their own cultural identity. Ask them to include any memories they have of growing up and becoming more aware of who they were in relation to their culture. Ask them to compare their own experiences with those of the author as they read.

POSTREADING

The following activities parallel the features with the same titles in the Student Edition.

Responses to Critical Thinking Questions

Possible responses are:

1. As a child, Wright was unaware of racial differences.
2. It is likely that as Wright matured he became more aware not only of his race, but of other people's attitudes toward it.
3. Guide students to compare how they perceive themselves with how they think other people react to their cultural identities.

☑ Guidelines for Writing Your Response to *Black Boy*

Have students share their journal entries with a partner. Or, as an alternative writing activity, have students work in pairs to collaborate on a list of differences between the attitudes of the young Wright in the selection and his mother. Before the students begin this activity, point out that the selection contains elements of both innocence and hardened realism. These elements are especially apparent in the differences in attitudes between Wright and his mother. Ask them to consider whether or not Wright's own perceptions of himself will change over the years.

Guidelines for Going Back Into the Text: Author's Craft

Ask students: "Could you write an *objective*

autobiography of yourself?" Discuss the pros and cons of reading autobiographies in order to learn about events and people.

Possible answers to the questions in the Student Edition are:

1. dialogue and personal remembrances
2. She comes from Irish, Scottish, and French heritages.
3. Sample answer: The mother's perspective might include her reasons for concealing her cultural background and reluctance to explain the "hard facts" of race to her son.

✔ FOLLOW-UP DISCUSSION

Use the questions that follow to continue your discussion of *Black Boy*. Possible answers are given in parentheses.

Recalling

1. To where were Wright and his mother traveling? (Arkansas)
2. What realization did Wright have concerning traveling accommodations on the train? (African Americans and whites were separated.)
3. What cultural background did Wright's father have? (African American, Native American, and European American)

Interpreting

4. Why do you think Wright's mother was irritated with his curiosity about race? (She did not have a simple answer for the young boy.)

5. What do you think Wright meant by this statement about Aunt Maggie's house: "I had no suspicion that I was to live here for but a short time and that the manner of my leaving would be my first baptism"? (Wright probably was forced to leave because of discrimination against African Americans.)
6. What symbolism is involved in the description of Wright and the bee sting? (Perhaps Wright is suggesting that the young boy's curiosity about race will one day "sting" him as well.)

Applying

7. What stereotypes do you think young African Americans must contend with today? Give some examples to support your answer. (Guide students to determine whether racial stereotypes continue to exist in the United States.)
8. In your opinion, can a history of discrimination cause a person to become disillusioned with his or her society? Use examples from the selection to support your answer. (Guide students to compare the young boy's innocence to his mother's hardened realism.)

ENRICHMENT

Students may benefit from viewing either of the two programs entitled *Black Americans*. The videos are available from Dallas County Community College District, Center for Telecommunication, 4343 North Highway 67, Mesquite, TX, 75150–2095, (214) 324–7988.

from **Four Directions**
from *The Joy Luck Club*
by Amy Tan
Saying Yes
by Diana Chang (pages 300–312)

OBJECTIVES

Literary
- to observe the author's use of conflicts in the characters' cultural identities
- to identify the elements of plot in the selections

Historical
- to evaluate the impact of World War II on the status of Chinese Americans
- to explain how immigration quotas and restrictions changed after World War II

Multicultural
- to discuss society's views on intergroup marriages
- to explain cross cultural views on marriage and courtship

SELECTION RESOURCES

Use Assessment Worksheet 39 to check students' vocabulary recognition, content comprehension, and appreciation of literary skills.

 Informal Assessment Opportuntiy

SELECTION OVERVIEW

In 1943, the United States Congress repealed the Chinese Exclusion Act. The decision was made, in part, because of a growing acknowledgment that the Chinese had been mistreated in the past. President Franklin D. Roosevelt stated that the repeal of the act was to "correct a historic mistake and silence the distorted Japanese propaganda." (Japan, during World War II, used allegations that the United States had discriminated against the Chinese for propaganda purposes.) In 1965, Congress abolished the national origins quota system and declared that immigration admission would be based only on skills or family relationships. This resulted in an increase of Chinese immigration, with many people coming from Hong Kong.

In these selections, the authors describe their experiences dealing with their ethnicity and cultural heritage.

ENGAGING THE STUDENTS

Allow students to write for five minutes on this topic:

> Why do some families react negatively when a family member marries someone from a different cultural background?

When they have finished writing, lead a guided discussion on this topic. Ask students who are comfortable with responding to share some of the thoughts they have written. Explain that in one of the selections, they will read about how one person's family dealt with her approaching marriage to a man from a different culture.

BRIDGING THE ESL/LEP LANGUAGE GAP

To prepare students to read the first half of "Four Directions," ask them to identify those actions by Richard that offended the author's parents. Ask: "What do these actions say about the values of the author's family? About Richard's values?"

As students prepare to read the second half of the story, ask them to think about what caused the conflict between the author and her mother. Ask: "How much of the conflict results from cultural differences? How much is due to a generation gap? How does the mother try to resolve the conflict?"

Have students read "Saying Yes" out loud and discuss as a class the conflict with which the speaker is struggling. Ask the students whether or not any questions posed in the selection have been difficult for them to answer for themselves. Suggest they write a poem in their journals about these kinds of conflicts.

☑ PRETEACHING SELECTION VOCABULARY

Write the following three lists of words on the chalkboard.

A	B	C
tense	strong	**slack**
despairing	tricky	frail
on edge	win	**guileless**
clenched	**ferocity**	innocent
amazed	authoritative	powerless
		defeated
		vulnerability

Explain that these words are used in the selection to develop characters through descriptions of their words and actions. Have students speculate on the type of person who would be described by each set of words.

After students have read the selection, discuss how a single person (the mother) could be described in terms of the words from *both* lists B and C.

Words printed in boldface type are tested on the selection assessment worksheet.

PREREADING

Reading the Author in Cultural Context

Help students to read for purpose by asking them to consider how parents and children might view the same cultural heritage differently. Before they read, ask: "Do you think cultural heritage is more important for first-generation immigrant parents than for their children being raised in the United States? Why or why not? Discuss how different views of cultural identity could create conflict between parents and children.

Focusing on the Selection

Ask students to consider some different cultural values that might exist in their own families. Point out that teenagers usually develop their own youth-oriented culture. Ask: "Why do teenagers often try to form identities separate from that of their parents? How can this cause conflict?" As they read, have students consider how differences in cultural values might make this generational conflict even more difficult.

POSTREADING

The following activities parallel the features with the same titles in the Student Edition.

Responses to Critical Thinking Questions

Possible responses are:

1. Both writers acknowledge the existence of conflicts caused by identifying with two cultures—Chinese and "American." Tan, however, seems more willing to embrace her American cultural identity.

2. It is likely that both writers have come to identify with American culture (with varying degrees of reluctance) and both have dealt with the conflict brought on by the merging of two cultures.

3. Guide students to compare cultural conflicts they are personally acquainted with to what they have read.

☑ Guidelines for Writing Your Response to "Four Directions" and "Saying Yes"

Have students share their journal entries with a partner. Or, as an alternative writing activity, have students work in pairs to describe the age group they believe this story would be most appropriate for. Point out to students that these selections contain both humorous and serious overtones. Discuss how the humor in "Four Directions" is bittersweet—Rich's remarks during the dinner scene provide humor, but they also distress the narrator. Call on each pair to explain and discuss their choice.

Guidelines for Going Back Into the Text: Author's Craft

Discuss with students the unique parameters of half hour television dramas that must resolve plots in about 22 minutes (allowing eight minutes for commercials). These television programs must in a short time create a conflict, develop the conflict, have a climax, and then provide resolution. Discuss with students whether or not television is well suited to the kind of in-depth character study found in "Four Directions."

Possible answers to the questions in the Student Edition are:

1. The speaker's conflict is over whether to be labeled "Chinese" or "American." The conflict is resolved by accepting both labels.

2. Her external conflict is with her mother and her mother's conceptions about who she should be and whom she should marry. Internally, the narrator has a conflict over her own Chinese cultural identity, her desire to please her mother, and her feelings for Rich.

3. The climax occurs when the narrator decides to confront her mother on the morning after the dinner with Rich. The resolution occurs later, when she tells her mother about the impending marriage.

☑ FOLLOW-UP DISCUSSION

Use the questions that follow to continue your discussion of "Four Directions" and "Saying Yes." Possible answers are given in parentheses.

Recalling

1. How does the narrator arrange for her mother to cook a meal for Rich? (by dropping in at her aunt's house for dinner)

2. According to the narrator, what do freckles symbolize? (good luck)

3. Who is Shoshana? (Rich's daughter)

Interpreting

4. Why is the narrator reluctant to have her mother meet Rich? (She was afraid her mother would spoil the relationship.)

5. How does Rich embarrass the narrator at the dinner party? (He fumbles with the chopsticks, drinks too much, drowns the food in soy sauce, and does not accept second helpings of food.)

6. How did the narrator's mother change the narrator's impression of Rich after the dinner? (The narrator begins to view Rich as "pathetic" and blind to her Chinese cultural background.)

Applying

7. Why do you think some parents might have reservations about their children marrying people from different cultures? (Guide students to understand why the narrator's mother was anxious about Rich.)

8. Do you think it is possible for anyone to describe himself or herself as just one type of person? Explain. (Guide students to understand the complexity of every person's identity.)

ENRICHMENT

Students may benefit from viewing *China: Land of My Father* which chronicles one filmmaker's search for her father's family roots. The video is available from New Day Films, 853 Broadway, Suite 1210, New York, New York, 10003, (212) 477–4304.

from *Stelmark: A Family Recollection*
by Harry Mark Petrakis (pages 313–320)

OBJECTIVES

Literary

- to identify main ideas in a selection
- to observe the ways an author conveys main ideas

Historical

- to describe the development of ethnic neighborhoods in American urban centers
- to explain why the traditional "melting pot" metaphor is no longer applicable to the cultural diversity in U.S. society

Multicultural

- to compare the development of ethnic neighborhoods among diverse cultural groups
- to explain why food is an important aspect of many cultures' heritage

SELECTION RESOURCES

Use Assessment Worksheet 40 to check students' vocabulary recognition, content comprehension, and appreciation of literary skills.

 Informal Assessment Opportunity

SELECTION OVERVIEW

Ethnic neighborhoods exist in almost every large U.S. city. Many of these neighborhoods have maintained their own languages (in addition to English), distinctive customs, ethnic food stores, and places of worship. For example, in an area of New York City known as the Lower East Side, a Puerto Rican neighborhood, an Italian neighborhood, and a Chinese neighborhood coexist within walking distance of one another. For many years, the dominant society believed that if ethnic cultures maintained their identities, U.S. culture would be fragmented. Now many people assert that cultural pluralism (allowing these cultures to maintain their ethnic identities) in fact reduces social and racial tension.

In this selection, the author describes his experiences as a youth confronting an older member of his ethnic group and what he learned from this confrontation.

ENGAGING THE STUDENTS

Ask students:

Is fast food destroying or creating cultural diversity in the United States?

Before they respond, call on a student to read aloud the first paragraph of the selection. Next, explain that rapid changes are taking place in peoples' food preferences. Foods once considered exotic in the United States are becoming commonplace because of new food processing techniques and more adventurous eating habits. Ask students if mass preparation changes the special qualities of certain cuisines, such as Mexican or Chinese. Tell them that in this selection they will read how food can link people to their cultural heritage.

BRIDGING THE ESL/LEP LANGUAGE GAP

Have students work in groups of three to four students to find the answers to these questions:

1. The author says that as a boy, he and his friends tried to be "really American." What does this say about how they felt about their culture and traditions?
2. What do you think the storekeeper was trying to teach the Greek boy?
3. Why did the storekeeper give the boy figs? Why do you think the author remembered those figs for so many years?
4. How does the author's attitude toward his culture and traditions change by the end of the selection? Why?

✔ PRETEACHING SELECTION VOCABULARY

Explain to students that in the selection they are about to read, the author gives the reader a sense of place by describing foods. Ask: "How is food related to our cultural identity?" Point out that what we eat is in many ways related to customs and traditions handed down from our parents. Tell students to skim the selection and the footnotes for names of foods—both familiar and unfamiliar. Challenge groups of four to six students to write the longest list. Be sure they find such words as **avgolemono, kokoretsi, feta, and lentils.** After students have engaged in this activity for five minutes, have them share their lists. Explain that they will learn from this selection that food names are not really important. Rather, the descriptions, associations, and traditions of foods are what matter to a culture. Ask students to write two food descriptions that evoke images of their own cultural heritage.

Words printed in boldface type are tested on the selection assessment worksheet.

PREREADING

Reading the Author in Cultural Context

Help students read for purpose by asking them to consider what ethnic neighborhoods exist in their community and whether these neighborhoods contain ethnic restaurants or food shops. Tell them to compare their own neighborhoods with the one they are about to read about in the selection.

Focusing on the Selection

On the board write:

> What is the most American food?
> Hamburgers? Tacos? Pizzas?

Tell students to write a short paragraph on one of their favorite meals. When they have finished writing, discuss how food is a part of our cultural identity. Tell students to compare their ideas on food with those of the author as they read.

POSTREADING

The following activities parallel the features with the same titles in the Student Edition.

Responses to Critical Thinking Questions

Possible responses are:

1. The author's attitude changes from embarrassment and withdrawal to pride and an eagerness to embrace his culture.
2. It is likely that Petrakis experienced first-hand the peer pressure a teenager feels to "fit in" and only later came to appreciate his cultural background.
3. Guide students to explore ways that people develop feelings and acquire knowledge about cultural identity.

✔ Guidelines for Writing Your Response to *Stelmark: A Family Recollection*

Have students share their journal entries with a partner. Or, as an alternative writing activity, ask each student to write in a few sentences what he or she believes Petrakis learned about his cultural identity. Next, have students form groups of three.

Tell group members to exchange their writing and to collaborate on a single response to this question: "In general, do teenagers embrace or distance themselves from their cultural heritage?" Call on student groups to read their responses. Discuss the different viewpoints.

Guidelines for Going Back Into the Text: Author's Craft

Discuss with students what a story would be like without a main idea, or with a main idea that was not supported and developed. Ask them to consider some of their favorite stories and to explain what the main ideas were. Challenge students to define the best ways to support a main idea. Ask: "Are facts and figures always the best way to support a main idea? How can an emotional appeal be equally effective?"

Possible answers to the questions in the Student Edition are:

1. As a youth, Petrakis ignored his Greek background so that he could fit in with his peers.

2. Petrakis explained the significance of his encounter with the storekeeper by suggesting that it changed his viewpoint on having a Greek heritage.

3. Sample answers: (second paragraph, page 314) The main idea is Petrakis' making fun of the Greek grocer and how his actions made him feel. (first paragraph, page 319) The main idea is Barba Nikos' connection to food and how food is a part of his heritage.

☑ FOLLOW-UP DISCUSSION

Use the questions that follow to continue your discussion of *Stelmark: A Family Recollection.* Possible answers are given in parentheses.

Recalling

1. What does Barba Nikos do when Petrakis threw the plum at him? (He swore in "ornamental Greek.")

2. How does Barba Nikos have Petrakis make up for throwing the fruit? (by working in the store)

3. What foods does Barba Nikos introduce to Petrakis? (olives, feta cheese, olive oil)

Interpreting

4. Why does Petrakis throw a plum at Barba Nikos? (He was being "tested" by his teenage peers.)

5. Why do you think Barba Nikos is surprised to discover that Petrakis is Greek? (He cannot understand why a Greek boy would throw something at a Greek grocer.)

6. What does Barba Nikos mean by this statement: "You don't understand that a whole nation and a people are in this store"? (Nikos means that his foods represent a way of life and the culture of Greek people.)

Applying

7. Do you think maintaining ethnic identity makes U.S. society more divisive or more cohesive? Use the selection to support your argument. (Guide students to understand how cultural pluralism is not divisive, but instead allows cultures to coexist and gives diversity to American society.)

8. What do you think Petrakis learns from the grocer? Why is this lesson important to him? (Petrakis learned not to reject or be embarrassed by his cultural identity.)

ENRICHMENT

For more on Greek culture, students may benefit from viewing *The Greeks.* The video is available from Films for the Humanities, 743 Alexander Road, Princeton, NJ, 08540, (800) 257–5126.

from *Fences*

by August Wilson (pages 321–331)

OBJECTIVES

Literary

- to understand conflicts portrayed in a selection
- to appreciate a playwright's use of dialogue to develop character

Historical

- to discuss the entry of African Americans into professional sports
- to explain the social and economic impact of U.S. civil rights legislation since the 1960s

Multicultural

- to compare the economic progress made by different cultural groups in the last half of the 20th century
- to discuss sports role models in different cultures

SELECTION RESOURCES

Use Assessment Worksheet 41 to check students' vocabulary recognition, content comprehension, and appreciation of literary skills.

☑ Informal Assessment Opportunity

SELECTION OVERVIEW

In 1947, Jackie Robinson joined the Brooklyn Dodgers and became the first African American player in modern major league baseball. This event opened the door to many other African American baseball players, who entered the major leagues after Robinson. Before then, African Americans played in the Negro Leagues, which were completely segregated. The Negro Leagues rarely got the publicity that the major leagues enjoyed, but they had many outstanding players, such as Satchel Paige, a pitcher, and Cool Papa Bell, an outfielder. In general, sports in the United States have represented a niche in the labor market where racial barriers have been difficult to maintain. Nevertheless, it was not until the 1980s that an African American was a quarterback on a professional football team or a manager of a major league baseball team.

In this selection, a young man argues with his father over whether or not he should play college football.

ENGAGING THE STUDENTS

On the board write:

National League   American League   Negro League

Ask students which of the leagues they recognize and which they do not. Ask: "Why do you think there once was a Negro League in the United States?" Explain to students that at one time African Americans were banned from playing in the major leagues.

BRIDGING THE ESL/LEP LANGUAGE GAP

Although the vocabulary in this selection is quite simple, students might be confused by some colloquialisms, such as: ain't ya, naw, whatnot, gonna, yessir, wanna, crackers. Review these words and their meanings with the students before they read the selection. Discuss how dialect is responsible for most of these unusual spellings.

☑ PRETEACHING SELECTION VOCABULARY

Sports—in particular, baseball—is an important topic in this selection. Have students skim the selection and make a list of all sports-related words. (Examples: **home runs, bench, timing, follow-through,** major league, **pitching,** strike-outs, recruited) Discuss how sports terminology has become important in our daily conversations. Challenge students to use each term on their lists in a non-sports-related way. Ask them to consider why sports and sports lingo are so important in U.S. society.

The words printed in boldface type are tested on the selection assessment worksheet.

PREREADING

Modeling Active Reading

Fences is annotated with the comments of an active reader. These sidenotes, prepared to promote critical reading, emphasize the cultural content of the piece, address author values, call attention to literary skills, invite personal response, and show how the selection is related to the theme. If you have time, read the entire selection aloud as a dialogue between reader and text. Encourage students to discuss and add their own responses to the ones printed in the margins. Model these skills by adding your own observations, too.

Reading the Author in Cultural Context

Help students to read for purpose by asking them this question:

Compared to your parents, are you more or less optimistic about the future?

Challenge students to determine which characters in the story are more optimistic about the future and which characters believe that cultural barriers are still a dominant factor in our society.

Focusing on the Selection

Tell students to weigh differences between their generation and the generation of their parents. Discuss why parents often have different perspectives than their children on topics such as work and leisure activities. Tell them to consider these differences as they read the selection.

POSTREADING

The following activities parallel the features with the same titles in the Student Edition.

Responses to Critical Thinking Questions

Possible responses are:

1. Guide students to compare the different perspectives of Troy and Cory. Troy seems to believe that racism was the reason he did not play baseball. Cory, on the other hand, seems more willing to accept other reasons, like Troy's age.
2. It is possible that Wilson experienced the same generational conflicts over his own aspirations to be a writer.
3. Sample responses: love, pride, success

☑ Guidelines for Writing Your Response to *Fences*

Have students share their journal entries with a partner. Or, as an alternative writing activity, have students work in groups of three to analyze characters' perspectives. Assign each student one of the characters in the story: Cory, Troy, or Rose. Have each student answer this question from his or her character's perspective: "Is Cory being unrealistic in his hopes to play college football?" When they have finished writing, have group members compare their writings and decide which perspective is the most accurate.

Guidelines for Going Back Into the Text: Author's Craft

Discuss with students whether or not good dialogue is becoming a "lost art." Have students write down the titles of the last three movies or plays they have seen and to rate the importance and quality of dialogue in each one. Suggest that the recent emphasis on action and special effects in film has lessened the importance of dialogue. Discuss with students their own opinions and have them support their ideas with their lists.

Possible answers to the questions in the Student Edition are:

1. Troy's feelings about cultural barriers, his ideas on work, and his life goals.
2. This dialogue reveals a deeper, more emotional side of Troy, including his love for Rose.

3. Guide students to examine whether or not teen-agers' dialogues accurately portray their feelings.

☑ FOLLOW-UP DISCUSSION

Use the questions that follow to continue your discussion of *Fences*. Possible answers are given in parentheses.

Recalling

1. Who is coming to visit Cory and his father? (a college football recruiter)
2. Where does Cory work? (the A&P)
3. How does Cory address his father? (yessir)

Interpreting

4. According to Troy, why should Cory forget about football? (Troy believes that the "white man" would not let Cory get anywhere in football.)
5. In your opinion, does Troy love his son? (Answers will vary, but students should note how Troy defines his relationship with Cory.)

Applying

6. Do you think Troy's bitterness is justified or not? Support your answer. (Students should weigh past patterns of discrimination and cultural barriers with Troy's beliefs about his son's future.)

ENRICHMENT

Students may enjoy viewing *The Black Athlete*, available from Pyramid Film and Video, Box 1048, Santa Monica, CA, 90405, (800) 421-2304.

RESPONSES TO REVIEWING THE THEME

1. Sample response: *Black Boy* and "Four Directions" both present interaction between parents and children. The mothers in both selections represent for their children sources of information about their cultures. The son in Black Boy is just beginning to become aware of the problems African Americans in general often face. He wants his mother's guidance in this area. The daughter in "Four Directions," although an adult herself, is just becoming aware of the personal issues of her Chinese American Heritage. She, too, looks to her mother for guidance. Although the childrens' experiences will be different from their mothers' because of changing times, the children are interested in gaining their mothers' perspectives.

2. Sample response: in her poem "Refugee Ship," Lorna Dee Cervantes focuses on the isolation and hardships encountered when Mexican Americans try to blend into the mainstream U. S. culture. She seems dismayed over her lack of identifiable heritage, and her tone is bleak. In "Saying Yes," Diana Chang focuses on her desire to be both Chinese and American. Her tone is hopeful and satisfied. The two poets express the same theme: a desire to fit into both their native and the mainstream United States cultures. However, the feelings evoked by the poems are quite different.

3. Encourage students to consider how enriching and educational the experience of reading selections from many different cultures has been. Being exposed to many different cultures in society can result in the same rewarding experiences.

☑ FOCUSING ON GENRE SKILLS

In his play *Fences,* August Wilson uses dialogue effectively to convey the personalities, values, and beliefs of his characters. The reader learns about the thoughts, beliefs, and desires of Troy and Cory through their words. Select, or have the students select, another drama (such as *Another Part of the Forest,* by Lillian Hellmann or *Dutchman* by Amiri Baraka [Leroi Jones]) and ask the students to read passages to illustrate the use of dialogue to convey personalities, values, or beliefs. Then ask them to evaluate the playwright's use of dialogue.

BIBLIOGRAPHY

Books Related to the Theme:

Simpson, George Eaton and Milton Yinger, *Racial and Cultural Minorities; An Analysis of Prejudice and Discrimination,* 5th Ed. New York: Plenum Press.

Weisbrot, Robert. *Father Divine and the Struggle for Racial Equality.* Urbana: University of Illinois Press, 1983.

Breaking Down Barriers and Building Communities

THEME PREVIEW

Breaking Down Barriers and Building Communities			
Selections	**Genre/Author's Craft**	**Literary Skills**	**Cultural Values & Focus**
Ishmael Reed, "America, The Multinational Society," pages 166–168.	essay	cause and effect	recognizing the cultural diversity in the United States
Margaret Walker, "For My People," Langston Hughes, "Let America Be America Again," pages 169–171.	poems	cataloging	maintaining cultural unity
MAKING CONNECTIONS: African American Traditions Linking Literature, Culture, and Theme, p. 172			
Henry Roth, "Petey and Yotsee and Mario," pages 173–175.	short story	point of view	learning to accept one's cultural heritage
Tato Laviera, "AmeRícan," Aurora Levins Morales and Rosario Morales, "Ending Poem," Heberto Padilla, "Instructions for joining a new society," pages 176–178.	poems	alliteration	breaking down prejudices
Hyemeyohsts Storm, *Seven Arrows*, pages 179–181.	tale	allegory	discovering one's culture and place in the universe

Assessment

Assessment opportunities are indicated by a ✓ next to the activity title. For guidelines see Teacher's Resource Manual, pg. xxxiii.

CROSS-DISCIPLINE TEACHING OPPORTUNITIES

Social Studies Work with the social studies teacher to examine how the workplace is changing in the United States and the implications these changes have for breaking down cultural barriers. Allow students to examine how the increased number of women working outside the home have changed societal perceptions of women's roles. Ask the social studies teacher to provide some historical perspective on how the Civil Rights Movement affected accessibility to education and higher paying positions for all cultural groups. Also, ask the social studies teacher to provide an overview of the legislative, social, and economic barriers that have traditionally kept some cultural groups out of high level work positions and how these barriers are changing.

Geography Remind students that geography is more than the study of place, it is also the study of how humans interact with their environment. Ask: "Is it possible that the growth of cities is a cause of cultural friction?" Point out that the large urban areas in the United States are the centers of cultural diversity but also the flash points of tensions. With the geography teacher, help students to list the characteristics of urban environments.

Music In the late 1980s and 1990s, a musical form known as rap became not only a musical phenomenon, but a cultural phenomenon as well. Contrary to more conservative perceptions, rap appeals to teenagers of all cultures. Some rap musicians have been criticized for conveying a message that is culturally divisive. Lead a discussion on this topic and on other forms of popular music with cultural messages. Ask students to justify opinions on the positive or negative effects popular music can have on cultural barriers. A good discussion of this topic can be found in the June 29, 1992, issue of *Newsweek* in the article "Rap and Race."

SUPPLEMENTING THE CORE CURRICULUM

The Student Edition provides humanities materials related to specific cultures in Making Connections on page 348–349 and in the art insert on pages A1–A16. You may wish to invite students to pursue the following activities outside of class.

- Since the 1980s, the creation of large quilts has become a popular means of expressing protest and political aspirations. These quilts go well beyond the traditional purposes of quilts as blankets or coverlets. Some of them are vast, covering acres, with patches contributed by many different people. Ask students to consider this recent trend. How does the quilt on page A16 express the theme, Breaking Down Barriers and Building Communities? Guide them to understand that big themes may require big art. Then discuss with students how other concepts and themes would neccessitate large artistic treatments. Also, have them suggest other artistic media that might express the themes of breaking down barriers and building communities. Have interested students bring illustrations of such works to class to show to the other students. (One traditonal medium for these grand themes is the mural, which many students will have seen in large public buildings.)

- In 1992, Los Angeles, California, was the setting for several days of rioting that for many people was evidence that cultural barriers continue to divide U.S. society. Have interested students research the causes and aftermath of the riots, including the public's perceptions of the Rodney King trial and verdict. National news magazines are a valuable source of information on the riots . The May 11, 1992, special issue of *Newsweek* gives a good overview.

- Challenge interested students to determine what causes barriers to form between groups of people. One theory is based on cultural elitism—the belief that one culture is better than another. An interesting perspective on this subject can be found in *The Third Wave*, by Ron Jones. Jones is a high school history teacher who conducted a class experiment in group behavior while studying the rise of Nazism. By using many of the same subtle tactics Hitler used, Ron Jones—to his surprise—was able to instill feelings of cultural superiority among his students. Before long, the students quickly formed a group that adhered to a policy of cultural elitism. Have interested students read *The Third Wave*. Ask them to report to the class on what they have read, including ideas on how social groups are formed and how cultural prejudice can be avoided. Guide students to understand that elitism is often an effect of some other cause, such as economic exploitation or religious prejudice.

- Many sociologists and historians believe that economics and equal opportunity in the workplace are the key to achieving intergroup harmony in U.S. society. Have interested students explore the economic system in the United States to find out how it might create barriers between cultures. Two excellent sources on this subject are Stephen Birmingham's *Certain People* (Little, Brown, 1977) and Michael Harrington's *The New American* (Holt Rinehart, and Winston, 1984).

INTRODUCING THEME 9: BREAKING DOWN BARRIERS AND BUILDING COMMUNITIES

Introduce the final theme by asking students to reflect for a moment on all that they have read and learned so far about relations among the diverse ethnic groups in the United States. Allow students a few minutes to write their ideas about how barriers between cultural groups are formed in the United States and whether these barriers are primarily the result of economic, social, or religious factors. While they write, put this question on the board:

> Why should you be concerned with breaking down barriers and building communities?

When they have finished writing, call on students to answer the question on the board. Tell them to support their answers with information from their writing.

Conclude your discussion by reading a British journalist's comment on the 200th birthday of the United States.

> For all its terrible faults, in one sense America still is the "last, best hope of mankind," because it spells out so vividly the kind of happiness which most people actually want . . . . We criticize, copy, patronize, idolize, insult, but we never doubt that the United States has a unique position in the history of human hopes.

Suggest to students that one reason why the United States is admired is its ongoing commitment to breaking down barriers and building communities.

✔ Developing Concept Vocabulary

Do a semantic mapping exercise on the board using the main words and concepts in the title of Theme 9. Write the theme title on the board: *Breaking Down Barriers and Building Communities*. Draw connecting lines from the words *Breaking Down, Barrier, Building,* and *Communities*. Lead a guided discussion on this theme and write the words students use to describe the ideas relating to the theme title. Encourage students to skim the selections in Theme 9 to participate in the class discussion.

Use the questions below to guide your discussion.

- Give an example of a barrier. What kinds of barriers keep people apart?
- Why is it important to break down these barriers? What can result from the breaking down of cultural barriers?
- Describe your community. What kinds of barriers does your community need to overcome?

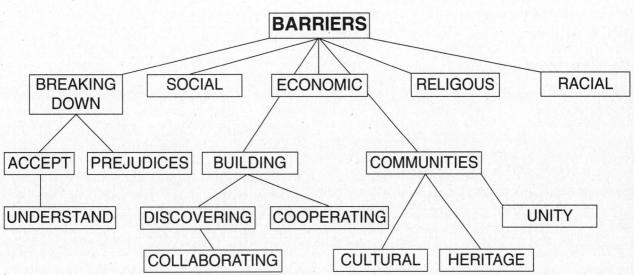

165

America the Multinational Society

by Ishmael Reed (pages 333–339)

OBJECTIVES

Literary

- to observe the author's attitudes concerning a multicultural society
- to recognize the use of cause and effect in an essay

Historical

- to explain how protest movements in the 1960s have affected views of cultural pluralism
- to discuss barriers to societal communication throughout U.S. history

Multicultural

- to compare different cultural views on assimilation and cultural identity
- to understand the traditional emphasis on western civilization in some segments of U.S. society

SELECTION RESOURCES

Use Assessment Worksheet 42 to check students' vocabulary recognition, content comprehension, and appreciation of literary skills.

✔ Informal Assessment Opportunity

SELECTION OVERVIEW

One of the most important aspects of multiculturalism is that it has made people more aware of literature outside the Western European tradition. Students, parents, and educators are challenging the exclusive focus traditionally placed on Western European accomplishments, influence, and history. (The term often used for this emphasis is *Eurocentrism*.) They point to the artistic, intellectual, and philosophical achievements of many cultures—Asian, African, and those of the pre-Columbian Western Hemisphere. There is a concurrent challenge to the value of the cultural "melting pot," the idea that the common good requires downplaying the unique aspects of different cultures. A multicultural approach to literature celebrates these cultural differences, while pointing out commonalities among all cultures.

In this selection, the author discusses the multicultural nature of U.S. society and reveals that it may be more prevalent than even its advocates are aware.

ENGAGING THE STUDENTS

On the board write two questions:

- Was Cleopatra black?
- Does it make any difference what race she was?

Explain to students that the first question is currently being debated among some historians and anthropologists and was recently asked (rhetorically) on the cover of a national news magazine. Discuss why someone would be interested in this question at all. (Point out that some people believe Western historians are unwilling to acknowledge an African heritage for Egyptian civilization.) Tell students that in the selection they are about to read, an African American author will shed some light on this subject as he reveals his feelings about cultural identity.

BRIDGING THE ESL/LEP LANGUAGE GAP

This selection may be difficult for some students because of its essay form and its many references to external events, places, people, and ideas. Assist students by explaining some of the main ideas before they read. Discuss what the term *multicultural* means and the difference between this concept and the melting pot concept. Define and discuss the following terms: *Puritan, the Elect* [educational and cultural], *Western civilization.* Then, ask students to read the opening paragraphs and discuss what makes Detroit, Houston, and Milwaukee examples of multiculturalism according to the author.

Next, give the students the following list of statements from the text. As they read, ask them to pay close attention to how these statements are used in the text and to determine whether the author agrees or disagrees with each statement and ask them to decide if he is using each statement only for the purpose of argument or illustration.

- Blurring of cultural styles occurs in everyday life in the United States to a greater extent than anyone can imagine.
- Western civilization was the greatest achievement of mankind.
- The Puritans were a daring lot, but they had a mean streak.
- The invasion of the American educational system by foreign curriculum . . . has already begun, because the world is here.

☑ PRETEACHING SELECTION VOCABULARY

Explain to students that the selection they are about to read makes many references to world cultures. On one side of the board list six of the cultures represented (Islamic, Jewish, African, Russian, American, Caribbean). On the other side of the board list words from the selection that are related to these cultures: **calypso, mosques, pastrami,** Puritans, **Yoruban,** Tsars. Ask students to match each word with the appropriate culture. Discuss the idea that American culture could actually be described by all of these words, as Ishmael Reed proposes.

The words printed in boldface type are tested on the selection assessment worksheet.

PREREADING

Reading the Author in Cultural Context

Help students read for purpose by asking them to consider their own education and what, if any, cultural bias they might have encountered. Tell them to use their own experiences to help them gauge what Reed is saying in the selection.

Focusing on the Selection

As they read, guide students to compare the author's viewpoint on U.S. culture with their own ideas. Ask them to consider what changes Reed would like to see in U.S. society and its educational system. Challenge them to determine if most Americans would agree with the author's interpretation or if they are aware of the biases the author brings up.

POSTREADING

The following activities parallel the features with the same titles in the Student Edition.

Responses to Critical Thinking Questions

Possible responses are:

1. Reed believes that a multicultural society is more interesting and equitable than one that is monocultural. The author points out that the United States is already a multicultural society—the problem stems from some citizens' unwillingness to acknowledge this.
2. It is likely that the author's own experiences as an African American have influenced his ideas on multiculturalism.
3. Guide students to evaluate Reed's ideas in relation to their own perspectives on U.S. culture.

☑ Guidelines for Writing Your Response to "America: A Multinational Society"

Have students share their journal entries with a partner. Or, as an alternative writing activity, have each student write a question he or she would ask Ishmael Reed if he or she could meet him. In groups of three, students can exchange their questions and discuss some possible responses Reed might make.

Guidelines for Going Back Into the Text: Author's Craft

Discuss why understanding cause-and-effect relationships can be helpful in all areas of study. Ask: "Why would historians want to know the causes and effects of warfare?" (to prevent future wars) "Why do scientists want to know the causes and effects of diseases?" (to find cures) "Why do writers want to know the causes and effects of prejudice and racism?" (Guide students to find their own answers to this question.)

Possible answers to the questions in the Student Edition are:

1. By providing examples of how cultures have blended in U.S. society, Reed demonstrates the effects and existence of multiculturalism.
2. These groups were persecuted and killed because of the belief that predominantly European Western civilization is superior to other civilizations.
3. Sample answer: Puritans' lack of understanding of other cultures caused them to kill Native Americans.

☑ FOLLOW-UP DISCUSSION

Use the questions that follow to continue your discussion of "America: A Multinational Society." Possible answers are given in parentheses.

Recalling

1. Who is Robert Thompson? (He is a Yale professor who refers to the United States as a cultural bouillabaisse.)
2. According to the article, where in U.S. society are Puritans idealized? (in school books)

3. What does Reed suggest about the origins of government in the United States? (He suggests that the U.S. system of government was influenced by the Iroquois system of government.)

Interpreting

4. Explain the final sentence in the selection: "The world is here." (The sentence reflects the author's belief that a multitude of cultures already lives in the United States; multiculturalism is not coming, it has always been here, but ignored.)
5. What is the significance of African painters having their art work shown in a McDonald's restaurant? (It demonstrates the blurring and intermingling of cultural styles in the United States. One could also argue, however, that since the paintings are not in a museum, there is a lack of appreciation for the African American painter.)
6. Why does Reed find it ironic that bilingual education is criticized in some Western states? (Spanish was spoken in much of this region before English.)

Applying

7. How do you think Reed's ideas apply to education? (Guide students to understand the connection between acknowledging the multiculturalism of U.S. society and the present cultural perspectives of textbooks and education.)
8. How does cultural bias lead to social tensions and conflicts? Give examples to support your answer. (Guide students to analyze how cultural bias has created an ethnic elite in U.S. society, as described by Reed, and the possible effects this might have on people who do not belong to the elite.)

ENRICHMENT

Students may benefit from viewing any of the twelve programs in the *Somebody Else's Place* series. The programs focus on exchange visits between pairs of young people from different backgrounds. The videos are available from Great Plains Instructional TV Library, University of Nebraska, P.O. Box 80669, Lincoln, NE, 68501 (402) 472–2007.

For My People
by Margaret Walker
Let America Be America Again
by Langston Hughes (pages 340–347)

OBJECTIVES

Literary
- to observe the poets' values and beliefs
- to appreciate the use of cataloging in free verse poetry

Historical
- to understand the position of minorities in the 1940s and 1950s, the period when the selections were written
- to examine the changes that have taken place since then

Multicultural
- to compare how different cultural groups have adapted to or rejected mainstream U.S. culture
- to examine how individuals of various cultural groups have balanced their commitment to their own particular heritage and to the broader United States
- to understand what has been done in recent years to help various cultural groups overcome the disadvantages from which they have suffered

SELECTION RESOURCES

Use Assessment Worksheet 43 to check students' vocabulary recognition, content comprehension, and appreciation of literary skills.

✔ Informal Assessment Opportunity

SELECTION OVERVIEW

As the United States emerged from World War II, satisfaction at having conquered the scourge of fascism could not hide the fact that for African Americans social and economic equality were still, in Langston Hughes's words, "a dream—still beckoning." Segregation and discrimination were still the rule in much of the United States. Hughes and Margaret Walker articulated the attitudes that

led to the Civil Rights Movement of the 1960s and 1970s. As a result of this movement many of the legal barriers that kept African Americans and others from participating equally in the political and social life of the nation were removed. Yet, as soon became clear, deeper economic and social barriers remained to be uprooted. At the same time that the African American middle class grew, with many more in the professions, in political office, and in white collar jobs, large numbers were still confined to crowded cities, attending inadequate schools, with little hope for adequate employment. Diverse African American voices were heard pressing for change. They ranged from Robert Woodson's appeal urging African Americans to shape their own destiny to the Reverend Joseph Lowery, who asked Americans to develop a national will to make change a priority.

In these selections, Walker and Hughes make a plea that is just as valid today as it was when they wrote, almost half a century ago.

ENGAGING THE STUDENTS

Tell students that the great African American leader W.E.B. Du Bois once (in 1903) wrote how difficult it was to be both "a Negro *and* an American." Ask them to discuss why he made this statement and to what extent they believe it to be true. Point out that two African American poets, Margaret Walker and Langston Hughes, about half a century ago also thought about this theme and elaborated on it in the two poems the students will now read.

BRIDGING THE ESL/LEP LANGUAGE GAP

Point out to the students that both poets use cataloging in their poems. Make sure students understand what cataloging as a poetic device means. Have each work with a partner to write a description of each list they find in Walker's "For My People." The description may establish a category or identity a theme that Walker is trying to communicate in that particular list.

Next, have students look for and describe the lists in Langston Hughes's "Let America Be America Again." Ask: "How does Hughes's cataloguing technique differ from Walker's? Does this create a different mood? What is the difference?"

✔ PRETEACHING SELECTION VOCABULARY

Tell students that the poems they are about to read have as a main theme the hope for a better future for all Americans. However, in order to call attention to the need for change, the poets must paint a true picture of the reality of their day. Therefore, many of the images in the poems are harsh, with descriptions full of bleak words. List some of these words on the board: disinherited, **dispossessed, blundering**, groping, **floundering, deceived, tangled,** worried. Ask students to look for words describing a happier alternative as they read. Have them list such words and compare them to the list on the board after they read the poems.

The words printed in boldface type are tested on the selection assessment worksheet.

PREREADING
Reading the Authors in Cultural Context

Help students read for purpose by asking them to consider how far the United States has come and how far it has to go to achieve racial equality. As they read, students should look for the authors' perspectives and ask themselves whether the conditions referred to in the poems still exist at the present time.

Focusing on the Selection

Tell the students to consider how the poems emphasize unity. Ask: How do good writers address cultural and other barriers?" Challenge students to find the themes in these poems that speak to all people.

POSTREADING

The following activities parallel the features with the same titles in the Student Edition.

Response to Critical Thinking Questions

Possible responses are:

1. Both poets tell the reader that equality has been difficult for African Americans to achieve and that many groups in the United States continue to face discrimination and economic hardships.

2. Both poets question the U.S. promise of freedom

and equality for its people. Hughes's tone could be interpreted as being more optimistic than Walker's. Hughes seems to suggest that the United States can still be a place where freedom and equality overbalance prejudice.

3. Help students evaluate the authors' "solutions." Both authors call for a rebirth or, in Hughes's poem, an allegiance to the Amrican ideals of equality and freedom for all people.

☑ Guidelines for Writing Your Response to "For My People" and "Let America Be America Again"

Have students share their journal entries with a partner. Or, as an alternative writing activity, have students choose one of the poems and write what they think the author of the poem would have to say about the Los Angeles riots of 1992 or another similarly disturbing incident. When they have finished writing, have students exchange or discuss their writing with a partner. Call on student pairs to summarize their analysis and discuss their interpretations.

Guidelines for Going Back into the Text: Author's Craft

Ask: "How do Walker and Hughes make their messages universal in scope?" Suggest that both poets have cataloged the breadth and scope of human feeling in their poems. Possible answers to the questions in the Student Edition are:

1. This stanza conveys the message that African Americans have worked hard for centuries but have not reaped the benefits of their work.
2. The catalogs include poor whites, Native Americans, and workers from all fields.
3. Guide students to discuss other examples of cataloging. Point out that this technique is often used in popular music.

☑ FOLLOW-UP DISCUSSION

Use the questions that follow to continue your discussion of "For my people" and "Let America Be America Again." Possible answers are given in parentheses.

Recalling

1. Where are the author's playmates from in "For My People?" (Alabama)
2. How does Langston Hughes characterize U.S. economic conditions? (He describes them as brutal, greedy, and unjust.)
3. What groups does Hughes mention in his poem? (Native Amricans, immigrants, poor people, African Americans)

Interpreting

4. Whom do you think Walker is referring to when she uses the term "My People?" (Guide students to determine whether Walker's message is exclusively for African Americans.)
5. What cultural groups do you think Hughes's poem would appeal to the most? Explain your answer. (Guide students to understand how the poem has appeal for many cultural and economic groups who have endured deprivation.)
6. How do you interpret the title of Hughes's poem. "Let America Be America Again?" (He is suggesting that we have not lived up to the ideals on which our country was founded.)

Applying

7. How are the economic themes in these poems applicable to conditions today? (Guide students to compare the authors' viewpoints on wages, working conditions, and economic mobility to today's economy.)
8. How different do you think the two poems would be if the authors wrote them today? (Guide students to see that while many of the specifics of the poems would be different, the overall themes would probably be the same.)

ENRICHMENT

Students may be interested in reading Langston Hughes's *The Best of Simple* (Hill & Wang, 1961), accounts of an ordinary African American, and comparing it to "Let America Be America Again." Another book they can profit from is W.E.B. Du Bois's *The Souls of Black Folk*, mentioned earlier.

MAKING CONNECTIONS AFRICAN AMERICAN TRADITIONS

OBJECTIVES

Overall

- to discover the similarities between African American art and music and the themes in Margaret Walker's "For My People" and Langston Hughes's "Let America be America Again"

Specific

- to compare and contrast the themes of dignity and pride in the selection "For My People" and the sculptures of Richmond Barthe
- to appreciate how Langston Hughes in "Let America Be America Again" and other African American artists embrace cultural diversity in their work
- To trace the roots of jazz music

ENGAGING THE STUDENTS

On the board write the following question:

> What is a legacy? What legacies have your cultural groups left for your generation?

Guide students to define legacy as something that is received from an ancestor or predecessor. Discuss their responses to the questions on the board. Next, have the students study the photographs on pages 348 and 349 to name some legacies African Americans have left for future generations. Finally, ask the students to name the legacies their generation may leave for future generations.

BRIDGING THE LEP/ESL LANGUAGE GAP

On the board write the following list: music, sculpture, dance and poetry. Have students work in pairs comprised of language proficient and LEP students. Have them discuss how each item in the list is in some way a form of communication. Ask them also to think of ways that the items communicate and pass on traditional attitudes and values. Finally, have the pairs collaborate on a few sentences that answer this question.: "If music, sculpture, dance and poetry can be forms of communication, can language be a barrier?"

EXPLORING ART

The influence of African Americans and the African culture has been far-reaching. For example, African American music influenced the development of rock and roll. In a similar manner, African masks influenced the innovative Cubist artworks of Picasso and Braque. Have students work in small groups to read the captions and examine the illustrations on pages 348 and 349. Ask: "How have the African traditions of music and dance influenced each art form or artists represented her? What things in your life have been influenced by music or dance, and more specifically by African American culture?"

Obtain recordings of blues and jazz artists from the 1920s and 1930s, such as Louis Armstrong, Bessie Smith. Robert Johnson, and Memphis Minnie, for students to enjoy and discuss. You might also obtain early recordings of Elvis Presley and the Beatles for students to hear the progression from blues to jazz to rock and roll. Also of interest are recordings of Langston Hughes reading his poetry. Invite interested students to research African American visual artists, such as Horace Pippin, Henry Ossowa Tanner, Faith Ringgold, Romare Bearden, Jacob Lawrence, and Barbara Chase-Riboud.

LINKING LITERATURE, HISTORY, AND THEME

Guidelines for evaluation

Walker mentions slave songs, dirges, ditties, blues, and jubilees. She believes that these songs unified and gave hope to the African American community.

Hughes's poem contains rhythm, rhyming words, and a call and response format. These elements give the poem a lilting, song-like feel that almost turns the selection into an anthem or spiritual.

Answers will vary. Sample answers: Music and art contain elements that transcend cultural barriers. Rhythm, paintings, sculptures, and melody appeal to all cultures and do not require a common cultural background to appreciate them. Jazz music best achieves the goal of breaking down barriers becuase music is the purest and most universal form of communication.

Petey and Yotsee and Mario
by Henry Roth (pages 350–355)

OBJECTIVES

Literary

- to observe how the author describes feelings about his heritage
- to understand point of view in a short story

Historical

- to explain the reasons why many Jews left Europe and relocated in the United States
- to locate and describe Jewish settlements in the United States in the early 1900s

Multicultural

- to describe how teenagers in diverse cultural groups reinforce or reject cultural stereotypes
- to compare how different cultural groups have fared economically in the United States

SELECTION RESOURCES

Use Assessment Worksheet 44 to check students' vocabulary recognition, content comprehension, and appreciation of literary skills.

✔ Informal Assessment Opportunity

SELECTION OVERVIEW

Persecution is an underlying current in the history of the Jewish people. At various times during the Middle Ages, Jews were expelled from England, France, and Spain. During the 18th century, Jews were tolerated in Europe, though prejudice against them was still widespread. In the late 1800s, a series of pogroms erupted in Russia and eastern European countries. Police and soldiers stood by as Jewish businesses and homes were destroyed. Jews who interfered were killed. Many Jewish people believed these attacks were condoned by the governments. As a result, between 1900 and 1914, 1.5 million Jews emigrated from Russia alone. Many of these people came to the United States. Jewish neighborhoods became cultural enclaves with their own associations, stores, theaters, restaurants, and cafes. The heart of the community was the synagogue.

In this selection, the author describes how, as a young boy, he was ambivalent toward his cultural identity. The episode he writes about helped to change his perception.

ENGAGING THE STUDENTS

On the board write:

Anti-Semitism

Ask students the following questions:

- What cultural group was expelled from England in the year 1290? From France in 1306? From Spain in 1492?

Tell students that the answer to all the questions is "the Jewish people." Have students consider this history of persecution as they read about the experiences of a young Jewish boy trying to "fit in" to U.S. society.

BRIDGING THE ESL/LEP LANGUAGE GAP

The vocabulary in the beginning of this selection may be difficult for some students. Help them by reading the opening paragraphs (through Fat's rescue) on page 00, and stopping to let students summarize what they understood of the passage.

Then ask students to read the remainder of the selection to themselves and to consider the following questions:

1. Why does Fat's mother decide to bake a cake for Petey, Yotsee, and Mario?

2. Why does Fat say, "Aw, Mom, they don't understand cakes like that"? What does this tell you about his attitude toward his culture?

3. Later in the story, when Fat's mother gives the boys the cake, Fat notices that they did in fact enjoy the cake. How do you think this realization made him feel?

4. What do you think Fat was afraid of?

✔ PRETEACHING SELECTION VOCABULARY

On the board write **Jewish cake.** Ask students to speculate on how a "Jewish cake" might be different from any other cake. Next on the board write **Gentile** cake. Explain to students that to Jews *Gentile* means "non-Jewish." Ask students to speculate on what the difference might be between a Jewish cake and a Gentile cake. Tell them to skim the selection for any clues to the meaning of these terms. (Examples: holiday, **spicecake, embossed** with walnuts, **crystallized** honey, raisins, Ward's, Tip-Top, Golden Queen)

The words printed in boldface type are tested on the selection assessment worksheet.

PREREADING

Reading the Author in Cultural Context

Help students to read for purpose by asking them to consider how they would feel if they were forced to move into a new neighborhood where they were looked upon as an "outsider." Tell them to examine how peer pressure for the main character led him to be ashamed of his cultural identity.

Focusing on the Selection

Ask each student to write a short paragraph that summarizes his or her own cultural identity. When they have finished, call on them to read their writing, and ask: "Would you be comfortable in any American neighborhood? Why or why not?" Discuss the student responses and tell them to think of the positive aspects of the neighborhood in Roth's story.

POSTREADING

The following activities parallel the features with the same titles in the Student Edition.

Responses to Critical Thinking Questions

Possible responses are:

1. As a teenager, Roth was ambivalent about his cultural heritage. The story suggests, however, that he learned to be proud of it.

2. It is likely that Roth had many experiences as a Jewish teenager living in a "Gentile" neighborhood that influenced his writing.

3. Guide students to examine their own attitudes toward people with different religious or cultural backgrounds.

✔ Guidelines for Writing Your Response to "Petey and Yotsee and Mario"

Have students share their journal entries with a partner. Or, as an alternative writing activity, have each student write in a few sentences what he or she believes Fat learned about his cultural identity. Next, have students exchange their writing with a partner. Ask them to collaborate on a single answer to this question: "Is it more or less difficult for teenagers to accept cultural differences than it is for adults?" Call on student pairs to read their answers and discuss the different viewpoints.

Guidelines for Going Back Into the Text: Author's Craft

Discuss with students the types of plot that would be better suited to a short story than to longer forms of fiction. Ask them why the plot of "Petey and Yotsee and Mario" is well-suited to short story form. (It has a simple plot and setting and only a single aspect of the character's personality is explored.) Possible answers to the questions in the Student Edition are:

1. first-person
2. It allows the reader to fully understand the thoughts and motivations of the main character.
3. Sample answer: It would not give as much insight into Fat's ambivalence about his cultural identity.

✔ FOLLOW-UP DISCUSSION

Use the questions that follow to continue your discussion of "Petey and Yotsee and Mario." Possible answers are given in parentheses.

Recalling

1. What river did Fat and his friends swim in? (the Harlem River)
2. What explanation did Mario give for saving Fat's life? (Fat was from "the block.")
3. How was the cake decorated? (with walnuts, crystallized honey, and raisins)

Interpreting

4. Do you think that Fat had a good understanding of his friends? Support your answer with details. (Students should note that Fat seemed to be more concerned with cultural differences than his friends, who seemed quite ready to accept him.)
5. Why didn't Fat want his mother to bake a cake for his friends? (He was embarrassed that it was a Jewish cake and not like the cakes boys on his block were used to.)
6. Explain Fat's mother's final statement in the story. (She is suggesting that anyone who would help her drowning son would certainly be "human" enough to like a cake that was different from those they were used to.)

Applying

7. Why are neighborhoods important for newly arrived immigrant groups? (They provide a feeling of home and security.)
8. Do you think Americans today are more or less tolerant of cultural differences? Use examples from the selection to support your answer. (Guide students to weigh the recent conflicts between cultural groups in the United States against the progress that has been made.)

ENRICHMENT

Students may benefit from viewing a video that deals with Jewish immigration. *West of Hester Street* is available from Ergo Media, Inc., P.O. Box 2037, Teaneck, NJ, 07666, (201) 692–0404.

AmeRícan
by Tato Laviera
Ending Poem
by Aurora Levins Morales and Rosario Morales
Instructions for joining a new society
by Heberto Padilla (pages 356–362)

OBJECTIVES

Literary

- to observe the poets' perspectives on cultural barriers
- to identify alliteration in the poems

Historical

- to describe the urban migration patterns of Puerto Rican immigrants
- to describe life in Hispanic neighborhoods in the United States

Multicultural

- to trace the diverse cultural roots of Hispanic Americans
- to explain how national origins remain an influence for cultural groups in the United States

SELECTION RESOURCES

Use Assessment Worksheet 45 to check students' vocabulary recognition, content comprehension, and appreciation of literary skills.

☑ Informal Assessment Opportunity

SELECTION OVERVIEW

In 1917, the Jones Act gave U.S. citizenship to Puerto Ricans. Most Puerto Ricans who came to this country before 1940 were recruited to do farm labor on a seasonal basis and then returned to Puerto Rico by their employers. A small percentage of these farm workers chose to stay on the mainland. By 1930, about 45,000 Puerto Ricans lived in East Harlem. During the 1940s Puerto Rican immigrants in New York were recruited by the garment industry after workers of Jewish and Italian backgrounds moved on to other neighborhoods and higher paying jobs. The 1950s saw a sharp rise in Puerto Rican migration to the U.S. mainland, as thousands of islanders moved to New York City and other large mainland cities in search of jobs. Many live in largely Puerto Rican neighborhoods and have retained a strong sense of cultural identity. The Spanish language is an important component of this identity, as it is for other Hispanic Amercan groups in the United States.

In these selections, the poets describe their experiences and thoughts on being Hispanics in U.S. society.

ENGAGING THE STUDENTS

Read the following remark by a member of the California state legislature:

If Hispanics have any hope of improving their lives and earning a decent living, they must learn to speak English—and speak it well.

Tell students that this remark was made during a debate on whether a law should be passed making English the official language of California. Then ask the students if all U.S. citizens should speak English. Ask them to justify their answers.

BRIDGING THE ESL/LEP LANGUAGE GAP

Have students work in groups of three to four. Ask them to answer to the following questions:

Poem 1
- How are the "AmeRícans" that Laveria talks about a "new generation"?
- Can you find examples of how the poet "plays" with language? What is the effect?

Poem 2
- How are the first two lines of this poem a conclusion?
- Why do you think this poem is called "Ending Poem"?

Poem 3
- What feelings does the poet communicate?
- Why does he have mixed feelings?

Ask each group to discuss what it thinks the term *new generation* means. Ask: "Do you think it could apply to the people described in all three poems? Why or why not?" Have each group select a spokesperson and share the group's ideas with the class.

☑ PRETEACHING SELECTION VOCABULARY

Explain that these poems talk about how immigrants blend into U.S. society. Write these lines from the first selection on the board:

. . . we blend/ and mix all that is good!

Suggest that the vocabulary of the poems shows the blending process. Some words combine two languages (**AmeRícan, spanglish, cascabelling, latinoamerica**), others describe people as coming

from a blended ethnicity (**mestiza**), while others are simply from a non-English language.

Encourage students to look for other references in the poems that indicate the process of cultural blending.

The words printed in boldface type are tested on the selection assessment worksheet.

PREREADING

Reading the Authors in Cultural Context

Help students to read for purpose by asking them to consider how most Americans have a cultural identity that is a blend of different backgrounds and national origins. Ask them to consider the authors' cultural identities as they read.

Focusing on the Selection

Before they read the poems, have students write a short paragraph on this topic: "If you had one opportunity to improve break down communication and cultural barriers among ethnic groups in the United States, where would you focus your actions?" When they have finished writing, call on students to read their plans. Tell them to refer to their writing after they have read the poems.

POSTREADING

The following activities parallel the features with the same titles in the Student Edition.

Responses to Critical Thinking Questions

Possible responses are:

1. All authors express pride in the diversity of their cultural backgrounds and acknowledge that their backgrounds are indeed diverse. All authors express a desire to break down cultural barriers.

2. It is likely that the authors experienced first-hand cultural and racial barriers.

3. Guide students to examine how they feel about accepting cultures other than their own.

☑ Guidelines for Writing Your Response to "AmeRícan," "Ending Poem," and "Instructions for joining a new society"

Have students share their journal entries with a partner. Or, as an alternative writing activity, have each student write some of his or her favorite lines from the poems. Suggest to students that poems in particular can capture direct, focused images in just a few words. Next, have students work in groups of four. Have each group exchange the lines they have chosen and collaborate on the single phrase or line they found to be the most effective. Call on groups to read the line they have chosen and compare the different groups' choices.

Guidelines for Going Back Into the Text: Author's Craft

Tell students that alliteration is a common technique used in lyrics written for popular music. Ask students to find examples of alliteration in their favorite music. Call on them to state their examples and discuss whether or not they fit the definition of alliteration in the student text. Possible answers to the questions in the Student Edition are:

1. Initial: "I am a late leaf" Internal: "mountain born, country-bred, homegrown Jibara child"

2. It links the diverse cultural roots of the author.

3. Example: "sweet soft spanish danzas gypsies" This alliteration creates a dreamy, lyrical effect.

☑ FOLLOW-UP DISCUSSION

Use the questions that follow to continue your discussion of "AmeRícan," "Ending Poem," and "Instructions for joining a new society." Possible answers are given in parentheses.

Recalling

1. What cultural roots are referred to in "AmeRícan"? (European, Native American, African American, Spanish)

2. What American city does Laviera claim is being integrated with a new way of life? (New York City)

3. What "two Americas" are the poets referring to in "Ending Poem"? (South America and North America)

Interpreting

4. What do you think Laviera intended by naming his poem "AmeRícan?" (Guide students to understand the author's intent of using one word to represent an identity rooted in two cultures.)

5. What "poison" do you think Laviera is referring to in the eighth stanza of his poem? (cultural prejudice)

6. Reread the last two stanzas of "Ending Poem." Whom do you think the poets are "speaking" to in the final stanza? (Help students to explore the likelihood that the poets are speaking to Hispanics who have worked for others over a long period of time.)

Applying

7. The selections you have read celebrate cultural diversity. What are some ways that your generation celebrates the diversity of cultures in U.S. society? (Guide students to explore how cultural diversity is celebrated through music, art, literature, and community activities.)

8. What are some cultural barriers the poets mention in the poems? Do these barriers continue to exist in U.S. society today? (Guide students to analyze the cultural barriers caused by prejudice and to determine whether they believe these barriers continue to plague U.S. society.)

ENRICHMENT

Students may benefit from viewing *Latino Profiles: Living in the U.S. Today*. The video is available from Phoenix/BFA Films, 468 Park Avenue South, New York, NY, 10016, (800) 221–1274.

from Seven Arrows
by Hyemeyohsts Storm (pages 363–368)

OBJECTIVES

Literary

- to observe the importance of traditional beliefs to the author
- to understand the use of symbols in the selection

Historical

- to appreciate the Plains nations' system of beliefs
- to discuss the relevance of traditional Native American beliefs to contemporary social problems

Multicultural

- to develop an awareness of a contemporary unheard voice
- to contrast how different cultural groups have worked to preserve their cultural heritage and beliefs

SELECTION RESOURCES

Use Assessment Worksheet 46 to check students' vocabulary recognition, content comprehension, and appreciation of literary skills.

 Informal Assessment Opportunity

SELECTION OVERVIEW

The Cheyenne were forced by European expansion and intertribal warfare to migrate first to present-day Minnesota from an area encompassing the Carolinas north to New York and then to the Plains during the latter half of the 1700s. Around the year 1775, a prophet named Sweet Medicine gave the Cheyenne a sacred bundle of arrows and established a code of sacred laws that he believed had been handed down from the Cheyenne Supreme Being, or the Creator. The Cheyenne believed that if they followed these laws, the Creator would ensure the cultural and national survival of the Cheyenne nation. The sacred arrows themselves are symbolic of unity and communication with the Creator. The teachings of Cheyenne prophets such as Sweet Medicine and the sacred arrow bundle remain an integral part of Cheyenne contemporary life.

In this selection, the main character, White Rabbit, explains the history and significance of some Native American traditional beliefs.

ENGAGING THE STUDENTS

On the board write:

- Is it better to give than to receive?
- Would you rather give away $10,000 or receive $10,000?
- Money is the root of all evil.

Discuss with students the materialistic aspects of U.S. society. Ask: "Whom do we honor most in the United States—people who have achieved financial success or people who dedicate their lives to helping others?" As they respond, challenge students to define their ideas on success, status, happiness, and their commitment to their community. Explain that in the selection they are about to read, they will examine Native American beliefs on giving, sharing, and social responsibility.

BRIDGING THE ESL/LEP LANGUAGE GAP

This selection is difficult because of its allegorical style and use of certain concepts that will be unfamiliar to most students.

First, review with the class the prereading introduction, particularly the passage about the Medicine Wheel. Ask: "What does the wheel represent? Why do the Native Americans use the wheel as a symbol? What does this tell you about their beliefs and values?"

Next, explain that the story is divided into two parts: the Medicine Wheel and the Animal Dance. Have students work in groups to focus on one of the sections and to discuss the following:

- What are some of the symbols used in the story, and what do they represent?
- What lessons do they think the author is trying to teach? Are these important lessons?

☑ PRETEACHING SELECTION VOCABULARY

Do a webbing exercise with students on the references to animals and nature in this selection. On the chalkboard draw a large wheel with several spokes and label it **Medicine Wheel.** First, tell students to skim the selection and write down four to five animals or references to nature from the selection. (Examples: **Medicine of the Eagle,** White Rabbit, Night Bear, **Clan** of the Buffalo, **Animal Dance,** Medicine Animals, Twin Medicine Lakes, **Sun Dance**) Call on students to state the words they have found and write them on the spokes of the Medicine Wheel.

Lead a guided discussion using the questions below:

- What do you already know about the importance of animals in Native American cultures? What are some symbolic meanings of animals in Native American cultures? How do you think the spokes of the Medicine Wheel might be connected?
- What do you think a Medicine Wheel might be? Is medicine only for curing physical illnesses? Is there such a thing as medicine for the spirit as well? How about medicine for social problems?

The words printed in boldface type are tested on the selection assessment worksheet.

PREREADING

Reading the Author in Cultural Context

Help students to read for purpose by asking them to consider all that has happened to Native Americans in the past three centuries, and how their ways of life have been transformed. Point out that despite these changes many Native Americans today continue to practice and maintain ancient cultural beliefs and ceremonies. Tell students to consider whether these beliefs are still applicable in today's world.

Focusing on the Selection

Tell students that some high schools in the United States require a number of hours of community service before graduation. Discuss with students the benefits of such a program. Tell them to consider the Native American perspective on community and social responsibility as they read the selection.

POSTREADING

The following activities parallel the features with the same titles in the Student Edition.

Responses to Critical Thinking Questions

Possible responses are:

1. Storm believes that the traditions and beliefs of his people continue to have meaning and/or application today.
2. He most likely used storytelling because this is how Native Americans have passed on knowledge for centuries.
3. The story suggests that all people, animals, and nature are related to one another.

☑ Guidelines for Writing Your Response to *Seven Arrows*

Have students share their journal entries with a partner. Or, as an alternative writing activity, have students, working in pairs, speak the dialogue between White Rabbit and Red Star on page 00. When they have finished, have each pair collaborate on a summary of the significance of peoples' names according to White Rabbit.

Guidelines for Going Back Into the Text: Author's Craft

Brainstorm with students on symbols that are universally understood. For example, love is often symbolized by hearts, tears symbolize sadness, and so on. Discuss how writers can also use symbolism to convey different levels of meaning. Possible answers to the questions in the Student Edition are:

1. Robes symbolize which way they come from.
2. Sample answer: Animals symbolize our relation to the Earth and represent peoples' nature.
3. The symbols tell us that people have both a spiritual and worldly nature, and that people, like animals, have different tendencies.

☑ FOLLOW-UP DISCUSSION

Use the questions that follow to continue your discussion of "Seven Arrows." Possible answers are given in parentheses.

Recalling

1. According to White Rabbit, where do all people pass through when they are born? (They enter one of the four Eagle Lodges at all Four of the Directions.)
2. Into what clan were White Rabbit and Red Star born? (Little Medicine Bird People)

Interpreting

3. Explain the concept of Give-Away in the Medicine Wheel. (All animals need one another and must Give-Away in order that they might grow.)
4. How might people of the same Animal Medicine have different perceptions? (Their perceptions also depend on where they were born on the Medicine Wheel—East, West, North, or South.)

Applying

5. How does understanding yourself help you to understand others? Use the selection to support your answer. (Each person is born with a special Gift of Perceiving. Understanding ourselves is the first step to understanding how all people are interrelated.)

ENRICHMENT

For more on Native American philosophy, spirituality, and prophecy, students may benefit from viewing *Red Road: Towards the Techno-Tribal*. The video is available from the Native American Public Broadcasting Consortium, Inc. (NAPBC), P.O.Box 83111, Lincoln, NE, 68501, (402) 472–3522.

RESPONSES TO REVIEWING THE THEME

1. Sample response: Although "America: The Multinational Society" is nonfiction and "Petey and Yotsee and Mario" is fiction, both selections point to the ernrichment of all cultures that results when people share their traditons within the common bond of humanity. Each author gives examples of how barriers are broken down when different cultures live side-by-side and share their heritages.
2. Sample response: Tato Laviera and Aurora Levins Morales were both born in Puerto Rico. Their poems are similar because they celebrate the mixing of cultures that resulted from immigration. However, whereas Laviera seems empowered by the experiences of his people, Morales dwells somewhat on the pain of the process. Both poets feel it is up to the victims of discrimination and injustice to stand strong and united in the face of it.
3. Student responses will depend upon their personal ideas and opinions.

FOCUSING ON GENRE SKILLS

Hyemeyohsts Storm weaves information about the Medicine Wheel into a tale, saying that learning is a continuous process of growth. The tale is symbolic; it is an allegory and can be read on a literal level and on other symbolic levels as well. Select, or have the students select, another tale (such as "Jack Beats the Devil" [Abrahams, Roger D. ed. *Afro-American Folktales: Stories from Black Traditions in the New World*. New York: Pantheon, 1985] or "The Man Who Knew the Language of Animals" [Maestas, Jose Griego and Rudolfo A. Anaya. *Cuentos: Tales from the Hispanic Southwest*. Santa Fe: The Museum of New Mexico Press, 1980]). Have students read the tale and identify the literal meaning. Then have them evaluate the tale for symbols and allegorical meanings.

BIBLIOGRAPHY

Book Related to the Theme:

Banks, James A. *Teaching Strategies for Ethnic Studies*. Boston: Allyn and Bacon, Inc., 1979.

COOPERATIVE/ COLLABORATIVE LEARNING

Individual Objective: to participate in a discussion about the theme of Struggle and Recognition.

Group Objective: to develop a five minute presentation inspired by viewpoints expressed in the discussion.

Setting Up the Activity

Have students work in heterogeneous groups of three. Stipulate that each group member is responsible for researching and expressing an individual viewpoint in the group discussion and for making a contribution to the group presentation. The individual topics are as follows: Student 1 expresses the overall theme of Unit 3; Student 2 explains the struggles cultural groups in the United States have had achieving equality; Student 3 gives specific examples from the selections to support the groups' viewpoint.

When they have finished their discussion, have each group elect a representative to summarize the group's discussion for the entire class. After each groups' presentation, use the questions below to discuss and evaluate their reports.

- What are some of the barriers to equality in the United States?
- In your groups' opinion, what are two ways that barriers to accepting cultural differences can be broken down?

Assisting ESL and LEP Students

Distribute or assign the ESL/LEP students among the groups. Help them choose viewpoints that they will be comfortable with. As you monitor each group's progress, be sure that the ESL/LEP students are being encouraged to participate fully.

Assessment

Before you begin the group activity, remind students that they will be graded both on an individual and group basis. Without individual contribution, the group presentation will not be complete. Monitor group progress to check that all three students are contributing and that a presenter has been selected.

Time Out to Reflect As students do the end-of-unit activities in the student edition, provide time to let them make a personal response to the content of the unit as whole. Invite them to respond to the following questions in their notebooks or journals. Encourage students to draw on these personal responses as they complete the activities.

1. What have I learned about equality among cultural groups in the United States? What are the barriers to equality in the United States? What cultural groups in my own community struggle with inequality?

2. What have I learned about cultural differences in the United States? What examples from the selections in Unit 3 had the most significance for me?

3. What have I learned about the barriers to cultural equality? How does cultural prejudice prevent communication between cultures?

4. How can I apply what I have learned to my own life and community? Am I accepting of cultural differences? Is my community a place of equal opportunity for all cultures?

WRITING PROCESS: PERSUASIVE ESSAY

Refer students to the model of a persuasive essay found on pages 400–402 in the Handbook section of the Student Edition. You may want to discuss and analyze the model essay if your are working with less experienced writers.

Guidelines for Evaluation

On a scale of 1–5: the writer has:

- clearly followed all the stages of the writing process.
- made clear and specific references to the chosen selections.
- made clear and specific references to the literary elements and techniques used in the selections.
- provided sufficient supporting details from the selections to prove his or her main points.
- clearly organized the paper so that the conclusion is easy to follow.
- written an opening or a closing paragraph that clearly summarizes the main points of the paper.
- made minor errors in grammar, mechanics, and usage.

Assisting ESL and LEP Students

You may want to provide a more limited assignment for these students so that they can complete their first drafts somewhat quickly. Then have them work at length with proficient writers in a peer-revision group to polish their drafts.

PROBLEM SOLVING

Encourage students to use the following problem-solving strategies to analyze one of the speeches, stories, autobiographies, or poems in Unit 3. Afterwards, have students reflect on their use of the strategies and think of ways they could have used the strategies more effectively.

Strategies

1. Use a semantic map, brainstorming list, or flow chart to identify the conflicts and struggles experienced by cultural groups in Unit 3 and the origins of these conflicts. Have groups of students choose the topic they most want to investigate further and the selection or selections that best shed light on the topic.

2. Use a second graphic organizer to explore in detail the topic they chose. Include information both from the selections and from personal experience. This graphic organizer should be used to show cause and effect relationships between conflicts and their causes.

3. Use the overlapping circles of a Venn diagram to organize information into similarities and differences. For example, how have the struggles of different cultural groups been similar? How have they been different? Then refine the diagram by putting check marks next to the most significant points.

4. Decide which topics need further research or discussion. Set up a plan that shares responsibility for doing the research.

Allow time for students to work on their presentations.

Guidelines for Evaluation

On a scale of 1-5, the writer has:

- provided adequate examples, facts, reasons, anecdotes or personal reflections to support his or her presentation.
- demonstrated appropriate effort.
- clearly organized the paper so that main ideas and supporting details are logical and consistent.
- demonstrated an understanding of challenges faced by many groups and individuals in the United States.

Abridgements

The Sheep of San Cristóbal

page 16, line 1 (beginning)

"No! No!" screamed Felipa. "Not Carlos. Not my Carlos! It cannot be!"

The young woman stood in the doorway of her little two-room house. Before her, on the dirt floor, lay the body of her husband. He had just been killed by Ute Indians in a sneak raid. The dust raised by their horses still hung in the air outside.

In tears, Felipa sank to her knees and covered her face with her hands. Why hadn't she been home? Why had she picked that hour—that minute even—to go for water? Together, she and Carlos might have driven the Utes away. . . . And where was Manuel?

"Manuel!" she cried. Jumping to her feet, she dashed outside to look for her seven-year-old son. Her nearest neighbors were coming on the run, and behind them, the village priest. Surprisingly, no one had seen the Utes until their damage had been done. The neighbors searched the small field next to the house for the boy, but before long everyone guessed the truth: He had been carried off by the Indians. That was the same as death.

The priest held Felipa's sobbing face in his old brown hands. "Come," he said. "Come, Felipa Sandoval. Come to the shrine of Our Lady of Light. And may the Lord have mercy."

Felipa followed the old man to the church. Prayer did not fill the void in her heart, but it did help her to go on living. And she knew what it would mean to go on living: She would have to work the little farm herself. She would have to plant all the squash and hoe all the beans and pick all the corn. There was no other way.

page 16, line 1 (end)

She could not forget the Ute raid.

page 22, line 29

Had Manual been kept alive by the Utes? Not carried off and killed?

The Man Who Had No Story

page 41, line 29

The big man who was in the company stood up and said that the dancing must stop now. "A couple of us must go for the priest, so that we can say Mass," said he, "for this corpse must go out of here before daybreak."

"Oh," said the girl with the curly dark hair, "there is no need to go for any priest tonight, the best priest in Ireland is sitting here beside me on the chair, Brian O Braonacháin from Barr an Ghaoith."

"Oh, I have nothing of a priest's power or holiness," said Brian, "and I do not know anything about a priest's work in any way."

"Come, come," said she, "you will do that just as well as you did the rest."

Before Brian knew he was standing at the altar with two clerks and with the vestments on him.

He started to say Mass and he gave out the prayers after Mass. And the whole congregation that was listening said that they never heard any priest in Ireland giving out prayers better than Brian O Braonacháin.

Bless Me, Ultima

page 46, line 10

Let me begin at the beginning. I do not mean the beginning that was in my dreams and the stories they whispered to me about my birth, and the people of my father and mother, and my three brothers—but the beginning that came with Ultima.

The attic of our home was partitioned into two small rooms. My sisters, Deborah and Theresa, slept in one and I slept in the small cubicle by the door. The wooden steps creaked down into a small hallway that led into the kitchen. From the top of the stairs I had a vantage point into the heart of our home, my mother's kitchen. From there I was to see the terrified face of Chávez when he brought the terrible news of the murder of the sheriff; I was to see the rebellion of my brothers against my father; and many times late at night I was to see Ultima returning from the llano where she gathered the herbs that can be harvested only in the light of the full moon by the careful hands of a curandera.

page 47, line 8

He went to work on the highway and on Saturdays after they collected their pay he drank with his crew at the Longhorn, but....

page 47, line 12

drank and

page 47, line 15

to drink

page 49, line 22

I could not make out the face of the mother who rested from the pains of birth, but I could see the old woman in black who tended the just-arrived, steaming baby. She nimbly tied a knot on the cord that had connected the baby to its mother's blood, then quickly she bent and with her teeth she bit off the loose end. She wrapped the squirming baby and laid it at the mother's side, then she returned to cleaning the bed. All linen was swept aside to be washed, but she carefully wrapped the useless cord and the afterbirth and laid the package at the feet of the Virgin on the small altar. I sensed that these things were yet to be delivered to someone.

page 50, line 5

We must return to our valley, the old man who led the farmers spoke. We must take with us the blood that comes after the birth. We will bury it in our fields to renew their fertility and to assure that the baby will follow our ways. He nodded for the old woman to deliver the package at the altar.

No! the llaneros protested, it will stay here! We will burn it and let the winds of the llano scatter the ashes.

It is blasphemy to scatter a man's blood on unholy ground, the farmers chanted. The new son must fulfill his mother's dream. He must come to El Puerto and rule over the Lunas of the valley. The blood of the Lunas is strong in him.

He is a Márez, the vaqueros shouted. His forefathers were conquistadores, men as restless as the seas they sailed and as free as the land they conquered. He is his father's blood!

page 50, line 8

Cease! she cried, and the men were quiet. I pulled this baby into the light of life, so I will bury the afterbirth and the cord that once linked him to eternity. Only I will know his destiny.

page 51, line 26

"Now run and sweep the room at the end of the hall. Eugene's room—" I heard her voice choke. She breathed a prayer and crossed her forehead. The flour left white stains on her, the four points of the cross. I knew it was because my three brothers were at war that she was sad, and Eugene was the youngest.

"Mamá." I wanted to speak to her. I wanted to know who the old woman was who cut the baby's cord.

page 52, line 17

A priest, I thought, that was her dream. I was to hold mass on Sundays like Father Byrnes did in the church in town. I was to hear the confessions of the silent people of the valley, and I was to administer the Holy Sacrament to them.

"Perhaps," I said.

"Yes," my mother smiled. She held me tenderly. The fragrance of her body was sweet.

"But then," I whispered, "who will hear my confession?"

"What?"

"Nothing," I answered.

page 52, line 18

"I am going to Jasón's house," I said hurriedly and slid past my mother. I ran out the kitchen door, past the animal pens, towards Jasón's house. The white sun and the fresh air cleansed me.

On this side of the river there were only three houses. The slope of the hill rose gradually into the hills of juniper and mesquite and cedar clumps. Jasón's house was farther away from the river than our house. On the path that led to the bridge lived huge, fat Fío and his beautiful wife. Fío and my father worked together on the highway. They were good drinking friends.

"¡Jasón!" I called at the kitchen door. I had run hard and was panting. His mother appeared at the door.

"Jason no está aquí," she said. All of the older people spoke only in Spanish, and I myself understood only Spanish. It was only after one went to school that one learned English.

¿Dónde está?" I asked.

She pointed towards the river, northwest, past the railroad tracks to the dark hills. The river came through those hills and there were old Indian grounds there, holy burial grounds Jasón told me. There in an old cave lived his Indian. At least everybody called him Jasón's Indian. He was the only Indian of the town, and he talked only to Jasón. Jasón's father had forbidden Jasón to talk to the Indian, he had beaten him, he had tried in every way to keep Jasón from the Indian.

But Jasón persisted. Jasón was not a bad boy, he was just Jasón. He was quiet and moody, and sometimes for no reason at all wild, loud sounds came exploding from his throat and lungs. Sometimes I felt like Jasón, like I wanted to shout and cry, but I never did.

I looked at his mother's eyes and I saw they were sad. "Thank you," I said, and returned home.

Roots

page 57, line 14

Once in a long while, all the women would be present—but only if a case held the promise of some juicy gossip.

The Way To Rainy Mountain

page 78, line 9

Yellowstone, it seemed to me, was the top of the world, a region of deep lakes and dark timber, canyons and waterfalls. But, beautiful as it is, one might have the sense of confinement there. The skyline in all directions is close at hand, the high wall of the woods and deep cleavages of shade. There is a perfect freedom in the mountains, but it belongs to the eagle and the elk, the badger and the bear. The Kiowas reckoned their stature by the distance they could see, and they were bent and blind in the wilderness.

Descending eastward, the highland meadows are a stairway to the plain. In July the inland slope of the Rockies is luxuriant with flax and buckwheat, stonecrop and larkspur. The earth unfolds and the limit of the land recedes. Clusters of trees, and animals grazing far in the distance, cause the vision to reach away and wonder to build upon the mind. The sun follows a longer course in the day, and the sky is immense beyond all comparison. The great billowing clouds that sail upon it are shadows that move upon the grain like water, dividing light. Farther down, in the land of the Crows and Blackfeet, the plain is yellow. Sweet clover takes hold of the hills and bends upon itself to cover and seal the soil. There the Kiowas paused on their way; they had come to the place where they must change their lives. The sun is at home on the plains. Precisely there does it have the certain character of a god. When the Kiowas came to the land of the Crows, they could see the dark lees of the hills at dawn across the Bighorn River, the profusion of light on the grain shelves, the oldest deity ranging after the solstices. Not yet would they veer southward to the caldron of the land that lay below; they must wean their blood from the northern winter and hold the mountains a while longer in their view. They bore Tai-me in procession to the east.

page 80, line 3

, naked to the waist, the light of a kerosene lamp moving upon her dark skin. Her long, black hair, always drawn and braided in the day, lay upon her shoulders and against her breasts like a shawl.

Seventeen Syllables

page 91, line 28

, an adventure both painful and attractive to Rosie. It was attractive because there were four Hayano girls, all lovely and each one named after a season of the year (Haru, Natsu, Aki, Fuyu), painful because something had been wrong with Mrs. Hayano ever since the birth of her first child. Rosie would sometimes watch Mrs. Hayano, reputed to have been the belle of her native village, making her way about a room, stooped, slowly shuffling, violently trembling (always trembling), and she would be reminded that this woman, in this same condition, had carried and given issue to three babies. She would look wonderingly at Mr. Hayano, handsome, tall, and strong, and she would look at her four pretty friends. But it was not a matter she could come to any decision about. On this visit, *page 91, line 31*

Too, Rosie spent most of it in the girls' room, because Haru, the garrulous one, said almost as soon as the bows and other greetings were over, "Oh, you must see my new coat!"

It was a pale plaid of grey, sand, and blue, with an enormous collar, and Rosie, seeing nothing special in it, said, "Gee, how nice."

"Nice?" said Haru, indignantly. "Is that all you can say about it? It's gorgeous! And so cheap, too. Only seventeen-ninety-eight, because it was a sale. The saleslady said it was twenty-five dollars regular."

"Gee," said Rosie. Natsu, who never said much and when she said anything said it shyly, fingered the coat covetously and Haru pulled it away.

"Mine," she said, putting it on. She minced in the aisle between the two large beds and smiled happily. "Let's see how your mother likes it."

She broke into the front room and the adult conversation and went to stand in front of Rosie's mother, while the rest watched from the door. Rosie's mother was properly envious. "May I inherit it when you're through with it?"

Haru, pleased, giggled and said yes, she could, but Natsu reminded gravely from the door, "You promised me, Haru."

Everyone laughed but Natsu, who shamefacedly retreated into the bedroom. Haru came in laughing, taking off the coat. "We were only kidding, Natsu," she said. "Here, you try it on now."

After Natsu buttoned herself into the coat, inspected herself solemnly in the bureau mirror, and reluctantly shed it, Rosie, Aki, and Fuyu got their turns, and Fuyu, who was eight, drowned in it while her sisters and Rosie doubled up in amusement. They all went into the front room later, because Haru's mother quaveringly called to her to fix the tea and rice cakes and open a can of sliced peaches for everybody.

page 91, line 35

Occasionally, her father would comment on a photograph, holding it toward Mrs. Hayano and speaking to her as he always did—loudly, as though he thought someone such as she must surely be at least a trifle deaf also.

The five girls had their refreshments at the kitchen table, and it was while Rosie was showing the sisters her trick of swallowing peach slices without chewing (she chased each slippery crescent down with a swig of tea) that her father brought his empty teacup and untouched saucer to the sink and said,

page 91, line 38

gulped one last yellow slice and

page 91, line 39

"We have to get up at five-thirty," he told them, going into the front room quickly, so that they did not have their usual chance to hang onto his hands and plead for an extension of time.

page 92, line 2

while Mrs. Hayano concentrated, quivering, on raising the handleless Japanese cup to her lips with both her hands and lowering it back to her lap.

page 92, line 8

Haru was not giving up yet. "May Rosie stay overnight?" she asked, and Natsu, Aki, and Fuyu came to reinforce their sister's plea by helping her make a circle around Rosie's mother. Rosie, for once having no desire to stay, was relieved when her mother, apologizing to the perturbed Mr. and Mrs. Hayano for her father's abruptness at the same time, managed to shake her head no at the quartet, kindly but adamant, so that they broke their circle and let her go.

page 92, line 15

I wish this old Ford would crash, right now, she thought, then immediately, no, no, I wish my father would laugh, but it was too late: already the vision had passed through her mind of the green pick-up crumpled in the dark against one of the mighty eucalyptus trees they were just riding past, of the three contorted, bleeding bodies, one of them hers.

page 92, line 24

—he always had a joke for her when he periodically drove the loaded pick-up up from the fields to the shed where she was usually sorting while her mother and father did the packing, and they laughed a great deal together over infinitesimal repartee during the afternoon break for chilled watermelon or ice cream in the shade of the shed.

page 94, line 17

But the terrible, beautiful sensation lasted no more than a second, and the reality of Jesus' lips and tongue and teeth and hands made her pull away with such strength that she nearly tumbled.

page 94, line 24

No one had missed her in the parlor, however, and Rosie walked in and through quickly, announcing that she was next going to take a bath. "Your father's in the bathhouse," her mother said, and Rosie, in her room, recalled that she had not seen him when she entered. There had been only Aunt Taka and Uncle Gimpachi with her mother at the table, drinking tea. She got her robe and straw sandals and crossed the parlor again to go outside. Her mother was telling them about the haiku competition in the Mainichi and the poem she had entered.

Rosie met her father coming out of the bathhouse. "Are you through, Father?" she asked. "I was going to ask you to scrub my back."

"Scrub your own back," he said shortly, going toward the main house.

"What have I done now?" she yelled after him. She suddenly felt like doing a lot of yelling. But he did not answer, and she went into the bathhouse. Turning on the dangling light, she removed her denims and T-shirt and threw them in the big carton for dirty clothes standing next to the washing machine. Her other things she took with her into the bath compartment to wash after her bath. After she had scooped a basin of hot water from the square wooden tub, she sat on the grey cement of the floor and soaped herself at exaggerated leisure, singing "Red Sails in the Sunset" at the top of her voice and using da-da-da where she suspected her words. Then, standing up, still singing, for she was possessed by the notion that any attempt now to analyze would result in spoilage and she believed that the larger her volume the less she would be able to hear herself think, she obtained more hot water and poured it on until she was free of lather. Only then did she allow herself to step into the steaming vat, one leg first, then the remainder of her body inch by inch until the water no longer stung and she could move around at will.

She took a long time soaking, afterwards remembering to go around outside to stoke the embers of the tin-lined fireplace beneath the tub and to throw on a few more sticks so that the water might keep its heat for her mother, and when she finally returned to the parlor, she found her mother still talking haiku with her aunt and uncle, the

three of them on another round of tea. Her father was nowhere in sight.

page 94, line 26

Preoccupied at her desk in the row for students on Book Eight, she made up for it at recess by performing wild mimicry for the benefit of her friend Chizuko. She held her nose and whined a witticism or two in what she considered was the manner of Fred Allen; she assumed intoxication and a British accent to go over the climax of the Rudy Vallee recording of the pub conversation about William Ewart Gladstone; she was the child Shirley Temple piping, "On the Good Ship Lollipop"; she was the gentleman soprano of the Four Inkspots trilling, "If I Didn't Care." And she felt reasonably satisfied when Chizuko wept and gasped, "Oh, Rosie, you ought to be in the movies!"

page 94, line 28

The lugs were piling up, he said, and the ripe tomatoes in them would probably have to be taken to the cannery tomorrow if they were not ready for the produce haulers tonight. "This heat's not doing them any good. And we've got no time for a break today."

page 96, line 9

Mr. Kuroda was in his shirtsleeves expounding some haiku theory as he munched a rice cake, and her mother was rapt. Abashed in the great man's presence, Rosie stood next to her mother's chair until her mother looked up inquiringly, and then she started to whisper the message, but her mother pushed her gently away and reproached, "You are not being very polite to our guest."

"Father says the tomatoes . . ." Rosie said aloud, smiling foolishly.

page 97, line 2

It was like a story out of the magazines illustrated in sepia, which she had consumed so greedily for a period until the information had somehow reached her that those wretchedly unhappy autobiographies, offered to her as the testimonials of living men and women, were largely inventions. Her mother, at nineteen, had come to America and married her father as an alternative to suicide.

page 97, line 21

"I had a brother then?" Rosie asked, for this was

what seemed to matter now; she would think about the other later, she assured herself, pushing back the illumination which threatened all that darkness that had hitherto been merely mysterious or even glamorous. "A half-brother?"

"Yes."

"I would have liked a brother," she said.

Cante Ishta—The Eye of the Heart

page 109, line 35

All I knew from childhood was that a menstruating woman had to keep away from all rituals, and the thought intimidated me.

page 110, line 15

until she no longer had her moon time.

page 114, line 7

As the hissing steam enveloped us there rose a chorus of cries: "Ow, ow, ow, Great Spirit, we thank you for making us suffer so. We are suffering for our poor brothers in jail. Make us suffer more!"

"Jesus Christmas," I thought, "these people don't sweat to purify themselves. They sweat to suffer."

Wasichus in the Hills

page 129, line 2

In the spring when I was twelve years old (1875), more soldiers with many wagons came up from the Soldiers' Town at the mouth of the Laramie River and went into the Hills.

There was much talk all summer, and in the Moon of Making Fat (June) there was a sun dance there at the Soldiers' Town to give the people strength, but not many took part; maybe because everybody was so excited talking about the Black Hills. I remember two men who danced together. One had lost a leg in the Battle of the Hundred Slain and one had lost an eye in the Attacking of the Wagons, so they had only three eyes and three legs between them to dance with. We boys went down to the creek while they were sun dancing and got some elm leaves that we chewed up and threw on the dancers while they were all dressed up and trying to look their best. We even did this to some of the older people, and nobody got angry, because everybody was supposed to be in a good humor and

to show their endurance in every kind of way; so they had to stand teasing too. I will tell about a big sun dance later when we come to it.

The Council of the Great Peace

page 133, line 19

When there is any business to be transacted and the Confederate Council is not in session, a messenger shall be dispatched either to Adodarhoh, Hononwirehtonh or Skanawatih, Fire Keepers, or to their War Chiefs with a full statement of the case desired to be considered. Then shall Adodarhoh call his cousin (associate) Lords together and consider whether or not the case is of sufficient importance to demand the attention of the Confederate Council. If so, Adodarhoh shall dispatch messengers to summon all the Confederate Lords to assemble beneath the Tree of the Long Leaves.

page 133, line 26

Adodarhoh and his cousin Lords are entrusted with the Keeping of the Council Fire.

4   You, Adodarhoh, and your thirteen cousin Lords, shall faithfully keep the space about the Council Fire clean and you shall allow neither dust nor dirt to accumulate. I lay a Long Wing before you as a broom. As a weapon against a crawling creature I lay a staff with you so that you may thrust it away from the Council Fire. If you fail to cast it out then call the rest of the United Lords to your aid.

5   The Council of the Mohawk shall be divided into three parties as follows: Tekarihoken, Ayonhwhathah and Shadekariwade are the first party; Sharenhowaneh, Deyoenhegwenh and Oghrenghrehgowah are the second party, and Dehennakrineh, Aghstawenserenthah and Shoskoharowaneh are the third party. The third party is to listen only to the discussion of the first and second parties and if an error is made or the proceeding is irregular they are to call attention to it, and when the case is right and properly decided by the two parties they shall confirm the decision of the two parties and refer the case to the Seneca Lords for their decision. When the Seneca Lords have decided in accord with the Mohawk Lords, the case or question shall be referred to the Cayuga and Oneida Lords on the opposite side of the house.

6   I, Dekanawidah, appoint the Mohawk Lords the heads and the leaders of the Five Nations Confederacy. The Mohawk Lords are the foundation of the Great Peace and it shall, therefore, be against the Great Binding Law to pass measures in the Confederate Council after the Mohawk Lords have protested against them.

No council of the Confederate Lords shall be legal unless all the Mohawk Lords are present.

page 134, line 18

10   In all cases the procedure must be as follows: when the Mohawk and Seneca Lords have unanimously agreed upon a question, they shall report their decision to the Cayuga and Oneida Lords who shall deliberate upon the question and report a unanimous decision to the Mohawk Lords. The Mohawk Lords will then report the standing of the case to the Firekeepers, who shall render a decision as they see fit in case of a disagreement by the two bodies, or confirm the decisions of the two bodies if they are identical. The Fire Keepers shall then report their decision to the Mohawk Lords who shall announce it to the open council.

11   If through any misunderstanding or obstinacy on the part of the Fire Keepers, they render a decision at variance with that of the Two Sides, the Two Sides shall reconsider the matter and if their decisions are jointly the same as before they shall report to the Fire Keepers who are then compelled to confirm their joint decision.

12   When a case comes before the Onondaga Lords (Fire Keepers) for discussion and decision, Adodarhoh shall introduce the matter to his comrade Lords who shall then discuss it in their two bodies. Every Onondaga Lord except Hononwiretonh shall deliberate and he shall listen only. When a unanimous decision shall have been reached by the two bodies of Fire Keepers, Adodarhoh shall notify Hononwiretonh of the fact when he shall confirm it. He shall refuse to confirm a decision if it is not unanimously agreed upon by both sides of the Fire Keepers.

13   No Lord shall ask a question of the body of Confederate Lords when they are discussing a case, question or proposition. He may only deliberate in a low tone with the separate body of which he is a member.

14   When the Council of the Five Nation Lords shall convene they shall appoint a speaker for the day. He shall be a Lord of either the Mohawk, Onondaga or Seneca Nation.

The next day the Council shall appoint another speaker, but the first speaker may be reappointed if there is no objection, but a speaker's term shall not be regarded more than for the day.

15   No individual or foreign nation interested in a case, question or proposition shall have any voice in the Confederate Council except to answer a question put to him or them by the speaker for the Lords.

16   If the conditions which shall arise at any future time call for an addition to or change of this law, the case shall be carefully considered and if a new beam seems necessary or beneficial, the proposed change shall be voted upon and if adopted it shall be called, "Added to the Rafters."

page 135, line 16

When a Lord is to be deposed, his War Chief shall address him as follows:

"So you, ————, disregard and set at naught the warnings of your women relatives. So you fling the warnings over your shoulder to cast them behind you.

"Behold the brightness of the Sun and in the brightness of the Sun's light I depose you of your title and remove the sacred emblem of your Lordship title. I remove from your brow the deer's antlers, which are the emblem of your position and token of your nobility. I now depose you and return the antlers to the women whose heritage they are."

The War Chief shall now address the women of the deposed Lord and say:

"Mothers, as I have now deposed your Lord, I now return to you the emblem and the title of Lordship, therefore repossess them."

Again addressing himself to the deposed Lord he shall say:

"As I have now deposed and discharged you so you are now no longer Lord. You shall now go your way alone, the rest of the people of the Confederacy will not go with you, for we know not the kind of mind that possesses you. As the Creator has nothing to do with wrong so he will not come to rescue you from the precipice of destruction in which you have cast yourself. You shall never be restored to the position which you once occupied."

Then shall the War Chief address himself to the Lords of the Nation to which the deposed Lord belongs and say:

"Know you, my Lords, that I have taken the deer's antlers from the brow of ————, the emblem of his position and token of his greatness."

The Lords of the Confederacy shall then have no other alternative than to sanction the discharge of the offending Lord.

20 If a Lord of the Confederacy of the Five Nations should commit murder the other Lords of the Nation shall assemble at the place where the corpse lies and prepare to depose the criminal Lord. If it is impossible to meet at the scene of the crime the Lords shall discuss the matter at the next Council of their nation and request their War Chief to depose the Lord guilty of crime, to "bury" his women relatives and to transfer the Lordship title to a sister family. The War Chief shall address the Lord guilty of murder and say:

"So you, ———— (giving his name) did kill ——— —— (naming the slain man), with your own hands! You have committed a grave sin in the eyes of the Creator. Behold the bright light of the Sun, and in the brightness of the Sun's light I depose you of your title and remove the horns, the sacred emblems of your Lordship title. I remove from your brow the deer's antlers, which was the emblem of your position and token of your nobility. I now depose you and expel you and you shall depart at once from the territory of the Five Nations Confederacy and nevermore return again. We, the Five Nations Confederacy, moreover, bury your women relatives because the ancient Lordship title was never intended to have any union with bloodshed. Henceforth it shall not be their heritage. By the evil deed that you have done they have forfeited it forever."

The War Chief shall then hand the title to a sister family and he shall address it and say:

"Our mothers, ————, listen attentively while I address you on a solemn and important subject. I hereby transfer to you an ancient Lordship title for a great calamity has befallen it in the hands of the family of a former Lord. We trust that you, our mothers, will always guard it, and that you will warn your Lord always to be dutiful and to advise his

people to ever live in love, peace and harmony that a great calamity may never happen again."

21 Certain physical defects in a Confederate Lord make him ineligible to sit in the Confederate Council. Such defects are infancy, idiocy, blindness, deafness, dumbness and impotency. When a Confederate Lord is restricted by any of these conditions, a deputy shall be appointed by his sponsors to act for him, but in case of extreme necessity the restricted Lord may exercise hiss rights.

22 If a Confederate Lord desires to resign his title he shall notify the Lords of the Nation of which he is a member of his intention. If his coactive Lords refuse to accept his resignation he may not resign his title.

A Lord in proposing to resign may recommend any proper candidate which recommendation shall be received by the Lords, but unless confirmed and nominated by the women who hold the title the candidate so named shall not be considered.

page 136, line 4

26 It shall be the duty of all of the Five Nations Confederate Lords, from time to time as occasion demands, to act as mentors and spiritual guides of their people and remind them of their Creator's will and words. They shall say:

"Hearken, that peace may continue unto future days!

"Always listen to the words of the Great Creator, for he has spoken.

"United People, let not evil find lodging in your minds.

"For the Great Creator has spoken and the cause of Peace shall not become old.

"The cause of peace shall not die if you remember the Great Creator."

Every Confederate Lord shall speak words such as these to promote peace.

They Are Taking the Holy Elements from Mother Earth

page 148, line 10

The mine workers do a lot of drinking and they take youngsters with them and give them liquor and wine. These are the children of my husband's grandchildren with whom we used to live.

The Slave Ship

page 159, line 13

When I looked round the ship too, and saw a large furnace or copper boiling and a multitude of black people, of every description, chained together, every one of their countenances expressing dejection and sorrow, I no longer doubted of my fate; and, quite overpowered with horror and anguish, I fell motionless on the deck, and fainted. When I recovered a little, I found some black people about me, who I believed were some of those who brought me on board, and had been receiving their pay: they talked to me in order to cheer me, but all in vain. I asked them if we were not to be eaten by those white men with horrible looks, red faces, and long hair. They told me I was not: and one of the crew brought me a small portion of spirituous liquor in a wine glass; but, being afraid of him, I would not take it out of his hand. One of the blacks therefore took it from him and gave it to me, and I took a little down my palate, which, instead of reviving me, as they thought it would, threw me into the greatest consternation at the strange feeling it produced, having never tasted any such liquor before. Soon after this the blacks who brought me on board went off, and left me abandoned to despair.

I now saw myself deprived of all chance of returning to my native country, or even the least glimpse of gaining the shore, which I now considered as friendly; and I even wished for my former slavery, in preference to my present situation, which was filled with horrors of every kind, still heightened by my ignorance of what I was to undergo. I was not long suffered to indulge my grief. I was soon put down under the decks, and there I received such a salutation to my nostrils as I had never experienced in my life; so that, with the loathsomeness of the stench, and with my crying together, I became so sick and low that I was not able to eat, nor had I the least desire to taste any thing. I now wished for the last friend, death, to relieve me, but soon, to my grief, two of the white men offered me eatables; and, on my refusing to eat, one of them held me fast by the hands, and laid me across, I think, the windlass, and tied my feet, while the other flogged me severely. I had never

experienced any thing of this kind before, and although, not being used to the water, I naturally feared that element the first time I saw it, yet nevertheless, could I have got over the nettings, I would have jumped over the side, but I could not; and besides the crew used to watch us very closely, who were not chained down to the decks, lest we should leap into the water. I have seen some of these poor African prisoners most severely cut for attempting to do so, and hourly whipped for not eating. This indeed was often the case with myself. In a little time after, amongst the poor chained men, I found some of my own nation, which in a small degree gave ease to my mind. I inquired of these what was to be done with us. They gave me to understand we were to be carried to these white people's country to work for them. I was then a little revived, and thought if it were no worse than working, my situation was not so desperate. But still I feared I should be put to death, the white people looked and acted, as I thought, in so savage a manner; for I had never seen among any people such instances of brutal cruelty: and this is not only shewn towards us blacks, but also to some of the whites themselves. One white man in particular I saw, when we were permitted to be on deck, flogged so unmercifully with a large rope near the foremast, that he died in consequence of it; and they tossed him over the side as they would have done a brute. This made me fear these people the more; and I expected nothing less than to be treated in the same manner. I could not help expressing my fearful apprehensions to some of my countrymen; I asked them if these people had no country, but lived in this hollow place, the ship. They told me they did not, but came from a distant one. "Then," said I, "how comes it, that in all our country we never heard of them?" They told me, because they lived so very far off. I then asked, where their women were: had they any like themselves. I was told they had. "And why," said I, "do we not see them?" They answered, because they were left behind. I asked how the vessel could go. They told me they could not tell; but that there was cloth put upon the masts by the help of the ropes I saw, and then the vessel went on; and the white men had some spell or magic they put in the water, when they liked, in order to stop the vessel. I was exceedingly amazed at this account, and really thought they were spirits. I therefore wished much to be from amongst them, for I expected they would sacrifice me; but my wishes were in vain, for we were so quartered that it was impossible for any of us to make our escape.

page 161, line 24

During our passage I first saw flying fishes, which surprised me very much: they used frequently to fly across the ship, and many of them fell on the deck. I also now first saw the use of the quadrant. I had often with astonishment seen the mariners make observations with it, and I could not think what it meant. They at last took notice of my surprise; and one of them, willing to increase it, as well as to gratify my curiosity, made me one day look through it. The clouds appeared to me to be land, which disappeared as they passed along. This heightened my wonder, and I was now more persuaded than ever that I was in another world, and that every thing about me was magic.

page 162, line 7

While I was in this astonishment one of my fellow prisoners spoke to a countryman of his about the horses, who said they were the same kind they had in their country. I understood them, though they were from a distant part of Africa, and I thought it odd I had not seen any horses there; but afterwards, when I came to converse with different Africans, I found they had many horses amongst them, and much larger than those I then saw.

page 163, line 3

O, ye nominal Christians! might not an African ask you, "learned you this from your God, who says unto you, Do unto all men as you would men should do unto you? Is it not enough that we are torn from our country and friends, to toil for your luxury and lust of gain? Must every tender feeling be likewise sacrificed to your avarice? Are the dearest friends and relations now rendered more dear by their separation from the rest of their kindred, still to be parted from each other, and thus prevented from cheering the gloom of slavery, with the small comfort of being together, and mingling their sufferings and sorrows? Why are parents to lose their children, brothers their sisters, or husbands their wives?

The Slave Who Dared to Feel Like a Man

page 167, line 4

So deeply was I absorbed in painful reflections afterwards, that I neither saw nor heard the entrance of any one, till the voice of William sounded close beside me. "Linda," he said, "what makes you look so sad? I love you. O, Linda, isn't this a bad world? Every body seems so cross and unhappy. I wish I had died when poor father did."

I told him that every body was not cross, or unhappy; that those who had pleasant homes, and kind friends, and who were not afraid to love them, were happy. But we, who were slave-children, without father or mother, could not expect to be happy. We must be good; perhaps that would bring us contentment.

"Yes," he said, "I try to be good; but what's the use? They are all the time troubling me." Then he proceeded to relate his afternoon's difficulty with young master Nicholas. It seemed that the brother of master Nicholas had pleased himself with making up stories about William. Master Nicholas said he should be flogged, and he would do it. Whereupon he went to work; but William fought bravely, and the young master, finding he was getting the better of him, undertook to tie his hands behind him. He failed in that likewise. By dint of kicking and fisting, William came out of the skirmish none the worse for a few scratches.

He continued to discourse on his young master's meanness; how he whipped the little boys, but was a perfect coward when a tussle ensued between him and white boys of his own size. On such occasions he always took to his legs. William had other charges to make against him. One was his rubbing up pennies with quicksilver, and passing them off for quarters of a dollar on an old man who kept a fruit stall. William was often sent to buy fruit; and he earnestly inquired of me what he ought to do under such circumstances. I told him it was certainly wrong to deceive the old man, and that it was his duty to tell him of the impositions practiced by his young master. I assured him the old man would not be slow to comprehend the whole, and there the matter would end. William thought it might with the old man, but not with him. He said he did not mind the smart of the whip, but he did not like the idea of being whipped.

While I advised him to be good and forgiving I was not unconscious of the beam in my own eye. It was the very knowledge of my own shortcomings that urged me to retain, if possible, some sparks of my brother's God-given nature. I had not lived fourteen years in slavery for nothing. I had felt, seen, and heard enough, to read the characters, and question the motives, of those around me. The war of my life had begun; and though one of God's most powerful creatures, I resolved never to be conquered. Alas, for me!

If there was one pure, sunny spot for me, I believed it to be in Benjamin's heart, and in another's, whom I loved with all the ardor of a girl's first love. My owner knew of it, and sought in every way to render me miserable. He did not resort to corporal punishment, but to all the petty, tyrannical ways that human ingenuity could devise.

page 168, line 28

"Go," said I, "and break your mother's heart."

I repented of my words ere they were out.

"Linda," said he, speaking as I had not heard him speak that evening, "how could you say that? Poor mother! be kind to her, Linda; and you, too, cousin Fanny."

Cousin Fanny was a friend who had lived some years with us.

Farewells were exchanged, and the bright, kind boy, endeared to us by so many acts of love, vanished from our sight.

page 169, line 12

We were not allowed to visit him; but we had known the jailer for years, and he was a kind-hearted man. At midnight he opened the jail door for my grandmother and myself to enter, in disguise. When we entered the cell not a sound broke the stillness. "Benjamin, Benjamin!" whispered my grandmother. No answer. "Benjamin!" she again faltered. There was a jingle of chains. The moon had just risen, and cast an uncertain light through the bars of the window. We knelt down and took Benjamin's cold hands in ours. We did not speak. Sobs were heard, and Benjamin's lips were unsealed; for his mother was weeping on his neck. How vividly does memory bring back that sad night! Mother and son talked together. He asked her pardon for the suffering he had caused her. She said she had nothing to forgive; she could not blame his desire for freedom. He told her that when he was captured, he broke away, and was about casting himself into the river, when thoughts of her came over him, and he desisted. She asked if he did not also think of God. I fancied I saw his face grow fierce in the moonlight. He answered, "No, I did not think of him. When a man is hunted like a wild beast he forgets there is a God, a heaven. He forgets

every thing in his struggle to get beyond the reach of the bloodhounds."

"Don't talk so, Benjamin," said she. "Put your trust in God. Be humble, my child, and your master will forgive you."

"Forgive me for what, mother? For not letting him treat me like a dog? No! I will never humble myself to him. I have worked for him for nothing all my life, and I am repaid with stripes and imprisonment. Here I will stay till I die, or till he sells me."

The poor mother shuddered at his words. I think he felt it; for when he next spoke, his voice was calmer. "Don't fret about me, mother. I ain't worth it," said he. "I wish I had some of your goodness. You bear every thing patiently, just as though you thought it was all right. I wish I could."

She told him she had not always been so; once, she was like him; but when sore troubles came upon her, and she had no arm to lean upon, she learned to call on God, and he lightened her burdens. She besought him to do likewise.

We overstaid our time, and were obliged to hurry from the jail.

page 170, line 20

With a strong arm and unvaried trust, my grandmother began her work of love. Benjamin must be free. If she succeeded, she knew they would still be separated; but the sacrifice was not too great. Day and night she labored. The trader's price would treble that he gave; but she was not discouraged.

She employed a lawyer to write to a gentleman, whom she knew, in New Orleans. She begged him to interest himself for Benjamin, and he willingly favored her request. When he saw Benjamin, and stated his business, he thanked him; but said he preferred to wait a while before making the trader an offer. He knew he had tried to obtain a high price for him, and had invariably failed. This encouraged him to make another effort for freedom. So

page 170, line 21

For once his white face did him a kindly service. They had no suspicion that it belonged to a slave; otherwise, the law would have been followed out to the letter, and the thing rendered back to slavery. The brightest skies are often overshadowed by the darkest clouds. Benjamin was taken sick, and compelled to remain in Baltimore three weeks. His strength was slow in returning; and his desire to continue his journey seemed to retard his recovery. How could he get strength without air and exercise? He resolved to venture on a short walk. A by-street was selected, where he thought himself secure of not being met by any one that knew him; but a voice, called out, "Halloo, Ben, my boy! what are you doing here?"

His first impulse was to run; but his legs trembled so that he could not stir. He turned to confront his antagonist, and behold, there stood his old master's next door neighbor! He thought it was all over with him now; but it proved otherwise. That man was a miracle. He possessed a goodly number of slaves, and yet was not quite deaf to that mystic clock, whose ticking is rarely heard in the slaveholder's breast.

"Ben, you are sick," said he. "Why, you look like a ghost. I guess I gave you something of a start. Never mind, Ben, I am not going to touch you. You had a pretty tough time of it, and you may go on your way rejoicing for all me. But I would advise you to get out of this place plaguy quick, for there are several gentlemen here from our town." He described the nearest and safety route to New York, and added, "I shall be glad to tell your mother I have seen you. Good by, Ben."

Benjamin turned away, filled with gratitude, and surprised that the town he hated contained such a gem—a gem worthy of a purer setting.

This gentleman was a Northerner by birth, and had married a southern lady. On his return, he told my grandmother that he had seen her son, and of the service he had rendered him.

page 170, line 35

He begged my uncle Phillip not to return south; but stay and work with him, till they earned enough to buy those at home. His brother told him it would kill their mother if he deserted her in her trouble. She had pledged her house, and with difficulty had raised money to buy him. Would he be bought?

"No, never!" he replied. "Do you suppose Phil, when I have got so far out of their clutches, I will give them one red cent? No! And do you suppose I would turn mother out of her home in her old age? That I would let her pay all those hard-earned dollars for me, and never to see me? For you know she will stay south as long as her other children are slaves. What a good mother! Tell her to buy you, Phil. You have been a comfort to her, and I have been a trouble. And Linda, poor Linda; what'll become of her? Phil, you don't know what a life they lead her. She has told me something about it, and I wish old Flint was dead, or a better man. When I was in jail, he asked her if she didn't want him to ask my master to forgive me, and take he home again. She told him, No; that I didn't want to go back. He got mad, and said we were all alike. I never despised my own master half as much as I do that man. There is many a worse slaveholder than my master; but for all that I would not be his slave."

While Benjamin was sick, he had parted with nearly all his clothes to pay necessary expenses. But he did not part with a little pin I fastened in his bosom when we parted. It was the most valuable thing I owned, and I thought none more worthy to wear it. He had it still.

His brother furnished him with clothes, and gave him what money he had.

They parted with moistened eyes, and as Benjamin turned away, he said, "Phil, I part with all my kindred." And so it proved. We never heard from him again.

Uncle Phillip came home; a;nd the first words he uttered when he entered the house were, "Mother, Ben is free! I have seen him in New York." She stood looking at him with a bewildered air. "Mother, don't you believe it?" he said, laying his hand softly upon her shoulder. She raised her hands, and exclaimed, "God be praised! Let us thank him." She dropped on her knees, and poured forth her heart in prayer. Then Phillip must sit down and repeat to her every word Benjamin had said. He told her all; only he forbode to mention how sick and pale her darling looked. Why should he distress her when she could do him no good?

The brave old women still toiled on, hoping to rescue some of her other children. After a while she succeeded in buying Phillip. She paid eight hundred dollars, and came home with the precious document that secured his freedom. The happy mother and son sat together by the old hearthstone that night, telling how proud they were of each other, and how they would prove to the world that they could take care of themselves, as they had long taken care of others. We all concluded by saying, "He that is willing to be a slave, let him be a slave."

I Leave South Africa

page 202, line 40

To them we will always be niggers.

page 204, line 19

"I would like to use the lavatory," I told Dr. Waller.

"There should be one over there." He pointed to a sign ahead which read RESTROOMS. "I'll wait for you at the newsstand over there."

When I reached the restroom I found it had the sign MEN in black and white on it. Just before I entered I instinctively scoured the walls to see if I had missed the other more important sign: BLACKS ONLY or WHITES ONLY, but there was none. I hesitated before entering: this freedom was too new, too strange, too unreal, and called for the utmost caution. Despite what I believed about America, there still lingered in the recesses of my mind the terror I had suffered in South Africa when I had advertently disobeyed the racial etiquette, like that time in Pretoria when I mistakenly boarded a white bus, and Granny had to grovel before the irate redneck driver, emphatically declare that it was an insanity "not of the normal kind" which had made me commit such a crime, and to appease him proceeded to wipe, with her lovely tribal dress, the steps where I had trod. In such moments of doubt such traumas made me mistrust my instincts. I saw a lanky black American with a mammoth Afro enter and I followed. I relieved myself next to a white man and he didn't die.

The black American washed his hands and began combing his Afro. I gazed at his hair with wonder.

page 205, line 15

I left the bathroom and rejoined Dr. Waller at the newsstand. I found him reading a magazine.

"There's so much to read here," I said, running my eyes over the newspapers, magazines, and books. Interestingly, almost all had white faces on the cover, just as in South Africa.

"Yes," replied Dr. Waller.

I was shocked to see pornography magazines, which are banned in South Africa, prominently displayed. The puritan and Calvinistic religion of the Afrikaners sought to purge South African society of "influences of the devil" and "materials subversive to the state and public morals" by routinely banning and censoring not only books by writers who challenged the status quo, but also publications like Playboy.

"So many black people fly in America," I said.

"A plane is like a car to many Americans," said Dr. Waller.

"To many of my people cars are what planes are to Americans."

I See the Promised Land

page 226, line 1

That's a strange statement.

page 226, line 10

And another reason that I'm happy to live in this period is that

page 229, line 3

And you know what's beautiful to me, is to see all of these ministers of the Gospel. It's a marvelous picture. Who is it that is supposed to articulate the longings and aspirations of the people more than the preacher? Somehow the preacher must be an Amos, and say, "Let justice roll down like waters and righteousness like a mighty stream." Somehow, the preacher must say with Jesus, "The spirit of the Lord is upon me, because he hath anointed me to deal with the problems of the poor."

And I want to commend the preachers, under the leadership of these noble men: James Lawson, one who has been in this struggle for many years; he's been to jail for struggling; but he's still going on, fighting for the rights of his people. Rev. Ralph Jackson, Billy Kiles; I could just go right on down the list, but time will not permit. But I want to thank them all. And I want you to thank them, because so often, preachers aren't concerned about anything but themselves. And I'm always happy to see a relevant ministry.

It's alright to talk about "long white robes over yonder," in all of its symbolism. But ultimately people want some suits and dresses and shoes to wear down here.

page 229, line 31

And so, as a result of this, we are asking you tonight, to go out and tell your neighbors not to buy Coca-Cola in Memphis. Go by and tell them not to buy Sealtest milk. Tell them not to buy—what is the other bread?—Wonder Bread. And what is the other bread company, Jesse? Tell them not to buy Hart's bread. As Jesse Jackson has said, up to now, only the garbage men have been feeling pain; now we must kind of redistribute the pain. We are choosing these companies because they haven't been fair in their hiring policies; and we are choosing them because they can begin the process of saying, they are going to support the needs and rights of these men who are on strike. And then they can move on downtown and tell Mayor Loeb to do what is right.

page 230, line 12

Let us develop a kind of dangerous unselfishness. One day a man came to Jesus; and he wanted to raise some questions about some vital matters in life. At points, he wanted to trick Jesus, and show him that he knew a little more than Jesus knew, and through this, throw him off base. Now that question could have easily ended up in a philosophical and theological debate. But Jesus immediately pulled that question from mid-air, and placed it on a dangerous curve between Jerusalem and Jericho. And he talked about a certain man, who fell among thieves. You remember that a Levite and a priest passed by on the other side. They didn't stop to help him. And finally a man of another race came by. He got down from his beast, decided not to be compassionate by proxy. But with him, administered first aid, and helped the man in need. Jesus ended up saying, this was the good man, this was the great man, because he had the capacity to project the "I" into the "thou," and to be concerned about his brother. Now you know, we use our imagination a great deal to try to determine why the priest and the Levite didn't stop. At times we say they were busy going to church meetings—an ecclesiastical gathering—and they had to get on down to Jerusalem so they wouldn't be late for their meeting. At other times we would speculate that there was a religious law that "One who was engaged in religious ceremonials was not to touch a human body twenty-four hours before the ceremony." And every now and then we begin to wonder whether maybe they were not going down to Jerusalem, or down to Jericho, rather to organize a "Jericho Road Improvement Association." That's a possibility. Maybe they felt that it was better to deal with the problem from the casual root, rather than to get bogged down with an individual effort.

But I'm going to tell you what my imagination tells me. It's possible that these men were afraid. You see, the Jericho road is a dangerous road. I remember when Mrs. King and I were first in Jerusalem. We rented a car and drove from Jerusalem down to Jericho. And as soon as we got on that road, I said to my wife, "I can see why Jesus used this as a setting for his parable." It's a winding, meandering road. It's really conducive for ambushing. You start out in Jerusalem, which is about 1200 miles, or rather 1200 feet above sea level. And by the time you get down to Jericho, fifteen or twenty minutes later, you're about 2200 feet below sea level. That's a dangerous road. In the days of Jesus it came to be known as the "Bloody Pass." And you know, it's possible that the priest and the Levite looked over that man on the ground and wondered if the robbers were still around. Or it's possible that they felt that the man on the ground was merely faking. And he was acting like he had been robbed and hurt, in order to seize them over there, lure them there for quick and easy seizure. And so the first question that the Levite asked was, "If I stop to help this man, what will happen to me?" But then the Good Samaritan came by. And he reversed the question: "If I do not stop to help this man, what will happen to him?"

page 230, line 23

You know, several years ago, I was in New York City autographing the first book that I had written. And while sitting there autographing books, a demented black woman came up. The only question I heard from her was, "Are you Martin Luther King?"

And I was looking down writing, and I said yes. And the next minute I felt something beating on my chest. Before I knew it I had been stabbed by this demented woman. I was rushed to Harlem Hospital. It was a dark Saturday afternoon. And that blade had gone through, and the X-rays revealed that the tip of the blade was on the edge of my aorta, the main artery. And once that's punctured, you drown in your own blood—that's the end of you.

It came out in the New York Times the next morning, that if I had sneezed, I would have died. Well, about four days later, they allowed me, after the operation, after my chest had been opened, and the blade had been taken out, to move around in the wheel chair in the hospital. They allowed me to read some of the mail that came in, and from all over the states, and the world, kind letters came in. I read a few, but one of them I will never forget. I had received one from the President and the Vice-President. I've forgotten what those telegrams said. I'd received a visit and a letter from the Governor of New York, but I've forgotten what the letter said. But there was another letter that came from a little girl, a young girl who was a student at the White Plains High School. And I looked at that letter, and I'll never forget it. It said simply, "Dear Dr. King: I am a ninth-grade student at the White Plains High School." She said, "While it should not matter, I would like to mention that I am a white girl. I read in the paper of your misfortune, and of your suffering. And I read that if you had sneezed, you would have died. And I'm simply writing you to say that I'm so happy that you didn't sneeze."

And I want to say tonight, I want to say that I am happy that I didn't sneeze. Because if I had sneezed, I wouldn't have been around here in 1960, when students all over the South started sitting-in at lunch counters. And I knew that as they were sitting in, they were really standing up for the best in the American dream. And taking the whole nation back to those great wells of democracy which were dug deep by the Founding Fathers in the Declaration of Independence and the Constitution. If I had sneezed, I wouldn't have been around in 1962, when Negroes in Albany, Georgia, decided to straighten their backs up. And whenever men and women straighten their backs up, they are going somewhere, because a man can't ride your back unless it is bent. If I had sneezed, I wouldn't have been here in 1963, when the black people of Birmingham, Alabama, aroused the conscience of this nation, and brought into being the Civil Rights Bill. If I had sneezed, I wouldn't have had a chance later that year, in August, to try to tell Americans about a dream that I had had. If I had sneezed, I wouldn't have been down in Selma, Alabama, to see the great movement there. If I had sneezed, I wouldn't have been in Memphis to see a community rally around those brothers and sisters who are suffering. I'm so happy that I didn't sneeze.

And they were telling me, now it doesn't matter now. It really doesn't matter what happens now. I left Atlanta this morning, and as we got started on the plane, there were six of us, the pilot said over the public address system, "We are sorry for the delay; but we have Dr. Martin Luther King on the plane. And to be sure that all of the bags were checked, and to be sure that nothing would be wrong with the plane, we had to check out everything carefully. And we've had the plane protected and guarded all night."

And then I got into Memphis. And some began to say the threats, or talk about the threats that were out. What would happen to me from some of our sick white brothers?

In the American Storm

page 237, line 10

My Charlestown barber friend took me in that first night with the distinct understanding that I could stay only one night. So the next morning bright and early, leaving all my belongings with the barber, I started out in search of a job.

Fences

page 322, line 1

The lights come up on the yard. It is four hours later. ROSE is taking down the clothes from the line. CORY enters carrying his football equipment.

ROSE: Your daddy like to had a fit with you running out of here this morning without doing your chores.

CORY: I told you I had to got practice.

ROSE: He say you were supposed to help him with this fence.

CORY: He been saying that the last four or five Saturdays, and then he don't never do nothing, but go down to Taylors'. Did you tell him about the recruiter?

ROSE: Yeah, I told him.

CORY: What he say?

ROSE: He ain't said nothing too much. You get in there and get started on your chores before he gets back. Go on and scrub down them steps before he gets back here hollering and carrying on.

CORY: I'm hungry. What you got to eat, Mama?

ROSE: Go on and get started on your chores. I got some meat loaf in there. Go on and make you a sandwich . . . and don't leave no mess in there.

(CORY *exits into the house.* ROSE *continues to take down the clothes.* TROY *enters the yard and sneaks up and grabs her from behind.*)

Troy! Go on, now. You liked to scared me to death. What was the score of the game? Lucille had me on the phone and I couldn't keep up with it.

TROY: What I care about the game? Come here, woman. (*He tries to kiss her.*)

ROSE: I thought you went down Taylors' to listen to the game. Go on, Troy! You supposed to be putting up this fence.

TROY: (*Attempting to kiss her again.*) I'll put it up when I finish with what is at hand.

ROSE: Go on, Troy. I ain't studying you.

TROY: (*Chasing after her.*) I'm studying you . . . fixing to do my homework!

ROSE: Troy, you better leave me alone.

TROY: Where's Cory? That boy brought his butt home yet?

ROSE: He's in the house doing his chores.

TROY: (*Calling.*) Cory! Get your butt out here, boy!

(ROSE *exits into the house with the laundry.* TROY *goes over to the pile of wood, picks up a board, and starts sawing.* CORY *enters from the house.*)

TROY: You just now coming in here from leaving this morning?

CORY: Yeah, I had to go to football practice.

TROY: Yeah, what?

CORY: Yessir.

TROY: I ain't but two seconds off you noway. The garbage sitting in there overflowing . . . you ain't done none of your chores . . . and you come in here talking about "Yeah."

CORY: I was just getting ready to do my chores now, Pop . . .

TROY: Your first chore is to help me with this fence on Saturday. Everything else come after that. Now get that saw and cut them boards.

page 327, line 12

the hell

page 327, line 15

the hell

page 327, line 16

damn fool-ass

page 327, line 21

goddammit

page 328, line 21

your black ass

page 328, line 26

the hell

[8] page 330, line 24

and I fall down on you and try to blast a hole into forever.

America: The Multinational Society

page 337, line 14

"Niggers and Spics Suck,"

ASSESSMENT WORKSHEETS
LITERARY SKILLS WORKSHEETS

There are forty-six assessment worksheets, one for each lesson plan in the
Teacher's Resource Manual. There are nine literary skills worksheets. Although
the literary skills worksheets are each based on a specific selection in the
Student Edition, these worksheets are generic in nature and can be used with
other selections to which they apply.

Name _____ Date _____

Assessment Worksheet 1
The Stars by Pablita Velarde (pages 7–14)

Vocabulary The sentences below are based on descriptions or events in the selection. Look for context clues to help you find the best word below to complete each one.

mole   destination   journey   **Milky Way**   coyote

1. The long, difficult _____ helped the people learn tolerance.

2. Telling of the future, the spider and the _____ were prophetic creatures.

3. Lost tribes still follow Long Sash over the Endless Trail, the _____.

4. The people hoped they would reach their _____ soon.

5. Long Sash said the people who remained in the hot dry land would prey on what

   they found, following the way of the _____.

Understanding the Selection Complete each sentence by adding appropriate information from the selection.

6. Long Sash and his people wanted to leave their homes because

   _____.

7. As the people traveled they learned to carry their belongings by

   _____.

8. Long Sash decided to leave the beautiful land with plentiful game because

   _____.

9. The turtle was the long–awaited sign because

   _____.

10. The storyteller said the children did not get to catch the stars because

   _____.

Author's Craft: The quest story contains the following elements. Match the element to its example. (One answer appears twice.)

   a. quest   b. hero   c. guidance by higher powers   d. obstacle

_____ 11. The goal of the people was to find their true home.

_____ 12. In time some of the people grew doubtful and violent.

_____ 13. Long Sash was a brave, persistent leader.

_____ 14. The spider spoke to them and showed them the direction to take.

_____ 15. In one land, wild beasts often pounced on the people.

Assessment Worksheet 2
The Sheep of San Cristóbal by Terri Bueno (pages 15–23)

Vocabulary The sentences below are based on descriptions or events in the selection. Look for context clues to help you find the best word below to complete each one.

Padre   penance   priest   indigent   prayed

1. Felipa had to pay _____ for uttering an evil prayer.

2. After hearing of Don José's death, Felipa _____ to Nuestra Señora de los Dolores.

3. The old _____ told Felipa she had to do penance.

4. Felipa called the priest _____.

5. Felipa had to give a sheep to the most _____ person in each village.

Understanding the Selection Complete each sentence by adding appropriate information from the selection.

6. The seven-year-old son had been _____ by the Utes.

7. On his way to the mesa with his sheep, Don José would torment _____ with his visits.

8. Felipa had to pay penance for herself and for _____.

9. The penance included giving to the poor because _____

_____.

10. The tall man who gave Felipa back her son was really _____

_____.

Author's Craft: The Moral Tale The moral tale contains the following elements. Match the element to its example.

a. hero   b. reward   c. theme   d. guiding force   e. villain

_____ 11. The way to peace and fulfillment is by giving away one's money and possessions.

_____ 12. Don José was a selfish, unkind man.

_____ 13. Felipa carried out the actions the priest commanded.

_____ 14. Felipa's religious faith gave her strength to carry on the difficult mission of giving away the sheep one by one.

_____ 15. Eventually Felipa's cloud of guilt finally left her and she got Manuel back.

Assessment Worksheet 3
from Things Fall Apart by Chinua Achebe (pages 24–28)

Vocabulary   The sentences below are based on descriptions or events in the
selection. Look for context clues to help you find the best word below to complete
each one.

orator   voluble   eloquent   delectable   cunning

1. The dishes prepared by the people of the sky were _____.

2. Tortoise was chosen to speak for the party because he was a great _____.

3. His speech on behalf of his bird friends was quite _____.

4. Tortoise was happy and _____.

5. Tortoise was clever and _____.

Understanding the Selection   Complete each sentence by adding appropriate
information from the selection.

6. Tortoise had not eaten a good meal for _____, which means two months.

7. The birds do not want Tortoise to go with them because he is so full of mischief, but Tortoise persuades

   them by saying _____.

8. Tortoise beguiles the birds into thinking they need to choose new names and names himself _____

   _____.

9. The birds were angry after Tortoise tricked them, so _____

   thought of a way to trick Tortoise.

10. The story explains why Tortoise's shell _____

    _____.

Author's Craft: A FolkTale   Folktales contain elements that are realistic and
elements that are fantasy. Write R to identify statements that are could be realistic.
Write F to identify statements that come from the writer's imagination.

_____ 11. There had been a terrible famine in the land.

_____ 12. A main food crop was yams.

_____ 13. Tortoise delivered eloquent speeches.

_____ 14. Speeches were popular during holidays and special occasions.

_____ 15. Tortoise was revived by a great medicine man.

Assessment Worksheet 4

In the Land of Small Dragon: A Vietnamese Folktale
by Dang Manh Kha (pages 29–35)

Vocabulary   The sentences below are based on descriptions or events in the selection. Look for context clues to help you find the best word below to complete each one.

rice paddy   indolent   scowling   revenge   monsoon

1. Cám's face was ugly, _____ with discontentment.

2. Number Two Wife was jealous of T'âm's beauty and she planned _____
   on the good child.

3. The father had a house, a garden, fish ponds, and a _____.

4. Cám was _____, slow and idle.

5. The _____ season brings rain and wind.

Understanding the Selection   Complete each sentence by adding appropriate information from the selection.

6. While Cám's mother only loved her, the father loved _____.

7. The father determined which would be Number One Daughter based on _____.

   _____.

8. Cám was able to get T'âm's fish by _____.

9. T'âm's beautiful dress and *hai* were a gift from _____.

10. T'âm's would never have been able to sort the rice in time without the help of _____.

    _____.

Author's Craft: A Narrative Poem   Narrative poem contains the following elements. Match the element to its example. One element appears twice.

a. central conflict   b. refrain   c. characters   d. stanzas

_____ 11. *Beauty is not painted on.*

   It is the spirit showing.

_____ 12. T'âm's step_mother mistreated her terribly and spoiled her own daughter.

_____ 13. Cám is ugly and unkind, while T'âm's is lovely and good.

_____ 14. *Real beauty mirrors goodness.*

   What is one is the other.

_____ 15. The narrative poem is divided into five sections.

Name _____   Date _____

Assessment Worksheet 5
The Man Who Had No Story
by Michael James Timoney and Séamus Ó Catháin (pages 38–43)

Vocabulary   The sentences below are based on descriptions or events in the selection.
Look for context clues to help you find the best word below to complete each one.

    fog   glen   fairy tale   wake-house   Fenian tale

1. Telling a _____ about the wee folk was something Brian had never done before in his
   life.

2. Brian thought that maybe if he sat down and ate his lunch the _____ would clear.

3. Brian had also never told a _____ dealing with the fight for Irish independence.

4. Although there were find rods growing in Alt an Torr, nobody dared cut them because the valley was a
   fairy _____.

5. Brian played the fiddle while the men danced in the _____.

Understanding the Selection   Complete each sentence by adding appropriate
information from the selection.

6. Brian ended up cutting rods in the fairy glen because _____.

7. After the fog and darkness descended, Brian followed the light, thinking _____.

8. Since he couldn't tell a fairy tale, Brian _____

   _____ to earn his keep.

9. The girl with the curly black hair persuaded Brian he could _____ and

   _____.

10. The solution to the problem of one pall bearer's being too tall was _____.

Author's Craft: A Story   Answer the following questions regarding elements used in the story.

11. What is one plot element that gets repeated?

    _____

12. What is one phrase that gets repeated?

    _____

13. What is one effect of the repetition in the story?

    _____

14. Why did Brian wake up in the fairy glen?

    _____

15. Why did the fairies play tricks on Brian?

    _____

Name _____ Date _____

Assessment Worksheet 6
from Bless Me, Ultima by Rudolfo A. Anaya (pages 45–55)

Vocabulary Match the Spanish words on the right with their English meanings on the left. Write the correct letter in the blank.

_____ 1. witches

_____ 2. feeling ill

_____ 3. a mill

_____ 4. healer

_____ 5. flat area, like a prairie

a. **curandera**

b. **llano**

c. **brujas**

d. **molino**

e. **crudo**

Understanding the Selection Complete each sentence by adding appropriate information from the selection.

6. Antonio's mother's people were _____.

7. His father's people were _____.

8. Antonio's mother called Ultima la Grande because _____

_____.

9. During Antonio's dream he sees _____

_____.

10. Antonio struggles to clear land for a garden because _____

_____.

Author's Craft: A Novel The story contains the following elements. Match the element to its example.

a. foreshadowing   b. realism   c. tension

d. spiritual qualities   e. sensory details

_____ 11. The war takes young boys overseas.

_____ 12. Antonio scattered grain for the hungry chickens and watched their mad scramble as the rooster called them to peck.

_____ 13. Antonio's mother wanted him to be like her people, but his father wanted him to be like his family.

_____ 14. Antonio's mother wants him to become a priest.

_____ 15. The summer Ultima came to live with his family, Antonio saw the beauty and magic of the earth for the first time.

Name _____ Date _____

Assessment Worksheet 7
from Roots by Alex Haley (pages 56–61)

Vocabulary   The sentences below are based on descriptions or events in the selection. Look for context clues to help you find the best word below to complete each one.

irate   decrease   succession   argumentative   irritable

1. When a family grows smaller, the family's farming rights also _____.

2. The _____ leader protested when the borrower lost a valuable tool.

3. An evil spell brought one woman a long _____ of troubles.

4. Two _____ people might raise quarrelsome children.

5. The impatient accuser became _____ when an Elder questioned him.

Understanding the Selection
Complete each sentence by adding appropriate information from the selection.

6. The Council of Elders met every month to _____

_____.

7. _____ usually attended the council meeting only when a

member of their immediate family was involved in a dispute.

8. Couples who wanted to marry had to undergo a _____

_____.

9. In the council, slaves sometimes accused _____

of taking more than their half-share of what was produced.

10. "New Men" were expected to attend council meetings because _____

_____.

Author's Craft: A Historical Novel   Historical novels contain elements that
are factual and other elements that come from the writer's imagination. Write *HF* to identify statements that are probably based on historical facts. Write *WI* to identify statements that probably come from the writer's imagination.

_____ 11. At Kunta Kinte's first council meeting, he witnessed a dispute about land.

_____ 12. A junior Elder named Omoro sat in front of Kunta at his first council

meeting.

_____ 13. People believed in the power of evil magic to produce bad fortune.

_____ 14. The Council of Elders had to give a couple its permission to marry.

_____ 15. The name of the next speaker at a meeting was announced by drumtalk.

Name _____   Date _____

Assessment Worksheet 8

To Da–duh, In Memoriam by Paule Marshall (pages 62–74)

Vocabulary   The sentences below are based on descriptions or events in the selection.
Look for context clues to help you find the best word below to complete each one.

> **truculent   unrelenting   formidable   dissonant   roguish**

1. The fresh white cloth wound as a turban gave Da–duh an almost _____ air.

2. Da–duh carried on an _____ struggle to hold her back straight.

3. The loud, raw voices of the villagers sounded _____.

4. Once in Da–duh's presence the previously _____ Ardy is reduced to

   the status of a child.

5. The granddaughter was a thin, _____ child.

Understanding the Selection   Complete each sentence by adding appropriate
information from the selection.

6. The grandchild's first impression of Da–duh's eyes was that _____

   _____.

7. The child thought she "won" the first encounter with her grandmother because _____

   _____.

8. The father had not made the trip because _____

   _____.

9. Da–duh wants to know about New York because _____

   _____.

10. The ultimate impact of learning about New York, culminating with the height of the Empire State

    Building on Da–duh is _____

    _____.

Author's Craft: A Novel   The author gives powerful descriptions of characters,
places, and encounters in the story. Fill in the blank with the person or place in the
following descriptions.

11. It was dark inside the crowded _____ in spite of daylight flooding in from outside.

12. There was an unrelenting struggle between her back which was beginning to bend and her will to keep

    it straight. _____

13. She was a thin truculent child, disturbing, even threatening. _____

14. The _____ plants leaves flapped like elephants ears in the wind.

15. England sent _____ as a show of force, flying so low the downdraft from them shook

    the mangoes from trees in the orchard.

Assessment Worksheet 9

from The Way to Rainy Mountain
by N. Scott Momaday (pages 75–82)

Vocabulary   The sentences below are based on descriptions or events in the selection.
Look for context clues to help you find the best word below to complete each one.

                    knoll   sacred   deicide   range   unrelenting

1. The Kiowa never understood the grim, _____ advance of the U.S. Cavalry.

2. Rainy Mountain is the name given to the _____ that rises out of the plain in Oklahoma.

3. The Kiowa gathered for their last Sun Dance, but there were no more buffalo and a group of soldiers

   dispersed them; always his grandmother carried, without bitterness this vision of _____.

4. The Kiowa had controlled the open _____ for more than a hundred years.

5. War was the Kiowa's _____ business.

Understanding the Selection   Complete each sentence by adding appropriate
information from the selection.

6. "The hardest weather in the world" (is at Rainy Mountain) means _____

   _____.

7. According to their origin myth, the Kiowa _____

   _____.

8. Aho's religious experiences included _____

   _____.

9. The narrator's feelings for his grandmother are _____

   _____.

10. The narrator believes a man, in his relationship to earth, should _____

   _____.

Author's Craft: A Novel   The selection contains the following elements. Match
the element to its example.

              a. plot   b. character   c. setting   d. description   e. tension

_____ 11. Great green and yellow grasshoppers are everywhere in the tall grass,

         popping up like corn to sting the flesh.

_____ 12. Aho had a reverence for the sun, an ancient awe.

_____ 13. Aho knew from birth the affliction of defeat, the dark brooding of old warriors.

_____ 14. The grandson makes a pilgrimage to visit his grandmother's grave.

_____ 15. Houses are like sentinels in the plain, old keepers of the weather watch.

Assessment Worksheet 10
i yearn by Ricardo Sánchez
[We Who Carry the Endless Seasons] by Virginia Cerenio
(pages 83–86)

Vocabulary   Poets often employ words in special or surprising ways to create an effect.
The sentences below are based on the special use of words in "i yearn" by Ricardo Sánchez.
Look for context clues to help you find the best word below to complete each one.

   chicano infinitum   calo   assail   ?que tal, hermano?   barrio

1. The poet yearns for his neighborhood and home, the _____ in El Paso.

2. He misses hearing _____, the special dialect of Spanish spoken in the barrio.

3. *Ad infinitum* is an expression that means (to infinity) "without end;" the poet makes a joke, by

    cataloging cities that have Chicano neighborhoods, labeling it _____.

4. The poet misses the lively foods that _____ a person's taste buds.

5. He yearns to hear the friendly greeting _____.

Understanding the Selection   Complete each sentence by adding appropriate
information from the selections.

6. The Filipino mothers want their daughters to _____

    _____.

7. The daughters, on the other hand, _____

    _____.

8. The ultimate problem for the daughters is _____

    _____.

9. In "i yearn" the poet misses _____.

10. The ultimate thing Sánchez yearns for is _____.

Author's Craft: Poems   The poets have used language and imagery to express
some of their feelings about their culture. Examine the meanings of the poems by
marking these statements T for true and F for false.

_____ 11. Cerenio feels the strong pull of her mother's dreams for her.

_____ 12. She uses powerful nature imagery including tropical rain in her blood,

            weeping her mother's tears, carrying the reminder of her mother's prayers

            on every breath in and out.

_____ 13. Cerenio hates the "shadows/attached to our feet."

_____ 14. Sánchez loves the sound and feel of speaking his dialect.

_____ 15. Sánchez misses the zesty food of his home.

Assessment Worksheet 11
Seventeen Syllables by Hisaye Yamamoto (pages 88–98)

Vocabulary   The sentences below are based on descriptions or events in the selection.
Look for context clues to help you find the best word below to complete each one.

delectable   giddy   grave   helplessness   dubious

1. Jesus told Rosie that he wanted to meet her alone, without the _____ glances of their

   parents.

2. After Jesus kissed Rosie, she felt a wonderful _____.

3. The feeling was too _____ to describe.

4. The next day Rosie felt serious and _____.

5. She also felt happy and _____.

Understanding the Selection   Complete each sentence by adding appropriate
information from the selection.

6. The title of the story refers to _____.

7. Rosie's father's feelings about poetry were probably _____

   _____.

8. Rosie's mother came to America and married her father because _____

   _____.

9. Jesus and Rosie have gotten to know each other because _____

   _____.

10. Her mother wanted Rosie not to marry because _____

    _____.

Author's Craft: A Story   The story contains the following elements. Match the
element to its example. One element appears twice.

a. climax   b. foreshadowing   c. plot   d. character

_____ 11. Early in the story it is apparent that Rosie sometimes says "yes" to her mother just

   because that is simpler than telling the truth.

_____ 12. Rosie's father explodes like cork from a bottle, stalks off to the house, axes the

   picture to pieces, and burns it.

_____ 13. Rosie's dreams are just beginning as her mother's are ending.

_____ 14. Rosie's father is a simple-minded man, threatened by his wife's new role as a poet.

_____ 15. Early in the story we learn that Ume Hanazono's life span as a poet only lasted

   about three months.

Name _____   Date _____

Assessment Worksheet 12
from Humaweepi, the Warrior Priest
by Leslie Marmon Silko (pages 99–106)

Vocabulary   The sentences below are based on descriptions or events in the selection.
Look for context clues to help you find the best word below to complete each one.

coral   yucca fiber   turquoise   succulent   obsidian

1. The black _____ arrowhead was shiny and sharp.

2. Some of Humaweepi's beads were sky-blue _____.

3. Some other beads were a dark red _____.

4. In the morning they ate tumbleweed sprouts that were _____ and tender.

5. The uncle's sandals were made of _____.

Understanding the Selection   Complete each sentence by adding appropriate
information from the selection.

6. In the opening of the story, Humaweepi does not realize he is being trained to be _____
_____.

7. When they stayed at the pueblo, Humaweepi usually stayed with clan members because _____
_____.

8. When Humaweepi realizes he has learned so much from his uncle he is _____
_____.

9. After Humaweepi sings the bear song he realizes _____
_____.

10. The ceremonial gesture that he makes is _____
_____.

Author's Craft: A Novel   The selection contains the following elements. Match
each element to its example.

a. dialogue   b. character   c. foreshadowing   d. climax   e. theme

_____ 11. Humaweepi's uncle is stern, self–assured, and dedicated.

_____ 12. It is important to learn the wisdom of one's culture, but the learning can take place indirectly,
simply by living beside someone.

_____ 13. The uncle teaches with few words: "Winter isn't easy. All the animals are hungry—not just you."

_____ 14. At sunset Humaweepi recognized the bear form in the rock, sang his prayer song to it, offered his
prized beads, and realized he had become a warrior priest.

_____ 15. As the two prepare for the journey, Humaweepi suddenly knows something unusual is going to
happen.

212

Name _____   Date _____

Assessment Worksheet 13
Cante Ishta—The Eye of the Heart
by Mary Crow Dog (pages 107–114)

Vocabulary   The sentences below are based on descriptions or events in the selection.
Look for context clues to help you find the best word below to complete each one.

purification   enraptured   peyote   sham   intellectualism

1. Leonard advised the narrator to look beneath the _____ realities to see things as they

   really are.

2. The narrator had been to some _____ meetings without really understanding them.

3. Mary Crow Dog felt _____ by the raw spiritual power of the Medicine man, Leonard.

4. The Medicine man's knowing is direct, rather than book–learning _____.

5. The goal of participating in a sweat is _____.

Understanding the Selection   Complete each sentence by adding appropriate
information from the selection.

6. Mary Crow Dog married Leonard who is a _____.

7. Mary learned by _____.

8. White limestone rocks are used for the sweat because _____

   _____.

9. After her first sweat Mary felt _____.

10. At the second sweat described, Mary did not cry out because _____

   _____.

Author's Craft: An Autobiography   The excerpt contains the following literary
elements. Match the element to its example. One element appears twice.

a. character   b. sensory detail   c. comparison   d. personification

_____ 11. From the lodge we can hear the murmur of the river's voice.

_____ 12. Crow Dog is an excellent teacher, and he speaks about harmony between

   humans and the earth.

_____ 13. The bent willow sticks of the sweat lodge are fastened together with strips of

   red cloth.

_____ 14. Green cedar is sprinkled over the hot rocks, filling the air with its aromatic odor.

_____ 15. Mary Crow Dog vows never to eat lobster again, knowing what they have to

   go through.

Name _____ Date _____

Assessment Worksheet 14
Black Hair by Gary Soto
ALONE/december/night by Victor Hernandez Cruz
(pages 115–118)

Vocabulary   Poets often use words in special or surprising ways to create an effect. The sentences below are based on the special use of words in "Black Hair" by Gary Soto. Explain something that is unusual about the words in their context.

1. July was a **ring of heat** we all jumped through.

   _____

2. Shade **rose** from our dirty feet.

   _____

3. Hector Moreno had hard, **turned** muscles.

   _____

4. Soto had an **altar** of worn baseball cards.

   _____

5. Hector **lined** balls deep into center.

   _____

Understanding the Selection   Complete each sentence by adding appropriate information from the selection.

6. To the boys, the game was more than baseball; it was _____

   _____.

7. "Gloves eating baseballs" shows _____.

8. The feelings in Soto's home are _____.

9. Cruz repeats one line three times to get across the idea that _____

   _____.

10. The people around Cruz think _____.

Author's Craft: Poems   The poets have used language and imagery to express some of their feelings about their varied cultures. Examine the meanings of the poems by marking these statements T for true and F for false.

_____ 11. Soto went to the baseball park because he loved to watch Hector Moreno and others who could play well.

_____ 12. He uses color, the brown light and the brown arms, and the black hair to convey a sense of oneness, of belonging.

_____ 13. Cruz may have missed going up on the roof to visit with his friends.

_____ 14. He has found new people to love.

_____ 15. Perhaps his music is his poetry and keeps him company.

Name _____ Date _____

Assessment Worksheet 15
Wasichus in the Hills from *Black Elk Speaks*
as told to John G. Neihardt (pages 125–130)

Vocabulary   Match the proper nouns from the selection with their significance. Write the correct letter in the blank.

_____ 1. **Crazy Horse**

_____ 2. **Red Cloud**

_____ 3. **Wasichus**

_____ 4. **Smoky Earth River**

_____ 5. **Plain of Pine Trees**

a. the Oglala Sioux chief

b. the great Lakota chief

c. the meeting place for the council between Native Americans and white people

d. "stealers of the fat"/white people

e. the place near the hills where Black Elk hoped for a vision

Understanding the Selection   Complete each sentence by adding appropriate information from the selection.

6. The Wasichus are _____.

7. Black Elk felt uncomfortable about throwing stones at the swallows because _____

_____.

8. The treaty between the Wasichus and White Cloud said that the Black Hills would belong to Black

Elk's people as long as _____.

9. Black Elk called the tarpaulin constructed in the middle of the circle for the council to meet under a

_____.

10. The Moon When the Calves Grow Hair is the month of _____

_____.

Author's Craft: Autobiography   The autobiography contains statements that are facts and others that are opinions. Write F to identify statements that are probably based on historical facts. Write O to identify statements that express Black Elk's opinions.

_____ 11. A cloud of white–tailed swallows flew over the heads of Black Elk's band.

_____ 12. The swallows seemed holy.

_____ 13. The yellow metal makes the Wasichus crazy.

_____ 14. Crazy Horse and Sitting Bull stayed away from the big council.

_____ 15. After the council "creeks" of white men flowed into the hills and began to make towns.

215

Name _____   Date _____

Assessment Worksheet 16
The Council of the Great Peace
from The Constitution of the Five Nations (pages 131–137)

Vocabulary   The sentences below are based on descriptions or events in the selection.
Look for context clues to help you find the best word below to complete each one.

candidate   contumacious   allies   royaneh   Confederacy

1. All Business of the five Nations was conducted by the two combined bodies of _____ Lords.

2. When a Confederate Lord is _____, his habit of disobeying the rules is a matter

   for the council of War Chiefs to consider.

3. True _____ are men possessing honorable qualities.

4. Friends, or _____ of the Five Nations will be able to see the smoke of the Council Fire.

5. When a _____ for a Lord position needs to be selected, the women of the Five

   Nations make the choice.

Understanding the Selection   Complete each sentence by adding appropriate
information from the selection.

6. The League of the Iroquois was the constitution of _____

   _____.

7. Peace and Strength are represented by the _____

   (which spread in all four directions) of the Tree of Great Peace.

8. If another nation wanted to join the Confederacy and live according to its beliefs, they could _____

   _____.

9. The Council Fire should not contain chestnut wood because _____

   _____.

10. Wampum belts may be constructed by _____.

Author's Craft: A Document   The Iroquois constitution was passed on orally for hundreds
of years and contains several elements found in poetry and narrative writing. Describe or give an
example of the following elements in the constitution; the article number is given to help you.

11. Sensory description (1) _____

12. Repetition (1) _____

13. Repetition (7) _____

14. Personification (2) _____

15. Repetition (24 and 28) _____

Assessment Worksheet 17
The Indians' Night Promises to be Dark
by Chief Seattle (pages 138–143)

Vocabulary   The sentences below are based on descriptions or events in the selection.
Look for context clues to help you find the best word below to complete each one.

wax   ebb   throng   thrill   somber

1. Seattle believes the White Man's God makes the White Man _____ in number and
   strength every day, while the Red Men's number wanes.

2. To Seattle, the rocks are not dumb and dead; rather they _____ with memories of
   important events.

3. Seattle says that the silent streets will _____ with his people's presence through their
   love of the land.

4. Seattle's people will welcome _____ silence because it will remind them of the beauty
   of their home.

5. Seattle says that his people will _____ away like a rapidly receding tide.

Understanding the Selection   Complete each sentence by adding appropriate
information from the selection.

6. Chief Seattle believes that the White Man's God is partial to _____.

7. The main condition Chief Seattle makes before accepting the White Man's proposition is _____
   _____.

8. The White Man's God has forsaken his Red children, and Chief Seattle fears his God, the Great Spirit,
   may also have _____.

9. Chief Seattle's people return to the tombs of their ancestors with reverence, so he is _____
   _____ that the White Men have abandoned the graves of their ancestors.

10. Chief Seattle warns the White Men that even if his people all die, _____.

Author's Craft: A Speech   Great orators often employ elements of poetry in
their speeches. Chief Seattle uses several, including metaphors and similes. Read the
following comparisons and label each one simile (S) or metaphor (M).

_____ 11. The Red children's multitudes once filled the continent as stars fill the firmament.

_____ 12. Chief Seattle fears that the Red Men are ebbing away like a rapidly receding tide.

_____ 13. The white Governor promises that his brave warriors will be a bristling wall of
         strength to protect the Red Men.

_____ 14. The Red Man has fled the approach of the White Man, as the morning mist flees
         before the morning sun.

_____ 15. Tribe follows tribe, and nation follows nation, like the waves of the sea.

Name _____   Date _____

Assessment Worksheet 18
They Are Taking the Holy Elements from Mother Earth
by Asa Bazhonoodah (pages 146–151)

Vocabulary   The sentences below are based on descriptions or events in the selection.
Look for context clues to help you find the best word below to complete each one.

pollen   **Holy Elements**   **particles**   **contaminate**   **herbs**

1. When Mother Earth needs rain, the author's people give prayers and _____.

2. Bazhonoodah says that _____ of coal dust get into the water and kill their animals.

3. The _____ for Bazhonoodah's people are the land, the air, and the water.

4. The _____ that Bazhonoodah has always gotten from Mother Earth are used to cure
   diseases that white medicine cannot cure.

5. As long as the coal mine is operating, it will _____ the Mother Earth.

Understanding the Selection   Complete each sentence by adding appropriate
information from the selection.

6. It is difficult to report dead sheep and other animals because _____
   _____.

7. The Navajo consider the earth to be _____.

8. When the earth needs rain, at the end of the ceremony, the pollen is _____
   _____.

9. The author believes going to the moon is bad because _____
   _____.

10. The author can no longer find her way around the mountain to collect herbs because _____
    _____.

Author's Craft: Essay   The author employs powerful cause and effect
relationships to make her point in this persuasive essay. Use the text to identify the
element that is missing in the following statements.

11. The sheep belonging to the author's children were killed because _____
    _____.

12. _____ scare the horses.

13. Because the coal mining is destroying grazing lands, _____
    _____.

14. They have trouble watering their livestock because _____
    _____.

15. Peabody Coal Company cannot replant the land because _____
    _____.

Assessment Worksheet 19
The Drinking Gourd
Steal Away (pages 153–156)

Vocabulary   Poets often use the same words over and over for effects of emphasis, rhythm, or rhyme. Often these words have been chosen or made up because they sound good. Sometimes the meanings of the words go beyond their usual dictionary definitions. Explain why any of these statements is true for each of the following words or phrases from "The Drinking Gourd" and "Steal Away."

  1. **drinking gourd** _____

  2. **old man** _____

  3. **a-waitin'** _____

  4. **steal away** _____

  5. **the trumpet sounds** _____

Understanding the Selections   Complete each sentence by adding appropriate information from the selections.

  6. "The Drinking Gourd" suggests that the slaves can find their way by following _____

  _____.

  7. Instead of looking for road signs, the slaves should look for _____

  _____.

  8. The old man in "The Drinking Gourd" represents _____.

  9. In "Steal Away" the slaves are called by the Lord through _____

  _____.

 10. The sounds the slaves "hear" inside are _____.

Author's Craft: Lyrics   The two songs use repetition and refrains to emphasize meaning, link ideas, and create rhythm. Answer the questions about these elements in the two songs.

 11. What line is repeated several times in "The Drinking Gourd"?

 _____

 12. Why is this line important enough to be repeated like this?

 _____

 13. How are the last three stanzas of "Steal Away" alike?

 _____

 14. How do these stanzas use repetition and refrain effectively in the three ways explained above?

 _____

 15. Why might repetition be especially effective in songs?

 _____

Assessment Worksheet 20

The Slave Auction by Frances E. W. Harper
The Slave Ship by Olaudah Equiano (pages 157–163)

Vocabulary   When talking about the horrors of slavery, authors often use words that have negative connotations (hidden meanings) for readers. The words below from "The Slave Ship" are examples. Write the dictionary meaning of each word to see if it matches any negative connotations of the word.

1. **slave ship** _____

2. **cruelty** _____

3. **fetters** _____

4. **avarice** _____

5. **pestilential** _____

Understanding the Selections   Complete each sentence by adding appropriate information from the selections.

6. "The Slave Auction" deals mostly with the emotional trauma associated with one result of slave

    auctions: _____.

7. Equiano talks most about one horror of the hold the slaves were placed in: _____

    _____.

8. Equiano was allowed to be on deck, without chains, because _____

    _____.

9. The slaves were shipped to Barbados because _____.

10. Equiano found the cruelest part of the slave auction to be _____.

Author's Craft: Autobiography   Autobiography contains facts as well as the author's feelings and opinions about what happened to him or her. Decide if each of the following sentences from "The Slave Ship" represents factual recall of events or an emotional interpretation of what happened. Write F for fact and E for emotion.

_____ 11. The first object that saluted my eyes when I arrived on the coast was the sea, and a slave ship

which was then riding at anchor and waiting for its cargo.

_____ 12. I was immediately handled and tossed up to see if I was sound by some of the crew, and I was now

persuaded that I had gotten into a world of bad spirits and that they were going to kill me.

_____ 13. The closeness of the place and the heat of the climate, added to the number in the ship, which

was so crowded that each had scarcely room to turn himself, almost suffocated us.

_____ 14. The shrieks of the woman and the groans of the dying rendered the whole a scene of horror

almost inconceivable.

_____ 15. On a signal given, (as the beat of a drum) the buyers rush at once into the yard where the slaves

are confined, and make a choice of that parcel they like best.

Name _____   Date _____

Assessment Worksheet 21
The Slave who Dared to Feel like a Man.
by Harriet A. Jacobs (pages 164–171)

Vocabulary   The words below are used in the selection to describe slave owners or
their attitudes toward slaves. Choose the word that best fits each sentence and write
it in the blank.

> master   overseer   chattel   yoke   indecorum   audacity

1. Benjamin's _____ often got him into trouble because slaves were not supposed

    to be bold.

2. The _____ told his slaves to obey him in every instance.

3. The _____ was instructed to discipline slaves who acted improperly.

4. The _____ of slavery was very heavy and owners did little to lighten the pressures.

5. Slaves were considered the owner's _____.

6. Any action that showed free spirit, even laughing, was considered to be an act of _____.

Understanding the Selection   Complete each sentence by adding appropriate
information from the selection.

7. The free grandmother in the narrative had her own house which the other family members liked to visit

    because _____.

8. The grandmother tried to explain slavery by saying _____

    _____.

9. Benjamin's personality was particularly unsuited to slavery because _____

    _____.

10. Benjamin's master tried to subdue him by _____

    _____.

Author's Craft: Personal Narrative   A narrative tells about an incident in
terms of a series of chronological events. Number the events listed below from "The
Slave who Dared to Feel like a Man." from 1-5 to show which happened first, second,
and so on.

_____ 11. A violent storm overtakes the ship Benjamin is on.

_____ 12. Benjamin meets Phil in New York.

_____ 13. Benjamin strikes down his master.

_____ 14. Benjamin is sold for three hundred dollars.

_____ 15. Benjamin is put in jail for many months.

Name _____   Date _____

Assessment Worksheet 22
The People Could Fly told by Virginia Hamilton
Runagate Runagate by Robert Hayden (pages 172–179)

Vocabulary   The sentences below are based on descriptions or events in the selections. Look for context clues to help you find the best word below to complete each one.

croon   thicketed   movering   underground   plausible   motif

1. The _____ of escaping from slavery is common in African American folktales.

2. When slaves went _____ they went into hiding as part of their escape attempts.

3. Toby's _____ restored the magic to the slaves in the field.

4. Folktales often present _____ explanations for seemingly supernatural events.

5. A _____ area would provide covering for a slave in hiding.

6. Slaves who traveled on the "underground train" were actually _____ from safe house to safe house on foot.

Understanding the Selections   Complete each sentence by adding appropriate information from the selections.

7. Toby knew it was time for Sarah to go when _____.

8. Toby laughed when the slave owner was going to kill him because _____.

9. The slaves in "The People Could Fly" fly away to _____.

10. The people in "Runagate Runagate" are running to _____.

11. The setting of "Runagate Runagate" is nighttime because _____.

Author's Craft: Folktales   Folktales contain the following elements. Match the element to its example from "The People Could Fly."

       a. character with desirable qualities   b. character with undesirable qualities

       c. oral tradition   d. problem resolution through magic

_____ 12. They say that the children of the ones who could not fly told their children. And now, me, I have told it to you.

_____ 13. Toby was there when there was no one to help her and the babe.

_____ 14. Then she felt the magic, the African mystery. Say she rose just as free as a bird. As light as a feather.

_____ 15. The owner of the slaves callin himself their Master. Say he was a hard lump of clay. A hard, glinty coal. A hard rock pile, wouldn't be moved.

Assessment Worksheet 23
Lali by Nicholasa Mohr (pages 182–189)

Vocabulary   The sentences below are based on descriptions or events in the selection. Look for context clues to help you find the best word below to complete each one.

> tenement   countryside   luncheonette   pollution   taciturn

1. Lali often daydreamed about the beautiful _____ of Puerto Rico.

2. The _____ of the city turned the snow an ugly brown color.

3. Because Rudi was _____, he failed to help Lali overcome her shyness.

4. Working at the _____ was difficult for Lali, but it helped her to forget her homesickness.

5. The _____ in which Rudi and Lali lived needed fresh paint.

Understanding the Selection   Complete each sentence by adding appropriate information from the selection.

6. Lali's work at the luncheonette includes _____.

7. Lali and Rudi's marriage has been a strain because _____.

8. Chiquitin is important to Lali because _____.

9. Lali is disappointed in New York because _____.

10. Rudi and Lali are both looking forward to visiting Puerto Rico because _____

_____.

Author's Craft: A Story   Setting is a very important element of fiction. In "Lali," Mohr uses both the actual setting of the story and a flashback setting effectively. Identify the setting of each example below as either from the time of the story (S) or from the flashback (F).

_____ 11. The old radiators hissed and clanked, sending out vaporous heat. In spite of all the noise and steam, the room remained cold.

_____ 12. Outside in the street, Lali heard the trucks changing gears; buses and cars honked their horns.

_____ 13. Lali inhaled the sweet and spicy fragrance of the flower gardens that sprinkled the countryside in abundance.

_____ 14. She saw the morning mist settling like puffs of smoke scattered over the range of mountains that surrounded the entire countryside.

_____ 15. ...as soon as the snow fell, traffic and pollution turned it to a brown and murky slush.

Name _____   Date _____

Assessment Worksheet 24
from Picture Bride by Yoshiko Uchida (pages 190–198)

Vocabulary   The sentences below are based on descriptions or events in the selection.
Look for context clues to help you find the best word below to complete each one.

kimono   pompadour   derby   sallow   flustered

1. Hana brought traditional Japanese items with her, including the _____ she wore

   on the ship.

2. Taro had been affected by American fashions, as the fact that he wore a _____ when

   he met Hana reveals.

3. Taro's _____ complexion indicates that he must work inside the store a lot.

4. Hana's _____ had the effect of making her look even smaller.

5. Hana was nervous and _____ when she first met Taro.

Understanding the Selection   Complete each sentence by adding appropriate
information from the selection.

6. The idea of going to America to wed Taro Takeda appealed to Hana because _____.

7. After receiving Taro's letters and picture, Hana knew _____ about him.

8. As she nears America, Hana is excited about _____

   and anxious about _____.

9. The immigrants were detained at Angel Island because _____.

10. Hana begins to feel better after she and Taro _____.

Author's Craft: A Novel   Characterization is an important element of fiction.
An author can achieve characterization in several ways. Match the techniques below
with the examples from *Picture Bride*.

a. telling directly about the character
b. describing what the character looks like
c. revealing a character's thought and feelings

d. showing what other characters say or think
   about the character
e. showing the character's actions

_____ 11. She was thin and small, her dark eyes shadowed in her pale face, her black hair

   piled high in a pompadour that seemed too heavy for so slight a woman.

_____ 12. In that same instant, Hana knew she wanted more for herself than her sisters

   had in their proper, arranged and loveless marriages.

_____ 13. "It has addled her brain—all that learning from those books," he said when he

   tired of arguing with Hana.

_____ 14. She had graduated from Women's High School in Kyoto which gave her five

   more years of schooling than her older sister.

_____ 15. Hana took a deep breath, lifted her head and walked slowly from the launch.

Name _____   Date _____

Assessment Worksheet 25
I Leave South Africa by Mark Mathabane (pages 199–206)

Vocabulary   Match each word on the right (from "I Leave South Africa") with the word on the left that has an opposite meaning. Write the letter in the correct blank.

_____ 1. truth   a. **railed**

_____ 2. apartheid   b. **freedom**

_____ 3. bondage   c. **dependent**

_____ 4. liberated   d. **fraud**

_____ 5. praised   e. **segregation**

Understanding the Selection   Complete each sentence by adding appropriate information from the selection.

6. Nkwame tells Mark that a black college would be best for him because he believes black colleges _____
_____.

7. Nkwame warns Mark about _____
_____.

8. Mark is held up in customs because _____
_____.

9. The incentive for South African blacks to copy black Americans was _____
_____.

10. Mark believes there is no apartheid in the United States when he sees _____
_____.

Author's Craft: Autobiography   Irony is an effective way for an author to contrast reality with expectations. Write Yes beside the examples from "I Leave South Africa" that use irony and No beside those that do not.

_____ 11. "Ever since I saw 'Roots' I have always wanted to know where my

homeland is."

_____ 12. The air seemed pervaded with freedom and hope and opportunity.

_____ 13. "There truly is no apartheid here."

_____ 14. "Integrated schools are the worst places for black folks."

_____ 15. Black Americans did indeed possess the sophistication to see through any

ruse an African puts up.

225

Name _____   Date _____

Assessment Worksheet 26
Ellis Island by Joseph Bruchac (pages 207–209)

Vocabulary   Write the words in the correct blanks to complete the story about immigration.

Ellis Island   old Empires   quarantine   decades   dreams   red brick

The immigrants coming from the 1. _____ of Europe had many

2. _____ as they sailed across the ocean toward America. Their excitement

peaked as the saw the 3. _____ buildings perched on 4. _____.

They patiently waited through the long days of 5. _____, until they were finally

free to go to the mainland. They hoped that the 6. _____ ahead of them would

bring fortune and happiness.

Understanding the Selection   Complete each sentence by adding appropriate information from the selection.

7. The setting of the poem as described in the first stanza is _____

_____.

8. The "tall woman" mentioned in the poem is _____

_____.

9. The "part of his blood" that loves Ellis Island is Bruchac's _____

_____.

10. According to the third stanza, the Native Americans' lifestyle was characterized by _____

_____.

11. As a result of immigration as described in "Ellis Island," the immigrants would _____

land whereas the Native Americans would _____ their land.

Author's Craft: A Poem   Poets use figurative language to help readers understand ideas and feelings. Read the literal descriptions below and find how the poet imaginatively expresses them in "Ellis Island." Write his words in the blank.

12. The Statue of Liberty is green in color. _____

13. For 90 years, Ellis Island represented the opportunity to be free and own land. _____

_____

14. Because the poet has a Native American heritage, he remembers the people who lived here before the

immigrants. _____

15. The Native Americans understood the seasons and moved around in order to take advantage of them.

_____

Name _____   Date _____

Assessment Worksheet 27
Those Who Don't, No Speak English, and The Three Sisters
by Sandra Cisneros (pages 212–217)

Vocabulary   Match the Spanish words on the right (from the selections) with
their English meanings on the left. Write the correct letter in each blank.

_____   1. Mother   a. **Las comadres**

_____   2. Close family friends   b. **Esperanza**

_____   3. Hope   c. **Cuando**

_____   4. When   d. **Mamacita**

Understanding the Selections   Complete each sentence by adding appropriate
information from the selections.

5. The people who live in the neighborhood described in "Those Who Don't" are mostly of a _____

_____ ethnic background.

6. The people of the neighborhood in "Those Who Don't" are uncomfortable when _____

_____.

7. Esperanza believes the reason Mamacita doesn't leave her apartment is because _____

_____.

8. To Mamacita, home is _____.

9. The three sisters describe Esperanza as _____.

10. Esperanza wishes for _____.

11. The three sisters tell Esperanza that she must _____.

Author's Craft: A Novel   Tone is the author's personal mark on a piece of
writing. Tone indicates the author's attitude toward the subject, the characters, or the
reader. Identify the tone of each excerpt below as angry (A), humorous (H), or
serious (S).

_____   12. They are stupid people who are lost and got here by mistake.

_____   13. Out stepped a tiny pink shoe, a foot soft as a rabbit's ear, then the thick

ankle, a flutter of hips, fuchsia roses and green perfume. The man had to

pull her, the taxicab driver had to push. Push, pull. Push, pull. Poof!

_____   14. And then to break her heart forever, the baby boy who has begun to talk,

starts to sing the Pepsi commercial he heard on T.V.

_____   15. I didn't understand everything they had told me. I turned around. They

smiled and waved in their smoky way.

Name _____   Date _____

Assessment Worksheet 28
I See the Promised Land
by Martin Luther King, Jr. (pages 225–231)

Vocabulary   The sentences below are based on descriptions or events in the selection.
Look for context clues to help you find the best word below to complete each one.

> Promised Land   human rights   injustice   grapple
>
> First Amendment   injunction   paddy wagons

1. When King refers in his speech to certain freedoms he is referring to the _____ of the
   U.S. Constitution.

2. King says that he has seen the _____ and that makes him happy, even if he might not
   reach it with the others.

3. King points out that the _____ movement is not about violence and nonviolence, but
   about nonviolence or nonexistence.

4. King and his followers were taken to jail in _____.

5. King labels the _____ in Memphis as illegal and unconstitutional.

6. King encourages the people to _____ with the problems they face rather than simply
   talking about them.

7. The main issue King and his followers are addressing is _____.

Understanding the Selection   Complete each sentence by adding appropriate
information from the selection.

8. King says that the members of the movement are not negative; rather, they are determined to _____
   _____.

9. King believes that the movement will have the most strength if _____.

10. King calls the injunction in Memphis unjust because it goes against certain constitutional rights _____
    _____.

11. King encourages his listeners to band together economically by _____.

12. King feels happy on the night of this speech because _____.

Author's Craft: A Speech   Allusions are references to familiar people, places, events,
literary works, or art works. By using allusions, speechmakers can say a lot through a few words.
An allusion and its associations can paint a vivid picture for the listeners. Martin Luther King,
Jr., often used allusions from the Bible in his speeches. Match the allusion on the left with the
main idea from the selection that it enriches. Write the correct letter in the blank.

_____ 13. promised land   a. civil rights

_____ 14. Pharaoh   b. equality

_____ 15. God's children   c. unity

Name _____   Date _____

Assessment Worksheet 29
See for yourself, listen for yourself, think for yourself
by Malcolm X (pages 232–237)

Vocabulary   The sentences below are based on descriptions or events in the selection. Look for context clues to help you find the best word to complete each one.

nonviolent   forgiving   violent   brutality   rumble

1. Hurting people in their own homes and hanging people are acts of _____ that Malcolm X mentions.

2. Malcolm X says that African Americans often fail to be _____ with one another.

3. Loving one another, being patient with one another, and _____ one another are actions of nonviolence.

4. Malcolm X says that a person only has to visit Harlem Hospital to see how _____ the people of Harlem are towards one another.

5. A _____ is a violent street fight.

Understanding the Selection   Complete each sentence by adding appropriate information from the selection.

6. Malcolm X tells about the incident on the plane because it shows _____.

7. Malcolm X says that young people can come to intelligent decisions for themselves only if they _____
_____.

8. Malcolm X points out that nonviolence among African Americans is only used with _____
_____.

9. Malcolm X says he can only accept the practice of nonviolence if _____.

10. Harlem is described in the speech as _____.

Author's Craft: A Speech   Restatement, repetition, and parallelism are used effectively in speeches to highlight important concepts. Decide which of the techniques is represented by each excerpt below. Write res (restatement), rep (repetition), or par (parallelism) in the blank.

_____ 11. The most important thing we can learn how to do today is think for ourselves ...it is very important to think out a situation for yourself.

_____ 12. ...the habit of going by what you hear others say about someone, or going by what others think about someone....

_____ 13. A person can come to your home...Or he can come put a rope around your neck....Or he can come to take your father out and put a rope around his neck....

_____ 14. So I myself would go for nonviolence if it was consistent....if everybody was going to be nonviolent, and if we were going to be nonviolent all the time.

_____ 15. But I don't go along—and I'm just telling you how I think—I don't go along with any kind of nonviolence....

Assessment Worksheet 30

from *China Men* by Maxine Hong Kingston
Immigration Blues from *The Gold Mountain Poems*
(pages 238–246)

Vocabulary   The sentences below are based on descriptions or events from *China Men*. Look for context clues to help you find the best word to complete each one.

pickaxed   **shoveled**   **sledgehammer**   **jarred**   **gouged**   **dynamite**   **diverted**

1. Once _____ was invented and used, it _____ the land so that the landscape became hard to recognize.

2. Streams were _____ by the force of the blasts.

3. When Ah Goong used the _____, his whole body was _____ by the impact of the tool striking rock.

4. When Ah Goong tunneled through dirt he _____ it.

5. The dirt was then _____ into mule carts.

Understanding the Selections   Complete each sentence by adding appropriate information from the selections.

6. Ah Goong's physical discomfort when tunneling included _____.

7. Time had little meaning to those digging tunnels because _____.

8. The changes dynamite brought to the tunneling efforts included _____.

9. Hardships created by snow included _____.

10. The Gold Mountain Poems all express anger at _____.

Author's Craft: A Story   Indirect characterization allows the author to reveal character in several ways, listed below. Write the letter of the correct technique next to the example from *China Men*.

   a. through the words, thoughts, or actions of the character

   b. through descriptions of the character's appearance or background

   c. by the ways in which other characters react to the character

_____ 11. This rock is what is real, he thought.

_____ 12. But he was a crazy man, and they didn't listen to him.

_____ 13. After tunneling into granite for about three years, Ah Goong understood the immovability of the earth.

_____ 14. "I felt time," he said. "I saw time. I saw world."

_____ 15. The steady banging reminded him of holidays and harvests; falling asleep, he heard the women chopping mincemeat and the millstones striking.

Assessment Worksheet 31
Napa, California by Ana Castillo
The Circuit by Francisco Jiménez (pages 247–257)

Vocabulary

harvest   **sweat**   **sun-beaten**   **dusk**   **sharecropper**   **braceros**   **corridos**

1. A _____ lived on a farm and worked the land, whereas _____ always had to move from farm to farm.

2. In "Napa, California" the _____ yields grapes.

3. Ana Castillo says that the workers' pride has been wiped away like the _____ from their _____ brows.

4. The workers labor daily from dawn until _____.

5. The boy in "The Circuit" says the music he most enjoys is _____

Understanding the Selections   Complete each sentence by adding appropriate information from the selections.

6. The people in "Napa, California" must work as hard as they do in order to _____
_____.

7. When the different picking seasons peak and the crops fall off, the family in "The Circuit" must _____
_____.

8. Carcanchita is _____.

9. Mama's favorite belonging is _____.

10. The children in the family are able to go to school when _____
_____.

11. The boy in "The Circuit" was going to start learning to play the trumpet, but _____
_____.

Author's Craft: A Story   Complete the following sentences to describe the setting in "The Circuit."

12. No matter which place the family is in, the weather is always _____
_____.

13. Although the family lives in different places, the building they call home is always _____
_____.

14. The seasons in "The Circuit" are not fall, winter, spring, and summer, but the seasons of _____
_____.

15. The presence of cardboard boxes always indicates _____
_____.

Name _____ Date _____

Assessment Worksheet 32
Speech by Sojourner Truth (pages 260–263)

Vocabulary   Match each religious name from the selection with the meaning it is intended to give to the reader. Write the correct letter in the blank.

_____ 1. **Eve**

_____ 2. **Jesus**

_____ 3. **Lazarus**

_____ 4. **Mary**

_____ 5. **God**

a. a man saved by women's pleas

b. responsible, with women, for men's existence

c. women as the cause of the world's problems

d. a man who never blamed women

e. responsible, with God, for men's existence

Understanding the Selection   Complete each sentence by adding appropriate information from the selection.

6. Sojourner Truth says that she can do these things as "much as any man": _____
_____.

7. Sojourner Truth says men shouldn't fear giving women their rights because _____
_____.

8. Sojourner Truth suggests that if men believe women upset the world, men should now _____
_____.

9. According to Sojourner Truth, man had no part in Jesus's coming into the world— _____
_____ created him and _____ bore him.

10. In the last sentence of the speech Sojourner Truth mentions another issue besides women's rights, the
issue of _____.

Author's Craft: A Speech   Humor, logic, and rhetorical questions are used by Sojourner Truth to make her persuasive speech effective. Find examples from the speech that fit the following descriptions.

11. Use of logic to point out women's equality _____
_____.

12. Use of humor to dismiss women's guilt as coming from Eve _____
_____.

13. A rhetorical question about how much man can do as compared to woman _____
_____.

14. A rhetorical question about man's responsibility for Jesus _____
_____.

15. Use of humor to describe the position man is in _____
_____.

Name _____ Date _____

Assessment Worksheet 33
from *I Am Joaquín* by Rodolfo Gonzales (pages 264–268)

Vocabulary   In *I Am Joaquín*, Gonzales uses the term *La Raza* to mean the
Mexican people. He also uses the Spanish words below for the different Spanish-
speaking people. Tell which group each word describes.

1. **Méjicano** _____   4. **Hispano** _____

2. **Epañol** _____   5. **Chicano** _____

3. **Latino** _____

Understanding the Selection   Complete each sentence by adding appropriate
information from the selection.

6. Joaquín believes that if he chooses to be true to his cultural heritage he will be _____

    _____; if he chooses to become "gringo" he will be _____.

7. Joaquín talks about two different groups whose oppression his people have overcome: _____

    _____.

8. Joaquín also describes the physical hardships his people have survived, including _____

    _____.

9. Finally, Joaquín refers to emotional hardships brought about by _____

    _____.

10. The two traits that Joaquín says will help his people survive are _____

    _____.

Author's Craft: Free Verse   Parallelism adds rhythm to free verse, while also
allowing poets to emphasize and unify their ideas. Decide whether each example of
parallelism from *I Am Joaquín* is used mostly for rhythm (R), emphasis (E), or unity
(U). Write the correct letter in the blank.

_____ 11. Part of the blood that runs deep in me....Part of the blood that is mine....

_____ 12. I have endured....I have survived....I have existed....

_____ 13. ...my spirit is strong, my faith unbreakable, my blood is pure.

_____ 14. ...confused by the rules,/ scorned by attitudes,/ suppressed by

      manipulation,/ and destroyed by modern society.

_____ 15. ...in the barrios of the city/ in the suburbs of bigotry/ in the mines of social

      snobbery/ in the prisons of dejection/ in the muck of exploitation/ and/ in

      the fierce heat of racial hatred.

Name _____   Date _____

Assessment Worksheet 34
Free to Go by Jeanne Wakatsuki Houston
and James D. Houston (pages 269–275)

Vocabulary   The sentences below are based on descriptions or events in the
selection. Look for context clues to help you find the best word below to complete
each one.

        nightriders   **assimilate**   **curfew**   **evacuation order**   **propaganda**   **internment camps**

1. In 1942 the army imposed a _____ on all west-coast Japanese; Gordon Hirabayashi

   protested this as racially biased.

2. The _____ of 1942 removed all Japanese and Japanese Americans living on the

   West Coast from their homes and placed them in _____.

3. Much of the _____ during the war was intended to turn people against the Japanese.

4. Japanese were often accused of refusing to _____.

5. Hostility against the Japanese was apparent in attitudes and in actions like the violence

   performed by _____.

Understanding the Selection   Complete each sentence by adding appropriate
information from the selection.

6. The family was not joyful about their release from the camp because _____.

7. While in the camp, the girl dreamed about _____.

8. The older children in the family decided to _____ after their release.

9. Papa would never leave the west coast because _____.

10. The family had felt safe in the camp because _____.

Author's Craft: Personal Narrative   Irony allows an author to contrast what is
stated and what is meant. Decide whether each example from "Free to Go" is
dramatic irony (D) or irony of situation (S). Write the correct letter in the blank.

_____ 11. In our family the response to this news was hardly joyful.

_____ 12. All the truly good things, it often seemed, the things we couldn't get, were outside....

_____ 13. The nuns expected us to ask for purity of soul, or a holy life. I asked God for some

      dried apricots.

_____ 14. "See you in New Jersey," we would wave, as the bus pulled out taking someone else to

      the train station in L.A.

_____ 15. She spent two and a half years awaiting the high court's decision, which was that she

      had been right: the government cannot detain loyal citizens against their will.

Name _____ Date _____

Assessment Worksheet 35
In the American Storm by Constantine M. Panunzio (pages 276-282)

Vocabulary   The sentences below are based on descriptions or events in the selection.
Look for context clues to help you find the best word below to complete each one.

mattock   pick and shovel   padrone   excavation   humiliating

1. Panunzio points out that the typical work available to immigrants like him was hard

   and _____.

2. For Italian immigrants the work in America that was most familiar to them was _____.

3. A _____ is a tool used to dig.

4. Panunzio and Louis found the work using _____ to be heavy and monotonous.

5. The _____ who found Panunzio and Louis their first job turned out to be a brute.

Understanding the Selection   Complete each sentence by adding appropriate
information from the selection.

6. During his first five days in America, Panunzio's activities consisted of _____

   _____.

7. Panunzio was happy to meet Louis because _____.

8. Panunzio and Louis were told that the only work available to Italians was _____

   _____.

9. The trouble Panunzio and Louis had with their first job was _____.

10. The trouble they had with their second job was _____.

Author's Craft: Autobiography   A writer's tone shows how he or she feels
toward the subject, characters, or audience. Write one descriptive word to label the
tone of each excerpt from "In the American Storm." (examples: informal, humorous,
excited, formal, restrained, serious, gloomy, optimistic)

_____ 11. Late in the evening of September 8, 1902, when the turmoil of the street traffic
was subsiding, and the silence of the night was slowly creeping over the city, I took my sea chest, my
sailor bag and all I had and set foot on American soil.

_____ 12. He drove me away, and I think I cried; I cried my first American cry. What
became of me that night I cannot say.

_____ 13. Just to look at Louis would make you laugh. He was over six feet tall, lank,
queer-shaped, freckle-faced, with small eyes and a crooked nose.

_____ 14. Now these were the first two English words I had heard and they possessed great
charm. Moreover, if I were to earn money to return home and this was the only work available for
Italians, they were very weighty words for me, and I must master them . . .

_____ 15. So we left. My bandaged hand hurt me, but my heart hurt more. This kind of
work was hard and humiliating enough, but what went deeper than all else was the first realization that
because of race I was being put on the road.

Name _____   Date _____

Assessment Worksheet 36
from Their Eyes Were Watching God
by Zora Neale Hurston (pages 285–288)

Vocabulary   Write a definition for each example of dialect from the selection.

1. speck _____

2. chile _____

3. useter _____

4. chillun _____

5. youngun _____

Understanding the Selection   Complete each sentence by adding appropriate information from the selection.

6. Janie was raised by _____.

7. Janie's playmates were _____.

8. Janie learned that she was African American when _____

_____.

9. People called Janie "Alphabet" because _____

_____.

10. Janie called her Grandma "Nanny" because _____

_____.

Author's Craft: Regional Writing   Dialect is used by writers to create realistic characters and to reveal the distinct characteristics of a region. Find the sentences from *Their Eyes Were Watching God* that say the same thing as the sentences below. Write the dialect sentences in the blanks. Think about how much more lively and colorful the dialect is.

11. I know what I need to tell you, but I don't know where to begin. _____

_____

12. Grandma used to catch us when we were misbehaving and give us all a spanking. _____

_____

13. I believe that we always deserved our punishment because we three boys and two girls were very

naughty. _____

_____

14. I had the nickname "Alphabet" because I had been called so many different names. _____

_____

15. Before I saw that picture I thought I was just like everyone else around me. _____

_____

Assessment Worksheet 37
Refugee Ship by Lorna Dee Cervantes
address by Alurista
Letter to America by Francisco X. Alarcón (pages 289–293)

Vocabulary   The sentences below are based on descriptions in the selections. Look for context clues to help you find the best word below to complete each one.

foreign   bronzed   lag   perdone   race

1. In "Refugee Ship," the Spanish name of the girl seems _____ to her.

2. At the beginning of "Letter to America" the poet apologizes for the _____ in writing.

3. Even though the person in "address" keeps politely saying _____, he is ignored.

4. When the girl in "Refugee Ship" looks in the mirror she sees black hair and _____ skin.

5. Ironically, the last piece of information that the interviewer asks for in "address" is_____ .

Understanding the Selections   Complete each sentence.

6. When the poet of "Refugee Ship" says that the ship will never dock, she means _____
_____.

7. The second stanza of "Refugee Ship" describes two things about the girl: _____
_____.

8. "Letter to America" names negative feelings whites have toward Mexican Americans, including _____
_____.

9. In "Letter to America" negative actions done by whites to Mexican Americans include _____
_____.

10. The last item on the list of information in "address" is _____.

11. The Spanish speaker in "address" is trying to tell his _____.

Author's Craft: Poetry   Word choice often greatly affects mood. Answer the following questions about the poets' choice of words for the poems in this lesson.

12. How do these words from "Refugee Ship" affect the poem's mood: wet cornstarch, orphaned, foreign,

    stumbling, captive? _____

13. How does the tone of these lines from "Letter to America" affect the mood of the poem: you fear us/ you

    yell at us/ you hate us/ you shoot us/ you mourn us/ your deny us? _____

_____

14. The Spanish speaker in "address" keeps saying "Excuse me." How does this affect the mood of the poem?

_____

15. "Refugee Ship" and "address" both include Spanish lines. How does this affect the mood of the poems?

_____

Name _____ Date _____

Assessment Worksheet 38
from Black Boy by Richard Wright (pages 294–299)

Vocabulary The sentences below are based on descriptions or events in the selection. Look for context clues to help you find the best word below to complete each one.

countered   evaded   taunting   peevishly   grudgingly

1. Early in the passage, Wright's mother's tone is described as _____, as she tries to make light of her son's serious questions.

2. Later in the passage, the mother begins to speak _____, as she becomes more and more irritated by the questions she is asked.

3. Several times when Wright asks a question, it is _____ by a question from his mother.

4. When Wright's mother doesn't want to answer a question directly, the question is _____ through a clever response.

5. When Wright's mother finally tells him about the ethnic background of his Granny, she does so _____.

Understanding the Selection Complete each sentence by adding appropriate information from the selection.

6. As Richard Wright boards the train for Arkansas, he realizes for the first time that _____ _____.

7. Wright is curious about Granny's _____.

8. Wright's mother explains that the people in their family got their names from _____ _____.

9. Wright is angered by his mother's response to his question regarding his own ethnic background because _____.

10. At Aunt Maggie's bungalow in Elaine, Richard enjoys _____ _____.

Author's Craft: Autobiography Choose one sentence from the excerpt that reveals Wright's attitude about each item listed below. Write the detail you select.

11. white people _____

12. his mother's feelings about Granny's background _____

13. his own ethnic heritage _____

14. Aunt Maggie's bungalow _____

15. his power over bees _____

Name _____ Date _____

Assessment Worksheet 39
from Four Directions by Amy Tan
Saying Yes by Diana Chang (pages 300–312)

Vocabulary   The sentences below are based on descriptions or events in the selections. Look for context clues to help you find the best word below to complete each one.

concocted   ferocity   slack   guileless   vulnerability

1. The mother in "Four Directions" reveals her personality even in the way she cooks—when she chops the food with _____.

2. The daughter in "Four Directions" has _____ a plan that will convince her mother to accept Rich into the family.

3. When the daughter finds her mother sleeping, she notices that her mother's mouth is _____ and all the lines on her face are gone.

4. Sleeping in this way, the mother looks like a _____ young girl.

5. The daughter realizes that she has never seen her mother's _____ before.

Understanding the Selections   Complete each sentence by adding appropriate information from the selections.

6. Rich makes several mistakes during dinner, including _____
_____.

7. When the daughter sees her mother sleeping, she is confused by _____
_____.

8. The mother tells her daughter that the traits she inherited from her father's side include: _____
_____.

9. From the Sun clan in Taiyuan the daughter receives other traits: _____
_____.

10. In "Saying Yes," the poet describes her dilemma in having to identify herself as either _____
_____ or _____; she thinks of herself as both.

Author's Craft: A Story   Plot is the sequence of events in a literary work. Conflict (internal and external), climax, and resolution are all components of plot. Decide which component of plot each excerpt from "Four Directions" involves. Write your response in the blank provided.

_____ 11. I'd never known love so pure, and I was afraid that it would become sullied ....

_____ 12. I put on my jogging clothes and headed out the door, got into the car, and drove to my parents' apartment.

_____ 13. My mother was doing it again, making me see black where I once saw white. In her hands, I always became the pawn.

_____ 14. And really, I did understand finally. Not what she had just said. But what had been true all along.

_____ 15. My mind was flying one way, my heart another.

Assessment Worksheet 40
from Stelmark: A Family Recollection
by Harry Mark Petrakis (pages 313–320)

Vocabulary   Match the Greek food on the right with the description of it on the left. Write the correct letter in the blank.

_____ 1. a roasted lamb dish

_____ 2. a rice cooked in poultry or meat broth

_____ 3. a soup made with lemon juice

_____ 4. a seed from the legume family of plants

_____ 5. a rich white cheese made from goat's milk

a. **avgolemono**

b. **kokoretsi**

c. **feta**

d. **lentils**

e. **pilaf**

Understanding the Selection   Complete each sentence by adding appropriate information from the selection.

6. Petrakis threw a plum at Barba Nikos because _____

_____.

7. When Barba Nikos describes the foods in his store, he talks about them in terms of _____

_____.

8. As Petrakis leaves the store after his day of work, Barba Nikos gives him a gift of _____

_____.

9. Petrakis believes Barba Nikos would be proud of him as an adult because _____

_____.

10. After 30 years, what Petrakis remembers most about his experience with Barba Nikos is _____

_____.

Author's Craft: Nonfiction   Decide whether each sentence below is a main idea or supporting detail from Petrakis' memoirs. Note that each quote is said by Barba Nikos. Write your answer in the blank provided.

_____ 11. "You are Greek!" he cried.

_____ 12. "Do you have 75 cents, boy?"

_____ 13. "You don't understand that a whole nation and a people are in this store."

_____ 14. "The men of Marathon carried small packets of these spices into battle, and the scents reminded them of their homes, their families, and their children."

_____ 15. "We are square now. Keep it that way."

Name _____ Date _____

Assessment Worksheet 41
from Fences by August Wilson (pages 321–330)

Vocabulary   The paragraph below could be considered as advice from Troy regarding how to play baseball. Fill in the blanks by writing each word below where it belongs.

>   bench   timing   follow-through   pitching   home runs

The key to being successful in baseball is being able to handle the 1. _____. You need to

play every day, working on your 2. _____ and 3. _____. You certainly won't

be able to hit any 4. _____ if you are sitting on the 5. _____ instead of

playing.

Understanding the Selection   Complete each sentence by adding appropriate information from the selection.

6. Cory wants his dad to buy a _____, but Troy says they

   need to buy a _____ instead.

7. Troy wants Cory to do two things besides practicing football: _____

   _____.

8. Troy is afraid that if Cory counts on playing football he will be disappointed because _____

   _____.

9. Troy says he takes care of Cory because _____

   _____.

10. Troy says that the best thing that ever happened to him was _____

   _____.

Author's Craft: Drama   A playwright depends on dialogue in order to develop character. Decide how Troy's character is developed by each example of dialogue below. Write the adjective that sums up the character trait that is revealed: practical, independent, bitter, or tired.

_____ 11. While you thinking about a TV, I got to be thinking about the roof...and
                       whatever else go wrong around here.

_____ 12. I don't want him to be like me! I want him to move as far away from my life as
                       he can get.

_____ 13. I give you the lint from my pockets. I give you my sweat and my blood. I ain't got
                       no tears. I done spent them.

_____ 14. I ain't gonna owe nobody nothing if I can help it.

_____ 15. The white man ain't gonna let you get nowhere with that football noway.

Name _____   Date _____

Assessment Worksheet 42
America: The Multinational Society
by Ishmael Reed (pages 333–339)

Vocabulary   The sentences below are based on descriptions or events in the selection.
Look for context clues to help you find the best word below to complete each one.

mosques   pastrami   Yoruban   calypso   bouillabaisse

1. A _____ sandwich often comes with cheese and mustard on Jewish rye bread.

2. _____ is spoken by one group of West Africans.

3. Robert Thompson refers to U.S. culture as a _____, because the mixture of different

   people and customs is like the many ingredients of a Russian fish soup.

4. The beautiful _____ often seen in areas with Islamic populations are centers of worship.

5. The _____ culture of Caribbean islands is typified by colorful fabrics.

Understanding the Selection   Complete each sentence.

6. As an example of how cultural styles are blurred, Reed describes the city of _____ in

   terms of its noticeable Islamic and Hispanic populations.

7. Reed disagrees with those educators who define the United States as a part of _____,

   because the nation is derived from much more than European roots.

8. Although Reed agrees that the Puritans were daring, he says that they also had a

   _____, often punishing and killing people in inhumane ways.

9. When Reed says, "Lady, they're already here," he means _____.

10. At the end of this selection, Reed says that the United States can become a place where

    _____.

Author's Craft: An Essay   Reed uses cause and effect in his essay to explain
aspects of history or society. Match each cause on the left with the resulting effect on
the right. Write the correct letter of the effect in the blank provided.

_____ 11. The Mexican American populace grows.

_____ 12. The Puritans encounter the
calypso culture of Barbados.

_____ 13. Artists of the early 20th century
discover African art.

_____ 14. A monoculturalist attitude is
prevalent in the United States.

_____ 15. The United States accepts the
heterogeneous world living
within its boundaries.

a. Cubism changes modern painting.

b. Japanese Americans, Chinese Americans,
Chicanos, and African Americans face
discrimination.

c. Directions at travel stations are given in
Spanish and English.

d. The cultures of the world crisscross in
the United States and enrich it.

e. Fear of what is not understood results in
Witchcraft Hysteria.

Assessment Worksheet 43
For My People by Margaret Walker
Let America Be America Again by Langston Hughes
(pages 340–347)

Vocabulary   The sentences below are based on descriptions in the selections. Look for context clues to help you find the best word below to complete each sentence.

dispossessed   floundering   tangled   deceived

blundering   consumption   anemia

1. Both Walker and Hughes describe African Americans as being _____ among themselves because of the chains placed on them by white people.

2. When Walker says that her people are _____, she means that they have nothing to call their own.

3. Walker uses alliteration effectively when she says her people are "distressed and disturbed and _____ and devoured."

4. When Walker talks about her people trying to fit into various associations, she describes them as _____ and _____.

5. _____ and _____ are two diseases Walker mentions as problematic for her people.

Understanding the Selections   Complete each sentence by filling in appropriate information from the selections.

6. Walker devotes the second stanza of her poem to the types of _____ her people do.

7. Walker says that what her people learn in school is that they are _____.

8. The new world that Walker asks for will be characterized by people who love _____.

9. Langston Hughes wishes for an America characterized by _____ and _____ for all.

10. Many of the vocabulary words footnoted in Hughes' poem refer to African Americans as

_____

11. The refrain that Hughes repeats throughout the poem is _____.

Author's Craft: Free Verse   By cataloging poets can suggest diversity and abundance, establish a setting, or develop a theme. Review the sections of the poems listed below and explain why cataloging is effective in each place.

12. the second stanza of "For My People" _____.

13. the seventh stanza of "For My People" _____.

14. the first stanza following italics in "Let America Be America Again" _____.

15. the last stanza of "Let America Be America Again" _____.

Name _____   Date _____

Assessment Worksheet 44
Petey and Yotsee and Mario by Henry Roth (pages 350–355)

Vocabulary   The sentences below are based on descriptions and events in the
selection. Look for context clues to help you find the best word below to complete
each sentence.

    Jewish cake   Gentile   embossed   crystallized   stoop   avid   spicecake

1. When Fat describes the rescue to his mother she refers to the _____ children who
   saved him as "blessed."

2. The _____, as Fat calls it, that Fat's mother bakes for the neighborhood boys is actually
   a holiday _____.

3. When the cake is finished, it is _____ with walnuts, dark with _____
   honey, and full of raisins.

4. Fat and his mother stand on the _____ and watch the boys devour the cake.

5. Everyone is excited about the cake, as is evident from their _____ cries for more.

Understanding the Selection   Complete each sentence by filling in appropriate
information from the selection.

6. The boys liked to swim at high tide in the _____.

7. Fat begins to "drown" after _____.

8. The boys' response following their rescue of Fat is to _____.

9. Fat doesn't want his mother to bake a cake for his friends because _____.

10. When Fat's mother gives the boys the cake, she apologizes for _____.

11. When the boys are given the cake, they _____.

12. Fat's mother says that if the boys didn't like her Jewish cake, they wouldn't have been the type to

   _____.

Author's Craft: A Short Story   An author's choice of point of view affects how
characters and events are developed. In "Petey and Yotsee and Mario," Roth chose to
use a first-person point of view. Tell why this point of view was effective in describing
the scenes listed below.

13. the "drowning" scene _____.

14. the discussion between Fat and his mother about her intention to bake a cake _____

   _____.

15. the presentation of the cake to the boys _____

   _____.

Name _____   Date _____

Assessment Worksheet 45

AmeRícan by Tato Laviera
Ending Poem by Aurora Levins Morales and Rosario Morales
Instructions for joining a new society by Heberto Padilla
(pages 356–362)

Vocabulary   Explain why each word below was created by the poets of the selections.

1. **AmeRícan** _____

2. **spanglish** _____

3. **cascabelling** _____

4. **latinoamerica** _____

Understanding the Selections   Complete each sentence by filling in appropriate
information from the selections.

5. As Laviera describes AmeRícan in the middle stanzas of the poem, he makes many, many references to

_____.

6. Laviera does not think of AmeRícan as separate from other cultures; rather he views it as a "_____

_____" of all that is good.

7. The woman in "Ending Poem," defines herself as a mestiza, a _____

_____.

8. She clearly states that she is *not* to be defined completely as a member of any of three cultures: _____

_____.

9. When she says, "We will not eat ourselves up inside anymore," she means _____

_____.

10. The three rules Padilla gives for joining a new society are: _____

_____

Author's Craft: Poetry   Poets use alliteration for the pleasing sound that is created,
but also to link and emphasize ideas. Decide whether each example from the selections
represents *initial* alliteration or *internal* alliteration. Write your answer in the blank.

_____   11. sweet soft spanish danzas

_____   12. strutting beautifully alert

_____   13. the language of garlic and mangoes

_____   14. talk being invented at the insistence of a smile

_____   15. I am a late leaf of that ancient tree

Name _____   Date _____

Assessment Worksheet 46
from *Seven Arrows* by Hyemeyohsts Storm (pages 363–367)

Vocabulary   Explain the significance of each term as it is used in the selection.

1. **Clan** _____

2. **Medicine of the Eagle** _____

3. **Animal Dance** _____

4. **Sun Dance** _____

5. **Medicine Wheel** _____

Understanding the Selection   Complete each sentence by filling in appropriate information from the selection.

6. The colors of the Eagle's robes indicate _____

_____.

7. People grow when they seek _____

_____.

8. People find their true Names through _____

_____.

9. The knowledge that animals have that people need to learn is _____

_____.

10. People must learn about themselves but also about the Ways of _____

_____.

Author's Craft: Allegory   Symbols in allegories help readers understand the meaning that goes beyond the literal level of the story. Match the symbols from *Seven Arrows* on the left with their literal meanings on the right. Write the correct letter in the blank provided.

_____ 11. True Names   a. those who are giving

_____ 12. Medicines   b. how people perceive things

_____ 13. Beginning Names   c. those who can see far

_____ 14. Buffalo   d. names given at birth

_____ 15. Eagles   e. names symbolizing a person's personality

Literary Skill: Metaphors

In the Land of Small Dragon: A Vietnamese Folktale
Told by Dang Manh Kha (pages 29–35)

A **metaphor** is an implied comparison of two dissimilar things.

A metaphor is a figure of speech used widely in speaking, narration, poetry, and many other types of writing. In some metaphors the dissimilar object is used in place of what it is compared with.

Metaphors   In the narrative poem "In the Land of Small Dragon," metaphors are used in two character descriptions, one of Elder Daughter and the other of the good fairy Nâng Tien. They are the two best characters in the poems terms of goodness and beauty. A similar figure of speech, the **simile,** is a comparison using the words *like* or *as*.

Find the two sections of description in "In the Land of Small Dragon." Write the metaphors and then interpret their meanings.

1. "Tâm's face _____ ,

   Her eyes dark as a storm cloud,

   Her feet _____ ,

   _____ ."

2. Interpret this description in prose and then compare the metaphor and the prose description.

   _____

   _____

   _____

   _____

3. "And from it rose Nâng Tien,

   A lovely _____ fairy.

   Her voice _____

   _____ ."

4. Interpret this description in prose and then compare the metaphor and the prose description.

   _____

   _____

   _____

   _____

Name _____   Date _____

Literary Skill: Point of View
To Da–duh, In Memoriam by Paule Marshall (pages 62–74)

Point of view is the perspective from which a story is told.

The person who tells the story is called the narrator. In *first–person point of view* the narrator is the character "I" or "me" refers to. In *third–person point of view* the narrator is outside the story and refers to the characters by name or by "he" or "she." An *omniscient* third–person narrator knows and tells what all the characters think and feel.

Point of View   In the short story "To Da–duh, In Memoriam," the narrator tells the story in *first–person* thereby allowing the reader to understand what she knows. It is also possible for the reader to see a broader picture than the nine–year old girl was able to understand. Complete the character analysis below, telling what the narrator is allowing the reader to know.

1. ". . . I finally made out the small, purposeful, painfully erect figure of the old woman headed our way."

   _____

   _____

2. "My mother, who was such a formidable figure in my eyes, had suddenly with a word been reduced to my status." _____

   _____

3. The granddaughter won the staring contest and reflected that "this was all I ever asked for of the adults in my life then." _____

   _____

4. Da–duh showed her the native trees and said "I know you don't have anything this nice where you come from . . . ." _____

   _____

5. The narrator persuades Da–duh that the Empire State Building is much taller than Bissex. "All the fight went out of her after that." _____

   _____

6. What did Da–duh learn during the narrator's visit? Did the narrator realize this? _____

   _____

   _____

Name _____ Date _____

Literary Skill: Cause and Effect
Seventeen Syllables by Hisaye Yamamoto (pages 88–98)

When one event comes before and brings about another event, the first is said to be the **cause** and the second the **effect.**

Cause and Effect   In fiction, the plot is a series of causes and effects. These may not appear in chronological order; primary causes may not be clear until the end of a story. The reader then must decide whether the cause–and–effect relationship in a plot is believable.

Answer the following questions regarding cause and effect relationships in the story.

1. What caused Rosie's mother to start writing *haiku?*

   _____

   _____.

2. What immediate effect does the writing have on Rosie's mother?

   _____

   _____.

3. What effect does the writing have on Rosie's father?

   _____

   _____.

4. What effect does the prize ultimately have on the poet?

   _____

   _____.

5. What effect does the kiss from Jesus have on Rosie's perception of her mother?

   _____

   _____.

6. What causes Rosie's mother to insist that Rosie never marry?

   _____

   _____.

7. What originally caused Rosie's mother to marry her father?

   _____

   _____.

8. Are the cause and effect relationships in the story believable? Explain your response.

   _____

   _____.

Name _____   Date _____

Literary Skill: Historical Document
from The Council of the Great Peace
by the League of the Iroquois (pages 131–137)

Historical documents may employ elements of literature such as repetition, parallelism, metaphors, and similes.

Historical Documents   "The Council of the Great Peace" is a record written in 1900 that reflects a Confederation formed around 1390. For many centuries the constitution had been kept in memory, and recalled in some cases by wampum that reflected individual laws and regulations. The written constitution reveals oral tradition, including elements of literature as well as the values and ideals of the Five Nations.

Use the text to identify elements listed below. Either quote from the material or summarize it.

1. Article 1—visual sensory descriptions that are repeated _____
_____

2. Article 1—first person narrator _____
_____

3. Article 2—symbolism to represent their willingness to welcome other individuals and nations into the
   Confederation _____
_____

4. Article 7—parallelism _____
_____

5. Article 19—cause and effect relationships _____
_____

6. Article 24—metaphor (The thickness of their skin shall be seven spans.) _____
_____
_____

7. Article 24—description of character of the members _____
_____

8. Article 28—a symbol (or emblem) and what it represents (The emblem of the deer's antlers are a
   symbol of the new person's "Lordship.")_____
_____
_____

Literary Skill: Sequencing
The People Could Fly
Told by Virginia Hamilton (pages 172–176)

Sequence is the order in which the major incidents of the plot take place.

In most stories, the sequence of incidents that develop the plot follow *chronological*, or time, order.

Words that show sequence   In the folktale "The People Could Fly," each major incident in the plot begins with the word *Say*. In other stories, the writer uses words like *next, suddenly, meanwhile, the following afternoon,* and *later* to make a transition to the next major event.

Trace the sequence of incidents in "The People Could Fly." Complete the outline below by summarizing what the writer tells you at each stage of the story.

I. **Introduction**—gives background information and catches the reader's interest

_____

_____

II. **Initiating Incident**—sets the plot in motion

_____

_____

III. **Major Incident**

_____

_____

IV. **Major Incident**

_____

_____

V. **Major Incident**—the high point, or *climax*

_____

_____

VI. **Resolution**—the main problem is resolved or settled

_____

_____

Literary Skill: Flashback
Lali by Nicholasa Mohr (pages 182–189)

A **flashback** is a change to a different setting.

In short stories and novels writers use flashbacks to give the reader a sense of having experienced the characters in another time and place. Flashbacks are a dramatic, direct way of giving information as well as deepening understanding of characters and of presenting contrasts.

Flashbacks   In "Lali" Nicholasa Mohr uses a flashback from winter in New York to the young woman's lush home in Puerto Rico. The writer presents entirely different types of details to describe the two settings. The cold, work-filled poverty in New York is a sad contrast to the sensory wealth and beauty of Puerto Rico.

Conduct a search through the story for sensory images (pictures, sounds, smells, tastes, feelings). Then draw a conclusion about the mood in the flashback compared with the mood in New York.

1. What visual images do you get of New York?

    _____   _____

    _____   _____

2. What visual images do you get of Puerto Rico?

    _____   _____

    _____   _____

    _____   _____

3. What sounds do you hear in New York?

    _____   _____

    _____   _____

4. What are the smells and sensory feelings in Puerto Rico?

    _____   _____

    _____   _____

    _____   _____

5. Describe the mood of the New York setting: the feeling created by the apartment, the diner, the exchanges between Lali and her husband.

    _____

6. Describe the mood of Puerto Rico as Lali remembers it. Note that her mother interrupts her daydreams.

    _____

Name _____ Date _____

Literary Skill: Speech
I See the Promised Land
by Martin Luther King, Jr. (pages 225–237)

Speeches can contain literary elements used in stories, poetry, and nonfiction writing.

Speeches, whether impromptu or written, often contain literary elements such as parallelism, allusion, repetition, and cause and effect relationships. Some orators speak with the power of great poets and authors.

Speeches   In his speech "I See the Promised Land," Martin Luther King, Jr. used many elements important in literature. Use the list below to help identify these elements. Some appear more than once.

a. allusion   b. transitions   c. parallelism   d. cause and effect   e. repetition

1. _____ (Paragraph) "Now, what does this mean in this great period of history?"

2. _____ "their long years of poverty, their long years of hurt and neglect…"

3. _____ He urges people to go around to stores and industries and say that they must treat all people fairly, but if they do not, "…Our agenda calls for withdrawing economic support from you."

4. _____ (Paragraph) "Secondly, let us keep the issues where they are."

5. _____ When Pharaoh wanted to prolong slavery in Egypt, he kept the slaves fighting with each other.

6. _____ "But somewhere I read of the freedom of assembly. Somewhere I read of the freedom of s Speech. Somewhere I read of the freedom of the press. Somewhere I read that the greatness of America is the right to protest for right."

7. _____ "I just want to do God's will. And He's allowed me to go to the mountain."

8. _____ (Paragraph) "Now, let me say as I move to my conclusion that we've got to give ourselves to this struggle until the end."

Literary Skill: Drama
Fences by August Wilson (pages 321–331)

A **drama** is a literary work written to be performed by actors on a stage.

The playwright develops the story through **dialogue,** or the conversations of the characters. **Stage directions** give instructions about props, lighting, costumes, scenery, and the movement and gestures of the actors.

Drama   In the play *Fences*, dialogue reveals the personalities of the characters and the conflicts in Troy's family.

Read the dialogue, and write one quality revealed about each character. Each segment of dialogue may reveal a different characteristic. The following list contains suggestions of qualities you may find

naive   (un)loving   (un)informed   (un)realistic   independent stubborn

single-minded   (un)sympathetic   (ir)responsible cynical

1. Troy:...While you thinking about a TV, I got to be thinking about the roof...and whatever else go wrong

   around here. Now if you had two hundred dollars, what would you do...fix the roof or buy a TV?

   Cory: I'd buy a TV. Then when the roof started to leak...when it needed fixing...I'd fix it.

   Troy _____   Cory _____

2. Troy:...And ain't no need for nobody coming around here to talk to me about signing nothing.

   Cory: Hey, Pop...you can't do that. He's coming all the way from North Carolina.

   Troy _____   Cory _____

3. Cory: Come on, Pop! I got to practice. I can't work after school and play football too. The team needs

   me. That's what Coach Zellman says...

   Troy: I don't care what nobody else say. I'm the boss...you understand? I'm the boss around here. I do the

   only saying what counts.

   Cory _____   Troy _____

4. Cory: How come you ain't never liked me?

   Troy: Liked you? Who say I got to like you? Wanna stand up in my face and ask a question like that.

   Talking about liking somebody.

   Cory _____   Troy _____

5. Rose: Why don't you let the boy go ahead and play football, Troy? Ain't no harm in that. He's just

   trying to be like you with the sports.

   Troy: I don't want him to be like me! I want him to move as far away from my life as he can get.

   Rose _____   Troy _____

Literary Skill: Short Story
Petey and Yotsee and Mario by Henry Roth (pages 350–355)

A **short story** is a brief work of fiction.

Short stories usually have a single setting and only a few characters who are involved in the **plot,** or central conflict. The conflict usually involves a series of complication until it reaches a **climax,** or point of highest tension. The climax is usually followed by the **resolution,** or end, of the conflict.

Short Stories   In the brief story "Petey and Yotsee and Mario," Henry Roth develops great tension in two important scenes. The reader learns a great deal about several characters through dialogue.

Complete a story outline of "Petey and Yotsee and Mario" by describing the following elements.

  I. Plot

    _____

    _____

 II. Characters

    A. Fats

    _____

    _____

    B. Fat's mother

    _____

    _____

III. Setting—The dock on Harlem River

    _____

    _____

IV. Major Incident—the near–drowning

    _____

    _____

  V. Climax

    _____

    _____

VI. Resolution

    _____

    _____

ANSWER KEYS

These answer keys are for the assessment worksheets.
The literary skills worksheets have open-ended questions that depend
on the student's responses to the selections.

Answer Keys

WORKSHEET 1   The Stars

Vocabulary 1. journey; 2. mole; 3. Milky Way, 4. destination; 5. coyote

Understanding the Selection 6. they had lived many years under a cruel leader. 7. working in pairs to drag their loads on poles. 8. he had not seen the signs of the mole and the spider. 9. its tracks were like the mole and it had the pattern of the spider on its back. 10. Long Sash is playing a game and catching them first.

Author's Craft: Quest Story 11. a; 12. d; 13. b; 14. c; 15. d

WORKSHEET 2   The Sheep of San Cristóbal

Vocabulary 1. penance; 2. prayed; 3. priest; 4. Padre; 5. indigent

Understanding the Selection 6. kidnaped, taken; 7. Felipa; 8. Don Jose; 9. Don Jose had been stingy, never giving to the poor; 10. San Cristobal

Author's Craft: A Moral Tale 11. c; 12. e; 13. a; 14. d; 15. b.

WORKSHEET 3   from *Things Fall Apart*

Vocabulary 1. delectable; 2. orator; 3. eloquent; 4. voluble; 5. cunning

Understanding the Selection 6. two moons; 7. he is a changed man; 8. All of You; 9. Parrot; 10. is not smooth

Author's Craft: A Folktale 11. R; 12. R; 13. F; 14. R; 15. F

WORKSHEET 4   In the Land of Small Dragon: A Vietnamese Folktale

Vocabulary 1. scowling; 2. revenge; 3. rice paddy; 4. indolent; 5. monsoon

Understanding the Selection 6. his daughters equally; 7. their work—which one brought the most fish; 8. tricking her into gathering flowers and stealing from her; 9. her beloved fish; 10. a flock of blackbirds.

Author's Craft: A Narrative Poem 11. b; 12. a; 13. c; 14. b; 15. d

WORKSHEET 5   The Man Who Had No Story

Vocabulary 1. fairy tale; 2. fog; 3. Fenian tale; 4. glen; 5. wake-house

Understanding the Selection 6. he had already cut all the others; 7. where there is light there must be people; 8. went for a bucket of water; 9. fiddle, perform surgery; 10. to trim off some of his legs, just below the knee

Author's Craft: A Story 11. Brian follows a light

to lead him to find people. 12. "That is something I never did in my life." 13. It provides a pleasurable sense of recognition; a rhythmic feeling, similar to a refrain. 14. The fairies brought him back there after his adventures. 15. He had intruded into their glen.

WORKSHEET 6   from *Bless Me, Ultima*

Vocabulary 1. c; 2. e; 3. d; 4. a; 5. b

Understanding the Selection 6. farmers; 7. cowboys; 8. she respected her so much; 9. the scene after his birth; 10. his mother wanted one and he wanted to make her happy.

Author's Craft: A Novel 11. b; 12. e; 13. c; 14. d; 15. a.

WORKSHEET 7   from *Roots*

Vocabulary 1. decline; 2. irate; 3. succession; 4. argumentative; 5. irritable

Understanding the Selection 6. solve disputes between people; 7. women; 8. a waiting period of one year; 9. their masters; 10. it would broaden their knowledge as they grew

Author's Craft: A Historical Novel 11. WI; 12. WI; 13. HF; 14. HF; 15. HF

WORKSHEET 8   To Da-duh, In Memoriam

Vocabulary 1. roguish; 2. unrelenting; 3. dissonant; 4. formidable; 5. truculent

Understanding the Selection 6. they were as keen as a young child's; 7. the grandmother looked away first; 8. he thought it was a foolish waste of money; 9. she has never been off the island and she wants to know it is finer than New York; 10. Da-duh's spirit is broken and she dies.

Author's Craft: A Novel 11. disembarkation shed; 12. Da-duh; 13. the narrator; 14. banana; 15. airplanes

WORKSHEET 9   from *The Way to Rainy Mountain*

Vocabulary 1. unrelenting; 2. knoll; 3. deicide; 4. range; 5. sacred

Understanding the Selection 6. that each season is utterly harsh; 7. emerged from a hollow log; 8. Kiowa practices and Christianity; 9. love, awe, respect; 10. think about, experience, really come to know one piece of land in every season.

Author's Craft: A Novel 11. c or d; 12. b; 13. e; 14. a; 15. c or d.

WORKSHEET 10   i yearn [We Who Carry the Endless Seasons]

Vocabulary 1. barrio; 2. calo; 3. chicano infinitum; 4. assail; 5. que tal, hermano.

Understanding the Selection 6. to marry handsome Filipino boys from good families; 7. have different dreams because they have grown up in the United States; 8. that they are pulled by both cultures; 9. his neighborhood, language, food, and friends; 10. to be with people like him who love that they exist.

Author's Craft: Poems 11. T; 12. T; 13. F; 14. T; 15. T

WORKSHEET 11 Seventeen Syllables

Vocabulary 1. dubious; 2. helplessness; 3. delectable; 4. grave; 5. giddy

Understanding the Selection 6. *Haiku*, a form of poetry that has seventeen syllables; 7. anger, confusion, fear, hatred; 8. she had disgraced her family by getting pregnant; 9. their families are working together to harvest tomatoes; 10. she thinks marriage may be the same kind of trap for her daughter.

Author's Craft: A Story 11. b; 12. b; 13. c; 14. d; 15. b.

WORKSHEET 12 from *Humaweepi, the Warrior Priest*

Vocabulary 1. obsidian; 2. turquoise; 3. coral; 4. succulent; 5. yucca fiber

Understanding the Selection 6. warrior priest; 7. his parents had died; 8. surprised; 9. that he has become a warrior priest; 10. to place the beads from his buckskin on the head of the bear.

Author's Craft: A Novel 11. b; 12. e; 13. a; 14. d; 15. c.

WORKSHEET 13 Cante Ishta—The Eye of the Heart

Vocabulary 1. sham; 2. peyote; 3. enraptured; 4. intellectualism 5. purification

Understanding the Selection 6. medicine man; 7. listening and by doing; 8. they do not break from the heat; 9. purified, amazed, relieved; 10. she represented Sioux women.

Author's Craft: Autobiography 11. d; 12. a; 13. b; 14. b; 15. c.

WORKSHEET 14 Black Hair ALONE/december/night

Vocabulary 1. The month of July is compared to a ring of fire all must jump through. 2. Dust rises from dirty feet; shade usually appears flat. 3. Pottery is turned on a wheel; Hector's muscles seem sculpted like a clay pot. 4. Soto compares his feelings for baseball and for Hector to religious worship. 5. Usually *line* is an adjective, as in *line drive*; here Soto uses it as a verb.

Understanding the Selection 6. a hero like themselves who had made it; 7. the quick finality of a clean catch; 8. tense, hurt, sad; 9. it's been a long time since he felt at home; life is too complex to try to understand.

Author's Craft: Poems 11. T; 12. T; 13. T; 14. F; 15 T.

WORKSHEET 15 Wasichus in the Hills from *Black Elk Speaks*

Vocabulary 1. b; 2. a; 3. d; 4. c; 5. e

Understanding the Selection 6. white people (stealers of the fat); 7. it reminded him of a vision he had; 8. grass grows and water flows; 9. shade made of canvas; 10. September

Author's Craft: Autobiography 11. F; 12. O; 13. O; 14. F; 15. F

WORKSHEET 16 The Council of the Great Peace from *The Constitution of the Five Nations*

Vocabulary 1. Confederate; 2. contumacious; 3. royaneh; 4. allies; 5. candidate

Understanding the Selection 6. the five Iroquois nations; 7. Tree of the Great Peace; 8. be admitted; 9. it makes sparks that would disturb the council; 10. any Lord of the Confederacy

Author's Craft: A Document 11. the soft white feathery down of the globe thistle seats under the shade of the Tree of Great Long Leaves (visual); 12. The shade and the thistle seats are described twice. 13. They offer thanks to the earth, to the streams, to the maize, the forest trees, and so on. 14. The tree roots have a nature of Peace and Strength. The eagle can warn people of threatening danger. 15. The passage "The thickness of their skin...shall be marked by calm deliberation" is repeated in both articles.

WORKSHEET 17 The Indians' Night Promises to be Dark

Vocabulary 1. wax; 2. thrill; 3. throng; 4. somber; 5 ebb

Understanding the Selection 6. White Men; 7. that his people shall be able to visit the graves of their ancestors; 8. has forsaken his Red children; 9. puzzled, disgusted, amazed; 10. their spirits will throng the land they love

Author's Craft: A Speech 11. S; 12. S; 13. M; 14. S; 15. S

WORKSHEET 18 They Are Taking the Holy Elements from Mother Earth

Vocabulary 1. pollen; 2. particles; 3. Holy

Elements; 4. herbs; 5. contaminate
Understanding the Selection 6. lack of
understanding, no trading post or police station to
report them; 7. their Great Mother; 8. thrown into
the water; 9. she prays to the moon—it is holy;
10. it is so disturbed by the mining
Author's Craft: Essay 11. The particles of coal dust
contaminate the water; 12. the explosions in the
mines; 13. the sheep are getting thin and not having
many lambs; 14. the mine destroys the springs and
water holes; 15. there is no soil left, just rocks

WORKSHEET 19 The Drinking Gourd, Steal Away
Vocabulary 1. the phrase is used because it sounds
more pleasing than *The Big Dipper* and because the
image is one of satisfying thirst, symbolic of the
thirst for freedom; 2. old man is used because it
sounds more pleasing than *God*; 3. *a-waitin'* is
dialect for waiting; again it sounds more rhythmic,
pleasing, and natural; 4. the "s" sound is very
prevalent in the poem, perhaps one reason why *steal
away* is used instead of *run away*; 5. the "s" sounds
contribute to the sound of the poem and the image
of a trumpet blowing helps the meaning of the poem
Understanding the Selection 6. the river; 7. signs
in nature; 8. the Lord; 9. sounds of nature;
10. trumpet sounds.
Author's Craft: Lyrics 11. "Follow the drinking
gourd;" 12. It is the theme of the song; 13. The last
two lines are exactly the same for all three stanzas;
14. Meaning is emphasized in the lines that are
different; linkages and rhythm are achieved through
repetition; 15. The same notes could be sung for all
repeated lines, emphasizing linkages and rhythm.

WORKSHEET 20 The Slave Auction, The Slave Ship
Vocabulary (Students should note that many words
used to define these vocabulary terms have negative
connotations too.) 1. a vessel used to carry captured
Africans to America to be sold as slaves; 2. actions
that inflict pain or suffering; 3. chains or shackles for
the feet; 4. greed; 5. carrying and spreading infection
Understanding the Selection 6. the separation of
families; 7. the horrible smell; 8. he became very ill
and was considered too young to make trouble if left
unchained; 9. the slaves were to be sold there as
workers; 10. the separation of families
Author's Craft: Autobiography 11. F; 12. E;
13. F; 14. E; 15. F

WORKSHEET 21 The Slave who Dared to Feel like a Man.

Vocabulary 1. audacity; 2. master; 3. overseer;
4. yoke; 5. chattel; 6. indecorum
Understanding the Selection 7. the grandmother
was kind and sympathetic and gave them good food;
8. it was the will of God; 9. he had a bold and
daring spirit and found any alternative preferable to
slavery; 10. whipping him, jailing him, keeping him
chained, and disallowing visitors
Author's Craft: Personal Narrative 11. 2; 12. 5;
13. 1; 14. 4; 15. 3

WORKSHEET 22 The People Could Fly, Runagate Runagate
Vocabulary 1. motif; 2. underground; 3. croon;
4. plausible; 5. thicketed; 6. movering
Understanding the Selection 7. she could no
longer get up; 8. Toby knew he could fly away;
9. freedom; 10. freedom; 11. the slaves had the best
chance of escaping in darkness
Author's Craft: Folk tales 12. c; 13. a; 14. d; 15. b

WORKSHEET 23 Lali
Vocabulary 1. countryside; 2. pollution; 3. taciturn;
4. luncheonette; 5. tenement
Understanding the Selection 6. handling the grill
during the early morning and late evening and
cooking the standard meals during the rest of the
day; 7. Lali, a shy person, was overwhelmed by New
York, and Rudi, a very practical but not overly
affectionate man, was unable to help draw her out;
8. she can talk honestly with him and has shared
school experiences with him; 9. everything,
including the weather, is unfamiliar and not as she
had imagined it might be; 10. Rudi hopes the visit
will help their marriage, and Lali hopes it will
relieve her homesickness
Author's Craft: A Story 11. S; 12. S; 13. F;
14. F; 15. S

WORKSHEET 24 from *Picture Bride*
Vocabulary 1. kimono; 2. derby; 3. sallow;
4. pompadour; 5. flustered
Understanding the Selection 6. she wanted to
escape the confinement of her family and village;
7. very little; 8. being in a new land, meeting Taro;
9. they were questioned about their plans and given
medical tests; 10. relax enough to share laughter
Author's Craft: A Novel 11. b; 12. c; 13. d;
14. a; 15. e

WORKSHEET 25 I Leave South Africa
Vocabulary 1. d; 2. e; 3. b; 4. c; 5. a
Understanding the Selection 6. allow blacks to
share their different backgrounds and help one

another; 7. prejudice, the Ku Klux Klan, and false integration; 8. he doesn't have a return ticket to South Africa; 9. they were often mistaken for black Americans and treated as honorary whites; 10. black and white people interacting

Author's Craft: Autobiography 11. Yes; 12. No; 13. No; 14. Yes; 15. No

WORKSHEET 26   Ellis Island

Vocabulary 1. old Empires; 2. dreams; 3. red brick; 4. Ellis Island

Understanding the Selection 7. Ellis Island; 8. the Statue of Liberty; 9. European heritage; 10. moving from place to place according to the seasons; 11. own, lose

Author's Craft: A Poem 12. the tall woman, green as dreams of forests and meadows; 13. this island, nine decades the answerer of dreams; 14. Yet only one part of my blood loves that memory. Another voice speaks of native lands within this nation; 15. Lands of those who followed the changing Moon, knowledge of the season in their veins.

WORKSHEET 27   Those Who Don't, No Speak English, The Three Sisters

Vocabulary 1. d; 2. a; 3. b; 4. c

Understanding the Selection   5. Mexican American; 6. they leave their neighborhood and go to one where other ethnic groups live; 7. Mamacita can't speak English; 8. her pink house back in Mexico; 9. very special; 10. a way to leave Mango Street; 11. come back to Mango Street to help those left behind

Author's Craft: A Novel 12. A; 13. H; 14. S; 15. S

WORKSHEET 28   I See the Promised Land

Vocabulary 1. First Amendment; 2. Promised Land; 3. human rights; 4. paddy wagons; 5. injunction; 6. grapple; 7. injustice

Understanding the Selection 8. to be treated like other people; 9. the people are united; 10. right to freedom of assembly, freedom of speech, freedom of the press; 11. withdrawing economic support from those who don't support civil rights and using only African American financial institutions; 12. he believes civil rights will be achieved

Author's Craft: A Speech 13. a; 14. c; 15. b

WORKSHEET 29   See for yourself, listen for yourself, think for yourself

Vocabulary 1. brutality; 2. nonviolent; 3. forgiving; 4. violent; 5. rumble

Understanding the Selection 6. how people can make judgments based only on the opinions of others; 7. see, listen, and think for themselves;

8. their "enemies"; 9. it is consistently practiced by all people; 10. a center of African American life

Author's Craft: A Speech 11. res; 12. par; 13. par; 14. res; 15. rep

WORKSHEET 30 from **China Men, Immigration Blues**

Vocabulary 1. dynamite, gouged; 2. diverted; 3. sledgehammer, jarred; 4. pickaxed; 5. shoveled

Understanding the Selection 6. dirt in his eyes, nose, and mouth; jarring of the pickax; loud sounds of metal on granite; 7. they were removed from light when working and they worked round-the-clock shifts; 8. terrain changes, more incidents of injury and death, faster tunneling, change in the men's jobs; 9. avalanches caused by dynamite, frostbite, snowblindness; 10. deportation and detention by immigration

Author's Craft: A Story 11. a; 12. c; 13. a; 14. a; 15. b

WORKSHEET 31   Napa, California The Circuit

Vocabulary 1. sharecropper, braceros; 2. harvest; 3. sweat, sun-beaten; 4. dusk; 5. corridos

Understanding the Selections 6. survive; 7. move to find more work; 8. Papa's car, a '38 Plymouth; 9. an old, large, galvanized pot; 10. the picking season falls off; 11. his family had to move again

Author's Craft: A Story 12. hot and sunny; 13. a one-room shack; 14. harvesting fruits and vegetables; 15. the family must move

WORKSHEET 32   Speech of Sojourner Truth

Vocabulary 1. c; 2. d; 3. a; 4. e; 5. b

Understanding the Selection 6. work: plow, reap, husk, chop, mow; carry; eat; 7. women won't take more than their share; 8. allow women to set it right; 9. God, woman; 10. slavery

Author's Craft: A Speech 11. her description of how much she can do as compared to what a man can do; 12. "Well if woman upset the world, do give her a chance to set it right side up again"; 13. ". . . and can any man do more than that?"; 14. "Man, where is your part?"; 15. ". . . and he is surely between a hawk and a buzzard."

WORKSHEET 33   from **I Am Joaquín**

Vocabulary 1. Mexican; 2. Spanish; 3. Latin; 4. Hispanic; 5. Mexican American

Understanding the Selection 6. proud but poor and hungry; unhappy but economically comfortable; 7. the Moors and the Europeans; 8. life in the rugged mountains and hard labor toiling in the fields; 9. bigotry, snobbery, exploitation, and racial

hatred; 10. strength of spirit and unbreakable faith
Author's Craft: Free Verse 11. R; 12. E; 13. E;
14. E; 15. U

WORKSHEET 34   Free to Go
Vocabulary 1. curfew; 2. evacuation order,
internment camps; 3. propaganda; 4. assimilate;
5. nightriders
Understanding the Selection 6. they had no home to
go to and they feared racism; 7. the outside world,
especially as represented by Sears, Roebuck; 8. move
to the East Coast; 9. he was too old, afraid, and proud;
10. they knew they were accepted by their neighbors
Author's Craft: Personal Narrative 11. D; 12. S;
13. S; 14. S; 15. S

WORKSHEET 35   In the American Storm
Vocabulary 1. humiliating; 2. excavation;
3. mattock; 4. pick and shovel; 5. padrone
Understanding the Selection 6. wandering around,
feeling lost; 7. they had things in common and
could communicate with each other; 8. pick and
shovel work; 9. being swindled by the "padrone";
10. being hated by the Russian workers
Author's Craft: Autobiography Students' exact
answers will vary, but they should use words that are
similar in meaning to those suggested here:
11. serious, formal; 12. serious, sad; 13. light,
humorous; 14. serious, hopeful; 15. serious, gloomy

WORKSHEET 36   from *Their Eyes Were Watching God*
Vocabulary 1. suspect; 2. child; 3. used to;
4. children; 5. young children
Understanding the Selection 6. her grandmother
and the Washburns; 7. the four Washburn children;
8. she saw a picture of herself at age six; 9. she had
been given so many nicknames; 10. all the
Washburn children called her that
Author's Craft: Regional Writing 11. Ah know
exactly what Ah got to tell yuh, but it's hard to
know where to start at; 12. Nanny used to ketch us
in our devilment and lick every youngun on de
place; 13. Ah reckon dey never hit us ah lick amiss
'cause dem three boys and us two girls wuz pretty
aggravatin', Ah speck; 14. Dey all useter call me
Alphabet 'cause so many people had done named
me different names; 15. But before Ah seen de
picture Ah thought Ah wuz just like de rest

WORKSHEET 37   Refugee Ship, Letter to America, address
Vocabulary 1. foreign; 2. lag; 3. perdone;

4. bronzed; 5. race
Understanding the Selections 6. she will never
stop feeling like a refugee without a homeland;
7. her language and her looks; 8. fear and hate;
9. using, shooting, and denying them; 10. race;
11. name
Author's Craft: Poetry 12. These words are all sad
and repressive, reflecting the confused and sad mood
of the poem; 13. The accusatory tone of these
phrases contributes to the angry mood of the poem;
14. The Spanish speaker is trying unsuccessfully to
communicate, which contributes to the cold,
uncaring mood of the poem; 15. It contributes to
the troubled feeling of the mood—revealing people
trying to combine different cultures.

WORKSHEET 38   from *Black Boy*
Vocabulary 1. taunting; 2. peevishly; 3. countered;
4. evaded; 5. grudgingly
Understanding the Selection 6. there are two
lines—one for African Americans and one for
whites; 7. ethnic heritage; 8. the white people who
owned them; 9. he feels she is hiding information
from him; 10. the feeling and smell of the dirt road
Author's Craft: Autobiography Note that student
answers will vary in this section; in parentheses
following each sentence suggested as a response is a
word describing Wright's attitude. Check to be sure
the sentence chosen expresses this attitude.
11. Naively I wanted to go and see how the whites
looked while sitting in their part of the train.
(curious); 12. She explained it all in a matter-of-
fact, offhand, neutral way; her emotions were not
involved at all. (serious); 13. All right, I would find
out someday. (determined); 14. It looked like home
and I was glad. (relaxed); 15. But I felt confident of
outwitting any bee. (powerful)

WORKSHEET 39   from **Four Directions, Saying Yes**
Vocabulary 1. ferocity; 2. concocted; 3. slack;
4. guileless; 5. vulnerability
Understanding the Selection 6. (Students may
include any of these responses.) drinking too much
wine, using the chopsticks incorrectly, taking large
portions of food before everyone else had been
served, not trying the most expensive dish, refusing
seconds, putting soy sauce on the mother's best dish;
7. the many different emotions she experiences;
8. goodness, honesty, occasional bad temper and
stinginess; 9. intelligence, strength, trickiness,
ability to win wars; 10. Chinese, American
Author's Craft: A Story 11. internal conflict;

12. climax; 13. external conflict; 14. resolution; 15. internal conflict

WORKSHEET 40 from *Stelmark: A Family Recollection*
Vocabulary 1. b; 2. e; 3. a; 4. d; 5. c
Understanding the Selection 6. his peer group expected him to "prove his mettle" by leading the attack on a storeowner of his own ethnicity; 7. their cultural and historic significance; 8. yellow figs; 9. he eats and enjoys all of the traditional Greek foods; 10. the sweet taste of those yellow figs
Author's Craft: Nonfiction 11. main idea; 12. detail; 13. main idea; 14. detail; 15. main idea

WORKSHEET 41 from *Fences*
Vocabulary 1. pitching; 2. timing (or follow-through); 3. follow-through (or timing); 4. home runs; 5. bench
Understanding the Selection 6. TV, new roof; 7. work at the A & P and do his chores; 8. Troy feels the white people will keep African Americans from being highly paid stars; 9. it is his duty and responsibility; 10. his relationship with Rose
Author's Craft: Drama 11. practical; 12. bitter; 13. tired; 14. independent; 15. bitter

WORKSHEET 42 America: The Multinational Society
Vocabulary 1. pastrami; 2. Yoruban; 3. bouillabaisse; 4. mosques; 5. calypso
Understanding the Selection 6. Detroit; 7. Western civilization; 8. mean streak; 9. the "foreigners" are already living in the United States; 10. the cultures of the world crisscross
Author's Craft: An Essay 11. c; 12. e; 13. a; 14. b; 15. d

WORKSHEET 43 For My People Let America Be America Again
Vocabulary 1. tangled; 2. dispossessed; 3. deceived; 4. floundering, blundering; 5. consumption, anemia
Understanding the Selection 6. work; 7. "black and poor and small and different"; 8. freedom; 9. freedom, equality; 10. slaves; 11. "America never was America to me"
Author's Craft: Free Verse 12. By cataloging the different jobs her people do, Walker shows variety but also adds emphasis to the fact that all of these jobs are in the service of others; 13. By cataloging the daily activities of her people, Walker brings in the theme that all people are the same—yet her people are laughed at; 14. By listing here the people who are "in the dark," Hughes emphasizes the fact that it is people of all cultures; 15. By listing the

natural wonders of the United States, Hughes tries to establish a new, beautiful setting.

WORKSHEET 44 Petey and Yotsee and Mario
Vocabulary 1. Gentile; 2. Jewish cake, spicecake; 3. embossed, crystallized; 4. stoop; 5. avid
Understanding the Selection 6. Harlem River; 7. a tug boat sends waves washing over him; 8. laugh and make jokes about it; 9. he is embarrassed that it is "different" from the cakes the boys eat; 10. her inability to speak English better; 11. eat it "with gusto"; 12. rescue Fat
Author's Craft: A Short Story 13. By experiencing the scene from Fat's viewpoint, the reader sees how close he actually was to drowning; 14. By knowing Fat's thoughts as he talks, the reader gains insight into the source of his discomfort; 15. We learn through first-person point-of-view that Fat is reluctant and embarrassed.

WORKSHEET 45 AmeRícan, Ending Poem, Instructions for joining a new society
Vocabulary 1. The poet wanted a word to represent the combination of American and Puerto Rican; 2. The poet wanted to name the "new" language that combined Spanish and English; 3. This is another attempt by the poet to combine the two languages and create a new word; 4. Morales is trying to give a new title to the United States that is more reflective of the Latino influence.
Understanding the Selection 5. music; 6. blend and mix; 7. woman of mixed race; 8. African, Taina, or European; 9. they will come to terms with their new identity; 10. be optimistic, be well turned out, and walk one step forward and two or three back.
Author's Craft: Poetry 11. initial; 12. internal; 13. internal; 14. initial; 15. initial

WORKSHEET 46 from *Seven Arrows*
Vocabulary 1. a division within a nation; 2. the ability to see far; 3. the sharing of traits among different animals; 4. the recognition of one's own strengths and the sharing of those strengths among all; 5. a symbolic way of showing a person's roots and personality
Understanding the Selection 6. which Way a person enters the world from; 7. the knowledge of Ways other than those from which a person is born; 8. a vision quest; 9. how to share their gifts; 10. their brothers and sisters
Author's Craft: Allegory 11. e; 12. b; 13. d; 14. a; 15. c